Hydrology and Water Resources Management in Arid, Semi-Arid, and Tropical Regions

Christopher Misati Ondieki
Kenyatta University, Kenya

Johnson Utu Kitheka
South Eastern Kenya University, Kenya

A volume in the Advances in Environmental Engineering and Green Technologies (AEEGT) Book Series

IGI Global
DISSEMINATOR OF KNOWLEDGE

Published in the United States of America by
IGI Global
Engineering Science Reference (an imprint of IGI Global)
701 E. Chocolate Avenue
Hershey PA, USA 17033
Tel: 717-533-8845
Fax: 717-533-8661
E-mail: cust@igi-global.com
Web site: http://www.igi-global.com

Library of Congress Cataloging-in-Publication Data

Names: Ondieki, C. M., editor. | Kitheka, Johnson U., editor.
Title: Hydrology and water resources management in arid, semi-arid, and tropical regions / Christopher Misati Ondieki and Johnson Utu Kitheka, editors.
Description: Hershey, PA : Engineering Science Reference, [2019] | Includes bibliographical references.
Identifiers: LCCN 2019016791| ISBN 9781799801634 (hardcover) | ISBN 9781799801658 (ebook) | ISBN 9781799801641 (softcover)
Subjects: LCSH: Water resources development--Africa. | Sustainable development--Africa. | Hydrology--Africa. | Arid regions--Africa.
Classification: LCC TD315 .H93 2019 | DDC 333.910096--dc23 LC record available at https://lccn.loc.gov/2019016791

This book is published in the IGI Global book series Advances in Environmental Engineering and Green Technologies (AEEGT) (ISSN: 2326-9162; eISSN: 2326-9170)

British Cataloguing in Publication Data
A Cataloguing in Publication record for this book is available from the British Library.

All work contributed to this book is new, previously-unpublished material.
The views expressed in this book are those of the authors, but not necessarily of the publisher.

For electronic access to this publication, please contact: eresources@igi-global.com.

Advances in Environmental Engineering and Green Technologies (AEEGT) Book Series

ISSN:2326-9162
EISSN:2326-9170

Editor-in-Chief: Sang-Bing Tsai, University of Electronic Science and Technology of China Zhongshan Institute, China & Ming-Lang Tseng, Lunghwa University of Science and Technology, Taiwan & Yuchi Wang, University of Electronic Science and Technology of China Zhongshan Institute, China

MISSION

Growing awareness and an increased focus on environmental issues such as climate change, energy use, and loss of non-renewable resources have brought about a greater need for research that provides potential solutions to these problems. Research in environmental science and engineering continues to play a vital role in uncovering new opportunities for a "green" future.

The **Advances in Environmental Engineering and Green Technologies (AEEGT)** book series is a mouthpiece for research in all aspects of environmental science, earth science, and green initiatives. This series supports the ongoing research in this field through publishing books that discuss topics within environmental engineering or that deal with the interdisciplinary field of green technologies.

COVERAGE

- Sustainable Communities
- Green Transportation
- Alternative Power Sources
- Air Quality
- Biofilters and Biofiltration
- Contaminated Site Remediation
- Waste Management
- Pollution Management
- Green Technology
- Policies Involving Green Technologies and Environmental Engineering

IGI Global is currently accepting manuscripts for publication within this series. To submit a proposal for a volume in this series, please contact our Acquisition Editors at Acquisitions@igi-global.com or visit: http://www.igi-global.com/publish/.

The Advances in Environmental Engineering and Green Technologies (AEEGT) Book Series (ISSN 2326-9162) is published by IGI Global, 701 E. Chocolate Avenue, Hershey, PA 17033-1240, USA, www.igi-global.com. This series is composed of titles available for purchase individually; each title is edited to be contextually exclusive from any other title within the series. For pricing and ordering information please visit http://www.igi-global.com/book-series/advances-environmental-engineering-green-technologies/73679. Postmaster: Send all address changes to above address.

Titles in this Series

For a list of additional titles in this series, please visit:
https://www.igi-global.com/book-series/advances-environmental-engineering-green-technologies/73679

Advanced Agro-Engineering Technologies for Rural Business Development
Valeriy Kharchenko (Federal Scientific Agroengineering Center VIM, Russia) and Pandian Vasant (Universiti Teknologi PETRONAS, Malaysia)
Engineering Science Reference • ©2019 • 484pp • H/C (ISBN: 9781522575733) • US $195.00

Spatial Planning in the Big Data Revolution
Angioletta Voghera (Politecnico di Torino, Italy) and Luigi La Riccia (Politecnico di Torino, Italy)
Engineering Science Reference • ©2019 • 359pp • H/C (ISBN: 9781522579274) • US $195.00

Global Initiatives for Waste Reduction and Cutting Food Loss
Aparna B. Gunjal (Asian Agri Food Consultancy Services Ltd, India) Meghmala S. Waghmode (Annasaheb Magar Mahavidyalaya, India) Neha N. Patil (Annasaheb Magar Mahavidyalaya, India) and Pankaj Bhatt (Dolphin (P.G) College of Biomedical and Natural Sciences Dehradun, India)
Engineering Science Reference • ©2019 • 328pp • H/C (ISBN: 9781522577065) • US $195.00

Green Public Procurement Strategies for Environmental Sustainability
Rajesh Kumar Shakya (The World Bank, USA)
Engineering Science Reference • ©2019 • 228pp • H/C (ISBN: 9781522570837) • US $185.00

Climate Change and Its Impact on Ecosystem Services and Biodiversity in Arid and ...
Ahmed Karmaoui (Southern Center for Culture & Sciences (SCCS), Morocco)
Engineering Science Reference • ©2019 • 408pp • H/C (ISBN: 9781522573876) • US $225.00

Bioenergy and the Advanced Application of Bio-Products and Microfluidic Devices
Mohammad Reza Rahimpour (Shiraz University, Iran) Reza Kamali (Shiraz University, Iran) Mohammad Amin Makarem (Shiraz University, Iran) and Mohammad Karim Dehghan Manshadi (Shiraz University, Iran)
Engineering Science Reference • ©2019 • 325pp • H/C (ISBN: 9781522575344) • US $215.00

For an entire list of titles in this series, please visit:
https://www.igi-global.com/book-series/advances-environmental-engineering-green-technologies/73679

IGI Global
DISSEMINATOR OF KNOWLEDGE

701 East Chocolate Avenue, Hershey, PA 17033, USA
Tel: 717-533-8845 x100 • Fax: 717-533-8661
E-Mail: cust@igi-global.com • www.igi-global.com

Editorial Advisory Board

Table of Contents

Detailed Table of Contents

ASALs in the tropics present challenges to sustainable livelihoods that include drought vagaries, changing land uses, population increase, water insecurity due to rising demands and quality deterioration, as well as environmental degradation leading to declining agricultural productivity and food security. Climate change and transboundary conflicts give rise to governance challenges in water resource management of the ASALs in the tropics, and knowledge of the hydrologic processes would be critical in policy and stakeholder engagement. This chapter discusses the various aspects pertaining to the pertinent hydrological processes crucial for sustainable resource planning and development in watersheds in the tropics and also proposes best management practices including rainwater harvesting that would ensure sustained information and data for improved livelihoods in ASALs.

This chapter examines how the availability and non-availability of river flows has affected benefits realized by communities residing along the Boteti River in Botswana. The Boteti River, which is about 350 km long, derives all its flows from outflows of the Okavango Delta and then discharges into the Makgadikgadi Pans. Peak outflows from the delta occur during the dry season, June to October, and during wet years such as from 1974 to 1982, water flows along the whole river from the delta to the Makgadikgadi Pans. Since 1991, outflows from the delta have only covered about 50 km, with the rest of the river being dry. There has been lack of flows on the downstream section in some years (e.g., 1929-39, 1941-47). Communities residing along the 50 km stretch that is annually flooded benefit from the river through livestock watering, flood recession crop cultivation, fishing, and harvesting of aquatic plants for food and construction. These benefits were not realized by those residing along the rest of the 300 km stretch that was not receiving flows from the delta during the 1991-2008 period.

This chapter presents the results of a study on the influence of streamflow variability on salt fluxes in a semi-arid Tiva River Basin located in Eastern Kenya. Measurements of salinity, total dissolved solids (TDS), conductivity, and river discharges were undertaken in sampling stations located within the main branch of the river as well as in the tributaries, namely Kauwi, Kalundu, Nzeeu, and Mwitasyano Rivers in the Upper Region of the Tiva River Basin. The mean and maximum river discharges for the sub-basins of the river ranged from 11 to 33 m3s-1 and from 118 to 210 m3s-1, respectively. The peak river discharge at the main Tiva branch was 270 m3s-1. The study shows that there is a significant relationship between the variability of streamflow and the variability of salt fluxes in the river as exemplified by variation of salinity, conductivity, and TDS in the river.

Lack of water allocation plans in semi-arid river basins increases the risks of hydrological droughts. The Mukurumudzi River Basin was experiencing extremely low flows which led to a water allocation assessment to quantify the available resource amidst competing demands from domestic, commercial, and environmental needs. An abstraction survey was conducted along the entire length of the river followed by a low flow study and water balance modelling using rainfall-runoff model. Observed flows at source and mouth were 0.028 m3/sand 0.043 m3/s, respectively. Simulated flows estimated an environmental flow (Q95) deficit of0.023 m3/s and 0.010 m3/s in July and August, respectively. Normal flow (Q80) of 0.190 m3/s and flood flow (Q50) of 0.520 m3/s were estimated. Compensating the environmental flow deficit in July and August was recommended. A more detailed hydrogeological study should be conducted to elucidate the potential of the groundwater resource in meeting current demands.

Kiladeda River in Pangani Basin, Tanzania plays a vital role of providing water for agricultural activities of the sub-catchment. However, it is experiencing a problem of inequitable distribution of irrigation water among farmers. Cross-sectional data was collected from farmers both in upstream and downstream using questionnaires, while river discharge measurements were conducted in referenced spot gauging stations. WEAP model was used to analyze water demand and allocation among farmers. Furrow irrigation (94%) and plastic buckets (6%) were the main irrigation water management practices. The model results revealed a water shortage of 46.4% of the total irrigation water required. The annual irrigation water demand and unmet demand were 13.93mm3 and 7.47mm3, respectively, and are both expected to increase twice in 2020. This high water demand for irrigation could be the main cause of excessive water abstraction. A partnering approach is recommended to improve irrigation water management, reviewing of laws, regulations, and water rights.

Juba River runs from its headwaters in Ethiopia through Somalia and discharges into the Indian Ocean near Kismayo. Severe droughts recur in Juba River Catchment. This chapter examines the time series of hydrologic droughts and the factors influencing their occurrence in the catchment. The author further demonstrates the application of probability distributions to compute the exceedance probabilities of the hydrologic drought experienced in the catchment in the past, from 2001 to 2014. The fits of the probability distributions are compared to show the best fitting of all distributions. The probability distribution that will reasonably fit well to the data set is adapted to generate graphical plots for forecasting the return periods of drought events in the catchment. Reconstructing past drought episodes and future drought predictions is useful in drought risk management in Baardheere region of Somalia.

A number of evolutions on data collection and sharing have been published. Countries have collected data, but lack of access and complexity to implement these technologies has limitations. HydroServer Lite, a web-based server for sharing water data, helps to address the need of data sharing and storing in a standard format. Namibia Hydrological Services has no common online system for storing and sharing of water data. This study extends the research on HSL features as data system linked to online ANN forecasting model. This is done by implementing a Namibian HSL using real-time connection to the database to operate in real-time tools developed to visualize and fill in missing data. Lastly, a model was build using Waikato Environment for Knowledge Analysis. Results of the best model obtained are coded in hypertext preprocessor with near real-time data to provide continuous forecast. Linking data system for water resource management in a standard format is practical and promising.

Advances in environmental remote sensing have provided an opportunity to monitor water resource systems in vulnerable regions with data scarcity. The spatial datasets can be used to build spatial models of reality to enable derivation of catchment-based characteristics, also often required by models in hydrology. The derived estimates can then be mapped and cartographically presented to support water resources planning within the concerned developing regions. This contribution presents a database of water resources information for an upstream catchment of the Mara River Basin of Kenya developed from freely available spatial datasets. Additionally, water quality parameters (pH, electrical conductivities, and total dissolved solids) selected as essential indicators of the suitability of the water resources for domestic applications were measured and mapped. The database, packaged as spatial maps, has been presented to the local stakeholders for developing appropriate catchment management strategies within the important watershed.

This chapter assessed the impact of human activities on deforestation and sustainability of water resources and livelihoods in the Congo Basin. It mainly aimed to assess forest degradation in the Yoko Reserve from 1976 to 2015 and investigate the compatibility of Landsat imagery for forest monitoring. Digital image processing for unsupervised classification was done using ENVI software while supervised classification was done by means of ArcGIS 10. Results show that forest landscape faced large-scale human-induced fragmentation over the last 40 years. If these trends continue, they will affect the sustainability of water resources and livelihoods in the Congo Basin of the Democratic Republic of Congo. Hence, policymakers need to look at key drivers and address impacts that may threaten the future of hydrological ecosystems services, including water and land resources in the Congo Basin. Authorities have to apply an integrated management of water, land, and ecosystems.

Climate change is causing great impact on water resources in Eastern Africa, and there is need to establish and implement effective adaptation and mitigation measures. According to IPCC, less rainfall during the months that are already dry could increase drought as well as precipitation, and this has great impact on both permanent and seasonal water resources. Increased sea surface temperature as a result of climate change could lead to increased drought cases in Eastern African and entire equatorial region. Climate change will also result in annual flow reduction in various river resources available within the region such as the Nile River. IPCC predicts that rainfall will decrease in the already arid areas of the Horn of Africa and that drought and desertification will become more widespread, and as a result, there will be an increased scarcity of freshwater even as groundwater aquifers are being mined. Wetland areas are also being used to obtain water for humans and livestock and as additional cultivation and grazing land. This chapter reviews the climate change impacts on water resources within the Eastern Africa Region. The climate change impacts on different water resources such as Ewao Ngiro have been highlighted and projection of future climate change on water resources examined. Stream flow for Ewaso Ngiro was found to have a significant increasing trend in 2030s of RCP4.5 and non-significant decreasing trend in stream flow in 2060s for RCP4.5.

This chapter provides a critical analysis and evaluation of the water issues relevant to Kenya's ASALs. This has particularly been centered on the current resource development and management, future resource demand, as well as extent of its vulnerability to climate variability and change. The water development policy interventions have also been evaluated. The analysis has revealed that water resources are being utilized to satisfy a myriad of demands and that in fact a water shortage already exists in these ASALs. A number of factors including population pressure, poor resource use and management, and other socio-economic activities have been noted to increase the vulnerability of the available water to the impacts of climate change. Recommendations for purposes of ensuring the sustainable utilization of the vital water resource have been proposed.

Management of water resources is at the heart of political discourse to raise awareness among local stakeholders for support in policy formulation and implementation of water sector development plans. The concept of integrated water resources management (IWRM) has been largely disseminated by the Global Water Partnership. Though theoretically appealing and sound, the process of implementation of participatory water resources management still has potential to yield results at local level. One reason is that the top-down approach used is too broad to be implemented and neither does it facilitate better understanding of the needs of each sector involved in the inter-sectoral collaboration to foster planning and benefit sharing of water resources. It is in favor of such practical action for water sector planning and development at small-scale catchment level that the concept of "light" IWRM or integrated watershed management (IWM) was developed to reduce various threats and severe water constraints affecting local stakeholders.

At the basin level, watershed resources (water, land, and ecosystem services) are often managed in "silos," whereby gender-sensitive data and analytics are employed by different stakeholders without regard to existence of similar systems in other parts of the same sub-catchment. These cases create missed opportunities and insights for harnessing better innovative water interventions from baseline scenarios by examining the relevant frameworks, protocols, and tools for data sharing as well as the means of disaggregating water and climate change data into gender-sensitive formats transferable to similar sub-catchments. Innovative water solutions would enable optimal water services provision and financing without farmers depending wholly on external organizations to provide solutions to their financial challenges through taking part in technological development of viable water infrastructure.

Foreword

It is now roughly 30 years ago that Christopher (Misati) Ondieki and myself have crossed the Laikipia district in Kenya to gain first insights into the hydrological conditions of this marvelous landscape in the vicinity of Mount Kenya burdened with serious problems related to water shortage in many surrounding regions. The work was done in the framework of the "Laikipia Research Project" of the University of Berne. The Laikipia project dealt with "Water and Land Management". The authors of the book have a wealth of experience in the arid, semi-arid, and tropical regions where they have sought to improve the management of the scarce water. In particular they have tried to develop best Practices for Managing Water Resources in the drylands and tropical regions through the studies in catchments in Kenya, Tanzania, Somalia, Namibia and Botswana.

The book presents concerns hydrology and water resources in arid and semi-arid lands (ASALs) and tropical regions where conflicts in resource use occurs due to poor knowledge of best practices that could be employed. Water allocation and abstractions have an important impact on the management of river basins undergoing land use/cover changes. The rising populations and climate change pose conflicts of water scarcity which is made worse in drought periods. An improved understanding of the drought events made possible by data sharing, assessment and dissemination to decision makers would be of value to sustainable resource use in the ASALs and tropical regions. This book has made a good attempt towards this purpose.

Food security in ASALs and tropical regions require adequate water availability. Yet the water resources have challenges ranging from data scarcity and climate change to inefficiency in water use and land use/cover changes over time. Management practices that minimize losses of the limited water have to be achieved by stakeholder involvement, especially through equitable sharing of irrigation water with reduced losses in farms.

This book has discussed problems commonly experienced in arid, semi-arid lands and tropical regions with potential solutions highlighted. Sustainable resource management in ASALs and tropical regions can be achieved by considering the approaches provided in the case studies. Concepts on hydrological processes and sustainable management concern many academic disciplines and professionals in industry. There is need for assured decision making for sustainable practices to support livelihoods in ASALs and tropical regions. The contents in this book are therefore useful in research, development and management of river basins in ASAL and tropical regions.

Christian Leibundgut
University of Freiburg, Germany

Christian Leibundgut *is a long serving professor of Hydrology in both Institute of Geography, University of Berne, Switzerland and Institute of Hydrology, University Freiburg, Germany (1988 -2008). He has served as Director of the Centre of Water Research (ZWF) of the University of Freiburg. His research topics include, Tracer hydrology, Solute transport, Impacts of land use change on hydrological systems, Runoff generation, Karst hydrology, Hydrology of arid and semiarid zones, Water protection, Water Resources Management, Hydrological maps/atlases/Information systems, Irrigation and History of Hydrology. In the '80s he has had intensive investigations on the hydrological conditions of the Mount Kenya region on problems related to water shortage in the framework of the "Laikipia-project" of the University of Berne. He has had cooperation in various scientific commissions: He was President of the International Association of Tracer hydrology (ATH, 1976-1981), President / Past President of the International Commission on Tracers of IAHS (International Association of Hydrological Sciences), 1991-1997, Secretary 1999-2001, Vice-President of the International Association of Hydrology (IAHS) 2003-2007. He was a member of the Hydrological Commission of the Swiss Academy of Sciences, 1978-90, member of the Working group 5.1 "Hydrological processes in arid and semi-arid zones" of IHP UNESCO, 1997-2002. He was Co-Editor "Hydrological Sciences Journal", 1993-1999, Co-Editor "Wasser- und Boden", 1998-2005. 2008-20013 he acted as President of the Swiss National Research Program NRP 61 "Sustainable Water Management."*

Foreword

Arid, Semi-Arid Lands (ASALs) and Tropical Regions are frontiers of growing natural resources extraction and socio-economic development. This book has presented the pertinent issues on the hydrology and water resources management in ASALs and Tropical regions; issues that often cause resource use conflicts either because they are poorly understood or due to lack of sharing of good practices. The book is divided into three sections. Key issues in ASALs and Tropical regions touch on water balance, water allocation and abstractions as well as overall watershed management based on land use/cover changes and their impacts. These issues are well captured in Section 1 which deals with hydrological processes and water allocation and abstractions showing that the various perspectives of the hydrology of ASAL and Tropical regions revolve around aridity and water security and challenges of measuring the same. These in addition to the impacts of climate change and transboundary nature of water resources hinder comprehensive planning and integrated water resources management. Our understanding of the relationship between stream flow and river water salinity, conductivity and TDS had been heretofore limited. Modeling approaches provide highlights on the importance of maintaining environmental flow as an important factor in equitable distribution of water resources and appropriate irrigation practices. The section concludes with assessment of the hydrologic droughts in the Juba river basin to demonstrate use of hydrological studies on risk management effects on the provision of water for domestic and irrigation use. The study calls for a better approach to irrigation water management through proper irrigation scheduling and sustainable water abstractions.

Solutions to data scarcity, data sharing, modeling and stakeholder engagement for sustainable management have been discussed in Section 2. Challenges in ASAL and Tropical regions worldwide can be pegged to the dominant hydrological processes whose response to the rising populations and climate change in watersheds threaten food security and create conflicts in society. The water scarcity is a critical issue and manifested in the severity of the drought vagaries. The important concern for the future is the availability of water and its efficient use for food security amidst challenges of climate change and land use change. Rainwater harvesting and equitable sharing of irrigation water with reduced losses in farms through collaboration are highlights for Best Management Practices that would ensure sustainability. Most approaches benefit from data monitoring and dissemination and improved understanding of process dynamics.

Overall the rest of the chapters describe key global problems and solutions to Water Resources Management and Policy leading to sustainable development in arid and semi-arid lands. The editors have put together a valuable volume with useful content. Water resources being central to socio-economic development, this book is a must read for individuals who want to know the critical issues that make for sustainable utilization of ASALs and Tropical regions resources. The need to meet current and future challenges in ASALs and Tropical regions should be the concern of many professionals in geography, environmental engineering and agriculture as well as decision makers, and other stakeholders with interest in sustainable practices especially in developing countries.

In academic institutions where concepts on hydrological processes and sustainable water resources management are sought through curricula of diverse departments, this book is useful for researchers, educators and practitioners in river basins especially in ASAL and Tropical regions and it provides potential for conflict resolutions through an integration of best management practices.

George Okoye Krhoda
University of Nairobi, Kenya

George Okoye Krhoda *is Associate Professor and the Managing Director of Research on Environment and Development Planning (REDPLAN) Consultants Ltd. Until December 2006, he was the Permanent Secretary, Ministry of Environment and Natural Resources and Chairman of the Negotiation Committee on the Nile Basin Cooperative Framework, and earlier Permanent Secretary in the Ministry of Water and Irrigation where most of the water sector reforms were carried under his watch. Prof. Krhoda is a Hydrologist/Water Scientist and worked for the Government of Kenya, has MA and PhD on river hydraulics and water resources planning. His current research concentrates on hydrology and water resources management, policy dialogue and conflict over use and access to natural resources, institutional reforms, floods and environmental management. Krhoda is a member of Kenya National Academy of Sciences and patron of Hydrological Society of Kenya.*

Preface

OVERVIEW

Arid and semi-arid lands (ASAL) are presently undergoing major transformations due to climate change and rapid increase of human population. ASALs are vulnerable environments which often have unsustainable water resources utilization practices. Trend in land use and cover changes lead to water scarcity that impacts the socio-economic livelihoods of people. Research in hydrology is important for determining the available water resources and their utilization. Most of the research is carried out in watersheds where monitoring of hydrological data is collected for instrumented sites. Analyses of the data and use of suitable models could yield information for managing watershed resources for future posterity. Managing the scarce water resources in the ASALs require appropriate practices derived from knowledge of spatial and temporal information on the state of watersheds. The use of GIS and remote sensing techniques is critical in most applications in river basins and these rely on data sharing among users as well as stakeholder involvement.

HYDROLOGY AND WATER RESOURCES TODAY/OBJECTIVES OF THE BOOK

Hydrology plays a central role in water resources. Hydrologic studies carried out in river basins with special attention to selected semi-arid watersheds of Africa are presented with a focus on water resources management in a number of case studies. The pertinent issues include hydrologic modeling, water scarcity, water quality, climate change and stakeholder involvement and hydrologic data management. These are important for integrated water resources management not only in ASAL areas but also in high potential areas now undergoing human induced changes and experiencing the threat of desertification. Good practices in the management of water resources are sought through research and review of documented records of by National agencies and global monitoring systems.

TARGET AUDIENCE

Sustainable water resources planning and management would improve livelihoods in ASALs. The book can be of use among professionals in environmental sciences and engineering. It has value to stakeholders in the business of water and livelihoods in vulnerable ASAL areas. Decision makers in government departments will derive immense benefits based on the need to understand the hydrologic processes affecting water resources availability and how the scarce water resources utilization can be planned appropriately in ASALs. The topical issues in this book are of relevancy in research, academics and postgraduate studies in universities, especially in the developing countries of the tropics. River basin management authorities and water resource management commissions that have to contend with the challenges of degradation of watersheds, water scarcity, conflicts and preservation of ecological integrity of Riverline and coastal systems draining ASAL areas will find this a useful reference text.

BOOK STRUCTURE AND CHAPTER SYNTHESIS

The book structure consists of three sections pertaining to the hydrology and water resources management of ASAL areas. These sections are:

- **Section 1**: Hydrological Processes and Water Allocation and Abstractions
- **Section 2:** Hydrological Data Systems and Land Use Change Dynamics and Impacts
- **Section 3:** Water Resources Management and Policy

Section 1: Hydrological Processes and Water Allocation and Abstractions

In Chapter 1, Christopher Ondieki presents perspectives of the hydrology of ASAL areas based on various studies. The areas have been characterized by aridity indices and challenges of water security which include; effects of climate change, transboundary issues and water use conflicts. The study notes the importance of rainwater harvesting and the use of information and sound databases in sustainable resource planning and development for productive ASALs through continuous research and dissemination of results.

In Chapter 2, Dominic and Motsholapheko examine the effect of availability and non-availability of river flows on communities along the Boteti River in Botswana. Communities residing along the 50 km stretch that is annually flooded benefit from the river through livestock watering, flood recession crop cultivation, fishing, and harvesting of aquatic plants for food and construction which is not the case for those along the rest of the 300 km stretch that is not receiving flows from the Delta due to excessive upstream abstractions, disruption of channel flows within the Delta by operators of tourist facilities. Most of the villages have to develop their own water sources, which are mostly hand-dug wells along the Boteti River bed.

In Chapter 3, Kitheka has presented the results of a study on the influence of streamflow variability on salinity, total dissolved solids (TDS) and conductivity in a semi-arid Tiva River Basin in Kenya. Measurements of salinity, TDS, conductivity and river discharges were undertaken in sampling stations by applying standard hydrologic methods. The study shows that there is a significant relationship between the variability of streamflow and the variability of salinity, conductivity and TDS in the river. The high salt concentrations were a result of high evapotranspiration and seepage of subterranean water from bank storage and groundwater aquifers. Inter-sub-basin variations in the levels of salinity were attributed to differences in land uses, lengths and sizes of the sub-basins. Sustainable irrigation and land use practices in the semi-arid Tiva River Basin requires construction of water reservoirs for control of salinity levels in the river.

In Chapter 4, Jacob M. Katuva, Christian T. Omuto and John P. Obiero have carried out a study titled 'Water Allocation Assessment and Hydrological Simulation on Mukurumudzi River Basin in Kenya.' The chapter provides highlights on the importance of managing water abstraction in rivers in order to maintain environmental flow. The chapter notes that water abstractions should be managed through issuance of permits for normal and flood flows to ensure consistent water balance without deficits. The authors show that the water deficits in ASAL areas could be compensated by constructing dams and that climate change has to be accounted for by use of holistic watershed modeling that includes groundwater dynamics and water quality.

In Chapter 5, Mwadini Khatib, Joy Obando and Shadrack Murimi have investigated the 'Effects of Irrigation Management Practices on Water Allocation Among Farmers in Kiladeda Sub-Catchment, Tanzania.' The chapter shows that inappropriate irrigation practices in upstream areas of ASAL river basins leads to conflicts in downstream. The chapter also shows that irrigation systems such as furrow irrigation and use of plastic buckets are unsustainable in ASALs leading to wastage of scarce water resources. The chapter shows that inappropriate irrigation practices will lead to significant water shortages by 2020. The study calls for a better approach to irrigation water management through proper irrigation scheduling and sustainable water abstractions.

In Chapter 6, Ann N. Njoroge has carried out a study on the hydrologic drought events in Juba River in Somalia. The study notes that hydrologic droughts in the Juba river basin have their underlying causes in the hydrological processes of the basin.

The study demonstrates how hydrologic studies can be used to promote drought risk management in ASALs paying special attention to Baardheere Region of Somalia.

Section 2: Hydrological Data Systems and Land Use Change Dynamics and Impacts

In Chapter 7, Jacobine Taukondjele Amutenya and Gerald (Augusto) Corzo Perez present a study titled 'Infrastructures for Data in the Context of Flow Forecasting Using Artificial Neural Network Model for Okavango River in Namibia.' The study has shown that the HydroServer Lite (HSL), a web based data storage system in Namibia Hydrological Services can promote sustainable water resources management in ASALs. Through linkages with to online ANN forecasting model, the system can run using the near real time data to provide continuous forecasts.

In Chapter 8, Luke O. Olang, Mathew Herrnegger, Doris Wimmer and Josef Fürst have presented a study titled 'A Spatial Database of Hydrological and Water Resources Information for the Nyangores Watershed of Kenya.' The Chapter has shown that spatial datasets including land cover, geomorphology and soil datasets when processed together with the observed hydrological and topological data, is useful in building a database that can support water resources planning in ASALs. The study notes that the preparation of products such as spatial maps can have application among various players involved in river basin management.

In Chapter 9, Jean-Fiston Mikwa Ngamba, Ewango Corneille Ekokinya, Cush Ngonzo Luwesi, Yves-Dady Botula Kahindo, Muhogwa Jean Marie and Hyppolite Nshimba Seya have presented a study on "Multi-Temporal Landsat Remote Sensing for Forest Landscape Fragmentation Analysis in the Yoko Forest, Kisangani, DRC." The study has shown that large scale human induced fragmentation in the Congo River basin over the last 40 years is affecting water resources and livelihoods in the Congo Basin. Emphasis has been placed on the need for Integrated Management of Water, Land and Ecosystems.

In Chapter 10, Opere, Omwoyo, Mueni and Arango have presented impact of climate change on water resources in Eastern Africa, pointing to the need for effective adaptation and mitigation measures. The decreasing rainfall during the dry months could increase drought and thereby impact the surface water resources and the drought cases could even increase as a result of climate change in Eastern African and entire equatorial region. Drought and desertification when widespread would adversely impact wetlands areas as well which are often used for additional cultivation and grazing land. The climate change impacts on water resources for Ewao Ngiro were highlighted with projections of future climate change on water resources examined. Stream flow for Ewaso Ngiro was found to have a significant increasing trend in 2030s of RCP4.5 and non-significant decreasing trend in stream flow in 2060s for RCP4.5.

Section 3: Water Resources Management and Policy

In Chapter 11, Samwel N. Marigi has given a contribution on 'Hydrology and Best Practices for Managing Water Resources in Arid and Semi-Arid Lands.' The chapter shows that water resources utilization especially for horticulture causes stress in the ASALs due to population pressure and poor resource use and management. The author has indicated that stringent management principles of IWRM are required to ensure sustainable water utilization and further notes the need for determination of vulnerability to climate change and policy interventions.

In Chapter 12, Joy Apiyo Obando, Cush Ngonzo Luwesi Nele Förch, Anthony Ogutu Opiyo, Chris Shisanya and Gerd Förch have contributed on the topic 'Kenya's Success Story in Water Resources Management: Participatory Capacity Building in Integrated Watershed Management.' The study points out the importance of political discourse in raising awareness among stakeholders and policy formulation. It emphasizes the need for integrated water resources management (IWRM) and integrated watershed management (IWM) in order to promote coordinated development and management of water, land, and related resources. The findings from this study indicate that IWRM is a key driver for strengthening stakeholder participation and capacity building.

In Chapter 13, Cush NgonzoLuwesi, Amos YesutanbulNkpeebo, Paa Kofi Osei-Owusu and Raphael MuambaTshimanga have presented a study titled 'Advancing New Pathways for Sustainable Deployment of Agricultural Ground Water With Gender-Sensitive Data Sharing.' The authors have shown that at basin level, watershed resources are often managed in "silos" with gender sensitive data and analytics are employed by different stakeholders without regard of existence of similar systems that might already be existing in other parts of the same sub-catchment, thereby creating missed opportunities for insights for harnessing better intelligence through data and analytics compiled from various technological applications in the basin. Innovative water interventions could develop baseline scenarios for examining the relevant frameworks that are usable in similar sub-catchments for optimal water services provision and financing of water infrastructure.

An integrated Holiyas, Grundfos and MPesa (HGM) technology is suggested to be among such solutions filling the gap in agricultural water supply through a groundwater recharge sub-system (Holiyas), groundwater pumping sub-system (GrundfosLifelink) and ICT based water management sub-system for both withdrawals and payments (MPesa mobile money).

IMPACT OF THE BOOK

This book addresses the role of hydrology in sustainable development of low potential economies of the tropics. Livelihoods must be supported in the face of extreme climate change impacts by improving water resources planning and management as well as ecological health of watersheds undergoing land use change. The chapters are focused on key challenges in planning for sustainable water resources management. The problem of poor data availability, common databases and data sharing opportunities globally are explored using advances in spatial technologies such as remote sensing and model applications in support of sustainable water allocations.

Acknowledgment

The editors would like to thank all who have contributed towards the realization of this book. We are sincerely grateful to all the Chapter authors some of whom gave of their time to improve the scientific quality of the Chapters through the blind Review process. We are greatly indebted to the IGI Global family for the guidance and insights they offered the editors throughout the preparation of this Enhanced book. The whole team was indeed helpful and supportive at every stage to completion of the publication which would be a valuable addition to efforts made towards research and development in ASALs and Tropical areas. In wishing God's blessings to all who have contributed to this product, we finally want to recognize that our families supported us throughout the process.

Christopher Misati Ondieki
Kenyatta University, Kenya

Johnson Utu Kitheka
South Eastern Kenya University, Kenya

Introduction

The arid and semi-arid lands (ASALs) and tropical regions are undergoing major environmental transformations. Hydrology research has shown that ASALs are vulnerable environments that undergo degradation characterized by declining water quality and diminishing water resources leading to unsustainable development. In Africa, these areas support a large segment of human, livestock and wildlife populations. The challenges in the ASALs include climatic change, transboundary waters, inadequate diffusion of innovative technologies and inadequate knowledge of hydrological processes in watersheds given the heterogeneity in watershed soils and rainfall characteristics. The future of the vulnerable environments is centered on sustainable growth through water resources monitoring, assessment, management and dissemination of information by sharing of standard data harmoniously. The problem of scale influences representativity of hydrological processes in watersheds. The hydrological processes affect the availability of water for the increasing human, livestock and wildlife populations, rain fed and irrigated agriculture, industrial production, recreation and maintenance of the environment.

Land use and land cover changes have occurred in selected catchments between certain periods with various factors contributing. The catchment responses have triggered negative effects that would require mitigation. Equitable water allocations with minimal losses, contamination and degradation of hydrological systems can only be achieved through best management practices. Insights on these have been provided through the Chapters on studies carried out in the African Region. Activities influencing stream flow will enable planners to formulate policies towards minimizing the undesired effects of the future land use and land cover changes on the hydrology of the rivers especially those undergoing threats of degradation resulting from rapid population increases. Modeling and spatial technologies are the key in taking advantage of the limitations in data availability and data sharing opportunities.

Trends in river flow during droughts and the rapid land use changes threatening the water availability in the future require building scenarios to predict changes that may occur to river flow regimes. Recent initiatives in watershed research in many river basins in the region are impacting the watershed resource management in

African through capacity development. Water demand projections and the realization of Vision 2030 are at stake. Water demand management and rainwater harvesting under IWRM strategies would support sustainable livelihoods. The IWRM strategy sets out a framework within which the management plan is formulated and requires the involvement and support of various stakeholders. An integrated watershed management plan is desirable and should involve all stakeholders in any one watershed.

Runoff harvesting in ASAL watersheds need to take into account the influence of man and climate change on the key hydrological processes. Scenarios in model applications must capture climate change aspects and land use change trends. To cope with the ASAL challenges, there is need for continuity of research programs to improve and enhance understanding of the hydrological process dynamics. Hydrology should influence policy at all scales via better understanding of hydrological systems behavior and adequate collaboration among scientists and policy makers. Although there are various initiatives for developing capacity for sound watershed management policies with policy instruments including; watershed management plan, master water plan, water allocation plans, EIA among others, there should be continuous policy reforms to accommodate the ensuing changes. The best practices for water resources management such as adaptation in water demand and usage, application of technology, data monitoring and policy initiatives for conservation are critical. Collaboration among National meteorological and hydrological services, research institutions and other stakeholders could provide solutions for the sustainable management of ASALs river basins.

Section 1

Hydrological Process and Water Allocation and Abstractions

Chapter 1
Aspects of Hydrology and Water Resource Management of ASALs in the Tropics

Christopher Misati Ondieki
Kenyatta University, Kenya

ABSTRACT

ASALs in the tropics present challenges to sustainable livelihoods that include drought vagaries, changing land uses, population increase, water insecurity due to rising demands and quality deterioration, as well as environmental degradation leading to declining agricultural productivity and food security. Climate change and transboundary conflicts give rise to governance challenges in water resource management of the ASALs in the tropics, and knowledge of the hydrologic processes would be critical in policy and stakeholder engagement. This chapter discusses the various aspects pertaining to the pertinent hydrological processes crucial for sustainable resource planning and development in watersheds in the tropics and also proposes best management practices including rainwater harvesting that would ensure sustained information and data for improved livelihoods in ASALs.

INTRODUCTION

Hydrology plays an important role in water resources planning, development and management. Hydrological processes in tropical Arid and Semi Arid Lands (ASALs) are characterized by low and erratic rainfall, periodic droughts, irregular agricultural productivity and high water scarcity. Water security, environmental sustainability, food security (crop and livestock production, fish production) and;

DOI: 10.4018/978-1-7998-0163-4.ch001

fuel wood energy resources are critical in tropical ASAL environments where land degradation threatens the livelihoods of poor communities whose rising poverty levels create conflicts for marginalized agro-pastoral communities now worsened by land use and climate change dynamics. Climate change manifesting in recurrent droughts in the Kenya's ASAL areas causing wildlife/ human conflicts. Sustainable agro-pastoral livelihoods require an understanding of the hydro-climatic patterns that would enable drought mitigation and coping mechanisms for food security.

Land use closely linked to demographic conditions has resulted in degradation and desertification in the Sahel with effects on landscape productivity and aridity (Amougu et al., 2010, Anyamba and Tucker, 2005, Descroix, et. al, 2011, Mahhanne et al, 2016). Unsustainable land use and climate change will cause increased desertification and landscape degradation. The increasing population and rapid economic development are key drivers of land use change that impact hydrological processes pertaining to water use which in turn determines water quantity and quality (Mundia & Ariya, 2006). High surface runoff, reduced lag time, reduced low flows, increased peak flows and channel erosion are a consequences of urbanization (Rose & Peter, 2001, Okoye and Kwamboka, 2016). Urbanization influences water quality and quantity and may cause transboundary conflicts in water use. Changes induced in the pre urban landscapes often make urban environments vulnerable to pollution and drainage/flood hazards yet for ASALs, the storm water can be managed by a strategy of increasing ground water recharge (Oyeabande, 1990, Pilgrim 1982). Ground water recharge for ASAL runoff is influenced by vegetation, landuse/cover, temperature and soil moisture gradient in a basin (Flint, 2001). The interaction between vegetation, soil moisture and surface runoff and; the subsequent erosion risk can be achieved by mapping the spatial and temporal variability of the processes. The recharge processes affect ground water quality and aquifer vulnerability to pollution and thus important in determining sustainable water abstraction levels and the ground water conservation zones in ASAL aquifer basins.

Hydrological networks equipped with suitable instrumentation with continuous monitoring of the variables would ensure adequacy of data for planning sustainable development and management. Effective rain fed agriculture and irrigation scheduling during droughts would reduce wastage and degradation of water quality. Research in hydrology has shown that ASALs are vulnerable environments and a classical example of hydrological processes study is the Walnut Gulch Experimental watershed in Arizona (Peter et al, 2000). In Africa, ASAL areas support large human, livestock and wildlife populations. Inadequate knowledge and diffusion of innovative technologies hinder sustainable development here due to inadequate understanding of the hydrological processes in the watersheds that are heterogeneous in soils and rainfall characteristics which would affect the availability of water for the increasing populations engaged in rain fed and irrigated agriculture, industrial production, recreation and maintenance of the environmental integrity (Chahine, 2007 ; Oki & Kanae, 2006; Gleick, 1998).

Perspective for sustainable livelihoods in the ASALs of the tropics and the underlying challenges for the vulnerable environments are explored. An understanding of hydrological characteristics of tropical ASAL watersheds is critical for planning and development of any country seeking to attain sustainable resource use and more recently, water stress indicators have been used to address water conflicts using water allocation mechanisms in an attempt to achieve sustainable development in the sub Saharan watersheds (Sungu, 2019).

PERSPECTIVES, KEY CONCEPTS, AND CHALLENGES

Characteristics of ASALs

ASALs may be characterized in various ways some of which depict the internal hydrological behavior of the catchments by way of flow pathways (Beven, 2001). Water flow in the vadose zone partitioning rainfall to runoff and infiltration influences the availability of water for evapo-transpiration in different soil layers and percolation to groundwater. Preferential flow process affects solute transport and contamination as well as soil water availability and catchment scale erosion and discharge generation (Edwards et al., 1993, Beven & Germann, 1982; Doerr et al., 2000, Schaik, 2010).

The areas are characterized by sporadic temporal and spatial rainfall patterns, low humidity, high incident radiation, shallow soils with low moisture storage and infiltration capacities and extreme flow variability, and; varying land use & vegetation cover patterns. Most streams are ephemeral and intermittent with quick rainfall/runoff response and high channel transmission losses. Their watersheds are prone to large sediment yields due to sparse vegetation cover and often rugged topography.

Rainy days are few and may occur once in a year or at irregular intervals and are highly localized. Daily rainfall maxima may exceed annual rainfall totals in some places. There is high inter annual variability and intensities of very short duration storms are common (Xiao, 1990). Interception rates are low and highly variable due to poor and irregular vegetation cover. Losses of evapo-transpiration are high with soil water availability controlled by profile characteristics as well as atmospheric moisture stress (Helling, 1973). Raindrop impact and soil crusting inhibits infiltration. Dry riverbed infiltration has an important effect on flood propagation and recharge to alluvial aquifers in ASAL watersheds and this infiltration is the dominant process of ground water recharge compared to distributed infiltration from rainfall on catchment soils which is limited.

Runoff occurs within an internal drainage basin in ASAL areas where fine and sand deposits can support widespread surface water ponding. Runoff generation process due to convective rainfall is highly localized in space reflecting the spottiness of the spatial rainfall fields in part of the catchment (McMahon, 1979). Overland flow is likely to occur during extreme precipitation events and where terrain is rough and; it is patchy in occurrence often terminating before reaching the channel due to re-infiltration. Ephemeral stream flows are significantly influenced by transmission losses to alluvium and the runoff is highly variable. The low annual precipitation in ASAL regions inevitably means low annual runoff with high intra annual variability. Flood wave transmission losses dissipate the flood and it is common to have no flood observed at a gauging station downstream when further upstream, a flood has been generated and lost to bed infiltration (Lane, 1977). Downstream flow volumes decrease in ephemeral rivers due to transmission losses resulting from infiltration of stream flow into the unconsolidated alluvium forming channel boundaries. Transmission losses are a source of ground water recharge. After a rainfall event, magnitude and rates of transmission losses are highly variable. The dependent factors are; storm sizes, position of storm track, drainage network, channel wetted perimeter, porosity, initial moisture content of channel bed and stratigraphy of the channel fill. In ASAL streams, hydrologic connections occur episodically during flood pulses and apart from providing substantial amounts of mass, momentum, energy and organisms to downstream perennial waters, they contribute to ground water recharge. Hydrologic connectivity in ASAL environments is important because of delivery of water, sediment, nutrients and compounds to downstream areas. The relationship between wadi flow transmission losses and ground water recharge depends on the underlying geology and alluvium wadi bed which minimizes evaporation loss through capillary rise (Parispoulos and Wheater, 1992). The resulting ground water resource may form within a few meters from the surface or recharge substantial alluvial aquifers depending on geology and the localized recharge from rainfall infiltration of catchment soils can vary greatly depending on local circumstances.

Hydrological Processes

The major hydrological processes; precipitation processes, interception, infiltration, recharge, ground water flow process, evapo-transpiration, runoff and; erosion & sedimentation processes are well known and discussed in many Hydrology and Water Resources texts (Miguel V., P., 1989; Ven Te Chow, et al.; 1988; Linsley R. K.et al.; 1975; Shaw, E.M. ; 1986; Oki, T.et al.; 2006; WMO, 2008; 2009). An understanding of the various aspects of hydrologic processes is essential to planning and management of water resources. Spatial-temporal precipitation patterns vary drastically in ASALs with highly localized storms occurring in few days and at irregular intervals. Storm intensities for short duration storms are common, varying from one location to another. The spatial variations are attributed to sources of the precipitation; with convective rainfall highly localized in space reflecting spottinesss of the spatial rainfall trends based on other factors such as the inter tropical convergence zone (ITCZ) and the rain shadow effects of mountains. The rainfall amount is usually low and plays a major role in all the other processes with its partinioning being critical in the different flow paths.

Interception is usually low due to the poor or low vegetation cover in the ASALs and the rates are highly variable. Infiltration also varies in space and time as a function of precipitation and the nature and density of the vegetation cover and soil types/characteristics, seasons, topography/slope and drainage patterns; thus, varies over the catchment space. Whereas, the dry riverbed infiltration recharges ground water aquifers, overland flow from extreme precipitation events may re-infiltrate over the land surface enhancing transmission losses before runoff is generated. The transmission losses are variable being dependent on storm sizes, position of storm track, drainage network, soil porosity and initial moisture conditions and nature of vegetation cover.

Runoff generation is dependent on precipitation and infiltration, topography and landuse/cover conditions. The runoff in ASALs is well reflected in the ephemeral channels resulting from rainfall responses and channel transmission losses with runoff episodes lasting short durations depending on storm characteristics, catchment size, land use cover conditions and season. Flash flood hydrographs produced by convectional rainfall event in a small watershed mostly due to dominance of overland flow in runoff generation are characterized by steep rising and receding limbs and short time base (Xiao, 1990). The duration of recession is generally much longer than the time required for peak flow and the resulting flood wave shape is such that almost the entire hydrograph is the recession curve. Flow ceases after some period of time causing the flow to be of finite duration and the shape of the curve can be compared to an exponential decay reference curves. The runoff will determine the flows in rivers as well as sediment and soil loss processes and water quality. The two forces experienced by channel water are; gravity moving water down slope and friction between water and channel boundaries that resist down slope movement determining the ability of the water to modify channel geometry and transport debris. Channel roughness, slope and depth determine velocity of the flowing water. Dissipation of energy in these channels occurs due to vegetation that normally grows on dry channel beds to exploit moisture in the subsurface sediment. Sediment movement in ephemeral streams occurs as a pulse in response to runoff generated by short duration, high intensity storms typical of ASAL regions. High velocity turbulent flow contains heavy sediment loads and pushes large amounts of coarse sediment through the system; normally sands and gravels. These deep sediments cause large transmission losses in all directions resulting in reduction in both flow volumes and velocity over the lengths of the stream and subsequent deposition of bed load materials and coarser suspended sediments in downstream segments (Fuller, 1974).

Aridity Index

Aridity index, the ratio of mean annual precipitation to potential evapo-transpiration indicates the degree of moisture deficit in an area. Four zones are discernible; hyper arid, arid, semi-arid and sub humid. The dry sub humid aridity index lies between 0.5 and 0.65, semi-arid Aridity index between 0.2 and 0.5 while arid Aridity index is between 0.05 and 0.2 and hyper arid is below 0.05. Since potential evapo-transpiration is higher than precipitation in all these areas, water availability is limited. According to FAO (1981), the Agro-ecological zones distinguished in ASALs are; Semi-arid (AEZ V), sub humid (AEZ IV), Arid (AEZ VI) and very arid (AEZ VII). Areas of humid lands with aridity index greater than 0.65 have potential threats of ASAL conditions given the impact of climate change and human influence.

Soil Degradation

Processes of soil degradation include water erosion by splash, sheet and gulley mass movement, wind erosion, excess salts by salinity through exchangeable sodium on soil colloids, chemical degradation by leaching, physical degradation and biological degradation. Degradation poses many challenges with poor natural resource management enhancing land degradation leading to severe rill, gully and soil erosion which pushes lots of silt into the river systems interfering with the low flow in the rivers. The land degradation characterized by fragile ASAL watersheds in an area of convective rainfall, Katuk-Odeyo in the Lake Victoria Basin, Kenya is illustrated in figure 1.

Such degradation results in clogged water pans that may require costly periodic de-silting so as to cater for the domestic and livestock needs of communities, especially during drought periods. Water related conflicts have to be averted and degraded areas conserved to ensure sustainable livelihoods and economic activities and this is possible using best management practices based on a good understanding of the hydrological processes key in the planning and development of the ASAL areas including averting of water resources use conflicts by involving all the Stakeholders for timely appropriate responses.

Figure 1. Severe gully erosion at Katuk- Odeyo and the mid- Nyando basin *(Sungu, 2019)*

Drought

Drought is a natural hazard that only becomes noticeable through its impacts over a region and it is not easily predictable (Pulwarty and Sivakumar, 2014, Wilhite and Glantz, 1985, Moorhead et al, 2015). By presenting drought as a single numeric value, drought indices greatly facilitate analysis and comparison over time and space. When used in combination with an appropriate threashold, all relevant drought characteristics; onset, drought magnitude, drought intensity and drought cessation can be derived (Okpara and Tarhule, 2015, Yevjevich, 1967, Dracup et al, 1980, Paulo and Pereira, 2006). Since each index calculates drought differently, the indices are often compared using same regional data with criteria such as; statistical properties, interpretability, data requirements and availability (Ntale and Gan, 2003). Historical time series of readily available data can easily be evaluated using qualitative criteria of robustness, tractability, scalability, sophistication, transparency and dimensionality (Quiring, 2009, Keyantash and Dracup, 2002). Robustness refers to the degree of sensitivity of the indices to any perturbation in the process as well as the ability of the indices to measure drought over a wide range of climatic conditions and; sensitive to drought impacts as well as discriminate among drought impacts. Tractability measures whether index can practically be calculated using easily available data. Transparency evaluates the rationale behind the index construction, its clarity and understandability to scientific community, decision makers, affected public and user community. Sophistication evaluates the indices for scientific merit; i.e accurate presentation of physical aspects of drought event. Extendability measures the degree to which an index can be extended across time. Dimensionality refers to the connection of drought index with physical world, e.g mm of soil moisture, percentage of normal precipitation rather than dimensionless unit. Okpara and Tarhule (2015) have evaluated and compared the performance of three drought indices in the Upper Niger watershed namely Standardized Rainfall Anomaly Index, Bhalme and Mooley Drought Index and; the Standardized Precipitation Index (SPI). The SPI ranked high in the basin resonance to its prior recommendation for adoption by the WMO (2012). This index can detect and monitor the early warning aspects of drought, hence for drought mitigation although there is complexity associated with aspects of climate change.

Drought has been shown to be most damaging of the natural hazards experienced in Kenya and elsewhere, especially in Turkana, Baringo and Tana River areas where a number of deaths were reported (Daily Newspapers March 2019) as well as drying of some perennial rivers including the Chania Falls in Thika whose touristic and environmental value have been adversely diminished. These would be averted by focusing on development of boreholes and rainwater harvesting in dams on the basis of drought studies in the basins and regulated river abstraction by pertinent allocation mechanisms. Drought dynamics can be used by local managers in the basin authorities such as WRMA and County governance structures on basis of periodic information. Monitoring, research and dissemination through multidisciplinary engagements via professional associations and stakeholder commitments should be enhanced and indigenous knowledge exploited. Linkages with institutes at global, national and local levels would be beneficial for decision making in complex hydrological systems amidst environmental and climate changes for reduction of uncertainty in water sustainability through transformative solutions within policy frameworks.

Climate Change Issues

Climate change accelerates the hydrologic cycle since a warming climate speeds up the cycle of evaporation, condensation and precipitation. This will make wet places wetter and dry places drier with rainfall patterns changing and the rate at which soil and some plants lose moisture increasing. Droughts will intensify adding to the pressure on water resources. Late or light rainy seasons will alter the speed at which reservoirs and aquifers refill. A warmer atmosphere holds more moisture since the water content of air rises by about 7% for every 1ºC of warming thereby increasing the likelihood of sudden heavy downpours that can cause flash flooding across parched ground. This will also add to sediments in rivers and reservoirs, affecting storage capacity and water quality. Altered weather patterns will mean that crops may wither where they once thrived. By 2050, even if temperature increases can be limited to 2°C, crop yields could slump by a fifth in Africa. Altered rainfall patterns could make conditions too dry and hot to grow some crops and precise forecasting for regions is difficult as past weather records are a less useful guide as the climate changes.

Many ASAL regions are experiencing landscape degradation and desertification due to water scarcity and poor management and climate change is expected to increase the water scarcity in the future. Countries need climate change adaptation strategies and this will influence how water is to be allocated more efficiently for domestic use, growing food, and industry needs. The basis for these strategies will be scenarios of regional climate change predictions (IPCC, 2007) which indicate an increase in temperatures for arid and semi-arid regions of at least 1.5 °C by the end of the 21st century. The inter annual and intra-annual variability is likely to increase, regardless of the total precipitation implying less events with more extreme rainfall intensities which will lead to higher runoff percentages. These changes in temperature and precipitation will probably decrease soil moisture availability and thereby influence the vegetation cover. A vegetation cover on one hand can increase infiltration due to stemflow and the creation of macropores around the root network and on the other hand slows down the runoff, through the higher surface roughness, thereby decreases erosivity of rainfall and decreases peak flow (Schaik, 2010). To assess the impact of climate change on water availability and vegetation, it is essential to fully understand the hydrological processes which influence infiltration and soil moisture distribution. ASALs are vulnerable to climate change with low adaptive capacity. Climate change aggravates water stress and it has been the subject of many river basin studies. River basins currently not experiencing water stress will become risky in time due to climate change impact. Climate change policy has to be determined through ASAL development policies on adaptation. Since climate change is a major development concern for ASALs, climate variability and extremes of weather patterns provide a basis for the design of climate adaptation initiatives. Integrating local knowledge and institutions of pastoral and agro-pastoral communities into planning can help achieve climate adaptation. *Adaptation* depicts 'changes in processes, practices and structures to moderate potential damages or to benefit from opportunities associated with climate change.' The adaptation strategies with scenarios capturing future expected changes require that capacity to cope be developed with collaboration for integrating policy and financial interventions. This should help realize sustainable development by building resilience of vulnerable sectors and ecosystems.

Trans-Boundary (TB) Challenges

Many river basins in the tropics are trans-boundary and transcend ASAL environments and there are examples in Africa, Asia and the USA. In the Middle East, Jordan, Israel and Palestine share a geographical area comprising shared water basins of the Jordan River and the Dead Sea which are subject to water stress given the region's dry climatic conditions. Any unsustainable exploitation of the common water resources can trigger geo-political conflicts leading to misallocation of resources and unregulated pollution. The Jordan River for example is threatened by extensive water diversion, pollution and inappropriate development. All tributaries feeding the Lower Jordan River are dammed leaving less than 2% of the 1.3 billion cubic meters that used to flow naturally (FOEME, 2010). Over 98% of this water is diverted for agriculture and domestic consumption with part of the Jordan River drying up completely for most of the year. In place of diverted fresh water, untreated sewage and saline water, agricultural and fish farm effluents are discharged into the Lower Jordan turning the vital ecological system into a drainage canal and the drying up of the Jordan River causing drastic shrinking of the Dead Sea.

Fixed rate of water transfer from Israel to Jordan fails to account for drought periods and reduced precipitation rates induced by climate change. This calls for better water allocation mechanisms that account for drought periods and reduced water flow due to diversions upstream. Resolving these issues requires transparency and equitable benefit sharing and rehabilitation of the Lower Jordan River, maintenance of recharge areas and control of aquifer extractions. Water allocation and management should make integrated planning and management of water resources possible through a mechanism based on principles of ecological sustainability, economic efficiency and equitable allocation of shared waters along with continuous monitoring of the shared water resources. Since water levels change from year to year, a dynamic allocation criteria including population, existing water use and socio-economic development levels should ensure the long term sustainability of the shared water resources within the recharge rates and natural carrying capacity.

Most trans-boundary river basins have management authorities for example; OMVS (Organisation pour la Mise en Valeur du fleuve SÈnÈgal) with defined water sharing agreement for the Senegal River between Mali, Mauritania and Senegal in West Africa. The regional water management commission for the Zambezi basin caters for eight Southern Africa countries to avoid future conflicts over water resources. Others include the Volta River and the Basin authorities in East Africa exemplified by water resource management authorities (WRMA) and the underlying Water Resource User Associations in Kenya.

The balance between the water needs of people and nature requires a water ethic based on best practices implemented in other parts of the world. This applies to the domestic, industrial and agricultural sectors of all the countries. Removal and treatment of pollutant sewage and solid waste need to ensure return of higher levels of fresh water to nature. Other initiatives include; changing water consumption practices in all the sectors and investments in sanitation solutions.

Policies addressing trans-boundary issues and other environmental & developmental concerns are evident in many basins including Pangani, Mara and Turkana basins in East Africa. Many foreign governments have been involved at district level on issues of integrated development programmes during the last three decades with the aim of achieving food security and improvement of income of the rural populations. These districts were in areas with trans-boundary river basins whose upstream/downstream water use would cause conflicts given the diversity of stakeholders as has been the case with Ewaso Ngiro (Ondieki, 1994). One major basin in Africa, the Nile Basin has most of its upstream catchment areas characterized by high population growth rate (over 3% per annum) with Ethiopia having over 56 million people in the upper catchments. Egypt relies almost entirely on the Nile waters, for its economic development and over time, there have been emerging issues causing conflicts with the upstream Nile riparian countries including Uganda, Kenya, Ethiopia and Sudan. With increasing population in upstream riparian countries, more water will be demanded for domestic, agriculture and power generation. This calls for a management approach that resolves the conflicting interests through collaboration at Basin scale. Bad practices yielding land degradation in the upper catches affecting river discharge and seasonal rhythm and pollutants emanating from agro-chemicals will need to be addressed. Poor practices have also contributed to increased drought occurrences; unfavorable changes in rainfall characteristics in terms of intensity, frequency and duration as well as low soil moisture retention and increased surface flow. Hurni, (1988) has lauded the serious land degradation in many parts of the Ethiopian highlands. This degradation in the Ethiopian Highlands alters the climate thus impacting on the volume and flow patterns of the Nile. The 15% of the average annual water flow of the Nile lost through evaporation at Aswan High Dam can be reduced by encouraging good practices of conservation such as; terracing, inter-cropping, afforestation, rain-water harvesting, agro-forestry practices and small scale dams/reservoirs in upstream riparian states. These conservation practices would enhance seepage and subsurface water flow that would make the Nile flow more regular and reliable. While this is true of many trans-boundary basins of the world, climate change has its toll as discussed above.

Land Use and Water Quality

Water quality problems are rife in hydrological systems world over and are especially critical in ASAL areas of tropical watersheds. Most of the impacts result from urban and agricultural pollution due to lack of adequate urban drainage and poor agricultural practices. About 75% of the African continent is ASAL and is prone to drought vagaries. Water access problem stems from a combination of overuse, poor management and adverse climatic conditions as well as growing water demands and declining ground water tables.

Various approaches to water pollution studies in Kenya's vulnerable hydrological systems reveal water quality to be related to set standards of compliance for various uses. Mining and disposal of mine waste influence surface and ground water quality and levels of organic pollution have been found to reach ten times maximum permissible limits for domestic water supply due to high erosion rates and leaching of associated bedrock for the Rift Valley area (Ondeki, 2015). The spatial-temporal variation of the physic-chemical water quality variables have been distinctly different for dry and wet seasons reflective of impacts of upstream agricultural activities for specified sites. The Nairobi River has been degraded and polluted by sewage and industrial effluents and also dump sites have the potential of polluting ground water thus threatening the future of sustainable water supply potential. This is true of most urbanizing watersheds in the tropics and environmental degradation is bound to undermine environmental integrity. Lake Naivasha in the Rift Valley is prone to pollution from point and non point sources due to growing population and economies, changing lifestyles and climate change that puts pressure on the water resources. The basin has a unique natural and business environment, hosting more than 70% of the country's cut flower business, geothermal power generation, tourism, fishing, pastoralism, small holder agriculture and expanding municipality (Hepworth et al, 2011, Ondieki, 2012).

SUSTAINABILITY AND BEST MANAGEMENT PRACTICES (BMPS) IN ASALS

Sustainability and best management practices (BMPs) in ASALs within the tropics are dependent on resource planning and development. The ASAL tropics are increasingly getting populated often by poor people with challenges that threaten their sustainable livelihoods; namely; droughts and water insecurity triggered by increasing demands and landuse changes inducing reduced agricultural productivity. Solutions must be pegged to applied research on problems of rain-fed and irrigated agriculture in addition to management of watersheds to sustain livelihoods in spite of any vulnerability associated to natural and social-economic factors. Collaborative studies in Kenya for example have focused on communities along rivers in search of solutions of cropping patterns, management practices, pests, irrigation water access and perceptions of weather and climate. The common challenges for these communities relate to; weather and climate, pests and irrigation frequency. The best management practices for incorporation pertain to; crop varieties, tilling practices, water harvesting, water rationing and legalities of water usage as has been shown by scientists seeking relationships between climate, water and farming, yet also there are concerns for tourism and wildlife critical for the region's economy. Most of the efforts made in this region are evidence by collaborations by American and European Universities with Kenyan Universities as may be exemplified by Bern and Nairobi who focused on hydro-ecology and socio-economy and there are lessons learned that could be useful for addressing challenges in other tropical ASAL areas for sustainable livelihoods.

Rainwater Harvesting (RWH)

Water scarcity has greatly affected many regions and climate continues to change, RWH is one approach to combat it (Peters,J.E. 2013). The drought menace occurs in varying levels of severity and frequency with its effect most felt in ASALs. The reduced water and moisture availability in areas of pastoralist and agro pastoral communities may trigger conflicts in resource use. The severity is made worse by changes in climate as rainfall is limited and unreliable. A sustainable development path critical for ASAL areas would depend on rainfall water harvesting technologies ranging from roof top, run-off irrigation to in-situ and sand/subsurface dams. These harnessing of water depends on; rainfall amounts and intensity, evapo-transpiration rates, soil infiltration rates, topography and socio-economic factors. In-situ RWH refers to rainwater harvesting stored within farmland to enhance soil moisture regime for crop production with storage mechanism being the soil itself thus focusing on soil conservation where infiltration is maximized and surface runoff minimized. The design and performance of the conservation structures are influenced by soil type and the climate. Generally, the storage component of a RWH system is based on affordability and there is potential to mitigate drought through storage. A study by Kooke *et al.*,(2012) attempted a design of an appropriate household rainwater harvesting system based on rainfall, catchment area and water requirements to ensure improved water and sanitation access in Magadi division of Kajiado County whose main socio-economic activity is pastoralism with increasing water demands by human and livestock. Interviews were conducted to assess attitude, knowledge and practice with regard to rainwater harvesting. Rainwater samples were collected from reservoirs and tested to determine their physical, chemical and biological integrity.

Simulation of water demand and reservoir size for a given catchment area was done by use of the Massrain and UNESCO-IHE computer rainwater harvesting models. Based on the rainfall, catchment area, and reservoir and water demand data and projections from the models, an appropriate household roof rainwater harvesting system was designed and constructed (Figure 2).

Results from the model analysis indicated that a roof of 100m^2 could supply 100 litres of water per day, 80% of the time if the tank size of 26m^3 (26,000 litres) were connected. Based on the minimum water requirement for a semi arid area such as Magadi, this water would be sufficient for a household of five people. The baseline survey showed that households in Magadi are aware that rain water can be an important and viable alternative source of water, except for the inadequate storage facilities that limit the amount of rainfall water that can be stored. Households that own large earth dams are able to have water available for domestic and livestock throughout the year.

A similar study carried out by Mwendwa (2014) for another semi arid area, Mikuyuni in Makueni County established the water sources there as; sand dams, riverbeds, springs, rainwater, dams and shallow wells. Most people utilised water from the multiple sources simultaneously based on its use for the various household activities as indicated in figure 3 below.

The area has a bimodal rainfall pattern with four months of dry spell each year varying from 480mm to 1180mm/annum with a mean annual amount being 500mm. The rainwater harvesting model showed that with a roof catchment of size of 122m^2 and storage of 30m^3, monthly domestic water demand of 4200 litres can be met for an household of five persons in the semi arid Mikuyuni and thus has potential for roof catchment rainwater harvesting. Water quality should however be considered as well.

Figure 2. Previous rainwater harvesting structure and current rainwater harvesting structure
(Kooke et al, 2012)

Figure 3. Water sources and uses in Mikuyuni
(Mwendwa, 2014)

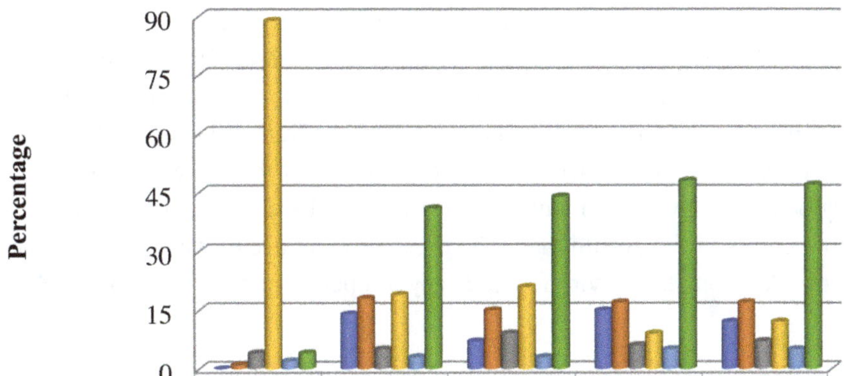

	Drinking	Washing	Cooking	Livestock	Cleaning
Dams	0	14	7	15	12
Springs	1	18	15	17	17
Shallow wells	4	5	9	6	7
Rainwater	89	19	21	9	12
Sanddams	2	3	3	5	5
Riverbeds	4	41	44	48	47

Sources Verses Use

Sand dams used to supplement local water supplies can recharge even with the unreliable rains. These are common in most ASAL countries including Kenya, Ethiopia, Tanzania, Namibia, Australia, U.S.A, Thailand, and New Mexico. The location of the suitable sites involves analyzing the underlying rock structures and soil properties with the ideal having impermeable or less permeable surface beneath the channel to enable water storage. Hydrological and geologic data are useful in computing potential storage capacity of a sand dam which can yield up to 6000m^3 of water after maturity in 2 to 7 years. Sand dams are environmentally friendly as they control erosion and manage silt deposition within river basins yielding clean water of good quality due to filtering effect of the sand. They also improve the ecosystems threatened by recurring droughts as a result of regeneration of indigenous vegetation and attraction of other biological resources. Hydrological data for most seasonal rivers are however unavailable or inadequate. Earth dams in ASALs cover large basins that may be trans-boundary with issues that must be resolved for sustainable development to be assured.

Sustainable Development and Management of ASALs

ASAL populations depend on certain activities for their livelihoods and in most cases, these cause conflicts that undermine the sustainability of the vulnerable environments. In Kenya, attempts have been made for climate resilient dry land development for the pastoral communities under the National Drought Management Authority seeking to enhance food security. These drylands can be food secure if policymakers and development agencies involve local stakeholders. Small scale farming communities would make the most efficient use of water by low-cost drip irrigation kits that are gravity fed and easy to operate and manage (figure 4).

Figure 4. Installation of Drip Irrigation System and application to vegetable farming at Nturukuma Village, Ewaso Nyiro Basin, Mount Kenya area
(Laikipia County Development Plan, Discussion Paper)

The average use in most poor communities is 29 liters per day although 100 liters per person per day is considered a minimum threshold (Falkenmark and Widstrand, 1992) for personal use. Additional uses for agricultural and industrial needs depicts countries with less than 1,700 cubic meters/ person/ year to be water stressed and clearly, water management presents conflicts of competing demands especially in ASALs whose hydrological information that could be used for planning are scarce. Hydrological measurements on ephemeral and seasonal streams are rare and have not been monitored effectively and the true water resource in these regions largely unknown. Data collection and research on hydrological processes need to be expanded to cover more rivers and aquifers in ASAL areas. Deforestation and overgrazing of marginal dry lands affect the hydrological cycle by reducing the infiltration and increasing the severity of droughts. The problem of sand extraction from seasonal dry riverbeds reduces the sub surface water storage capacity of the rivers. The vulnerability of environmental conditions can be reduced and stakeholders can be engaged in risk spreading strategies which would include water harvesting to improve crop and livestock production. The indigenous ways in which populations in the ASALs have lived with climate variability and adapted to extremes of weather patterns provide basis for the design of climate adaptation initiatives so that the full potential of these lands are exploited.

An initiative for addressing development challenges facing the Kenyan ASALs (GOK, 2012) has sought to enable self-sufficiency in these areas to cater for loss of livestock due droughts, water scarcity and land degradation. Security issues that create conflicts over resources should be a thing of the past if investment in both crop agriculture and livestock production through infrastructure development and water rationing as best practices are implemented. Trends in river flow during droughts and the rapid land use changes threatening the water availability in the future require building scenarios to predict changes that may occur to river flow regimes. Recent initiatives in watershed research in many river basins in the region are impacting the watershed resource management in Africa through capacity development (www.iwmnet.eu). The EU eco-village approach strives to reduce poverty and vulnerability to climate change by integrating interventions for adaptation strategies. The communities are transformed to adapt to climate change in relation to water resources management, farmer managed natural re-vegetation as best approach to re-green semi arid areas, and improving the health of rangelands to increase the resilience of pastoralists in Northern Tanzania. The EU global climate change alliance initiatives help the world's most vulnerable countries to adapt to climate change using indicators covering social, economic and environmental aspects critical for 'climate resilience' development, thus providing useful information on a country's vulnerability to climate change to inform the planning process (http://www.gcca.eu/). Water demand projections for the realization of Vision 2030 and water demand management require integrated water resource management (IWRM) strategies through involvement and support of various stakeholders with runoff harvesting taking into account the influence of man and climate change on the key hydrological processes (Ondieki C., 2013).

CONCLUSION AND FURTHER RESEARCH DIRECTION

The challenges that hinder sustainable livelihoods in the ASAL areas of the tropics include; droughts, land use changes and associated environmental degradation, climate change impacts and insufficient stakeholder engagements. Best practices at diverse scales require coordination of ASAL water resources research and adequate funding to overcome implementation challenges.

Water Scarcity and Quality

Climate change adaptation and mitigation strategies are desirable in the tropical ASAL watersheds. There should be a possibility of insurances and other innovations for the vulnerable ASAL communities. Land use change management has to be affected to reduce pollutants through buffer zones or river riparian areas to cater for upstream degradation. Investments in water storage would increase water access in pursuit of Vision 2030 objectives. Performance of institutions such as WRMA, NEMA through strategic development plans benefiting from climate change initiatives and input from learned societies/associations and proper agricultural practices on watersheds would help promote water quality standards. Climate modification in urban areas by way of open spaces would improve hydrological processes of infiltration and ensure ground water recharge.

River Basin Management

Water resources management challenges would be addressed using solutions that cater for increasing water demand for agriculture, industry and urban growth. There should be a focus on rainwater harvesting and drip irrigation and involvement of all stakeholders in the resource governance of the watersheds. Crop insurance schemes and improved agricultural practices are essential. Risk management systems that predict likely shortages and other policy initiatives could best be practiced in effective IWRM governance systems at various levels. The challenges pertaining to trans-boundary issues should be addressed through viable and harmonious geopolitical engagements so as to reduce the risks of vulnerability through appropriate and suitable innovative techniques of best management practices.

Research Needs

Research in all hydrological systems has to be integrated into overall planning and development for timely action using BMPs and suitable model approaches. A deeper understanding of the characteristics of tropical ASALs hydrological processes should be carried out using technological advances in monitoring and modeling that captures climate change and associated impacts. The capacity of tropical ASAL research and development should be enhanced through collaborative efforts of relevant institutions focusing on key areas. The institutional programs should be enriched to address knowledge gaps and follow-ups made to ensure solutions are availed to key problems where necessary. There should be fruitful long-term collaboration of the East and South African Universities already focusing on integrated water resources management and these should incorporate thematic areas of agencies such as IAHS, FAO, WMO, Global Institute of Water security and; the National Science Foundation. The various themes should provide impetus for focused outputs of products that address the needs of the tropical ASALs and thereby benefit the policy mandates of tropical countries. This is especially true in watershed conservation measures that would require greater sound hydrological research given the increasing drought severity, upstream downstream conflicts, ground water depletion and; water quality problems.

Assessment and mitigation of water pollution should be investigated using tracers in both subsurface and surface runoff patterns in ASAL catchments via pollutant signatures. Water quality measurements, modeling and process understanding should inform water resource management decisions. The data scarcity challenge need to be addressed to cater for risk associated to scales in time and space in drought risk management and water use planning of best practices. Hydrological approaches should blend with understanding of global development processes, communication and practice incorporating change and innovation for meeting the challenge of data measurement procedures (best practices) that are consistent as regards to standards in tandem with the 2018-2028 Decade for action on water.

ACKNOWLEDGMENT

I thank the reviewers of this chapter for their valuable comments.

REFERENCES

Amougu, O., Descroix, L., Yero, K. S., Le Brreton, E., Mamadou, I., Ali, A., ... Belleudy, P. (2010). Increasing Riverflow in the Sahel. *Water (Basel)*, *2*(2), 170–199. doi:10.3390/w2020170

Anyamba, A., & Tucker, C. J. (2005). Analysis of Sahelian Vegetation Dynamics using NOAA-AVHRR NDVI data from 1981 -2003. *Journal of Arid Environments*, *63*(3), 596–614. doi:10.1016/j.jaridenv.2005.03.007

Beven, K. (2001). *Rainfall-runoff modelling: The Primer*. John Wiley & Sons Ltd.

Beven, K., & Germann, P. (1982). Macropores and water flow in soils. *Water Resources Research*, *18*(5), 1311–1325. doi:10.1029/WR018i005p01311

Chahine, M. T. (2007). The hydrological cycle and its influence on climate. *Nature*, *359*(6394), 373–380. doi:10.1038/359373a0

Chow, Maidement, & Mays. (1988). *Applied Hydrology*. McGraw Hill.

Descroix, L., Esteves, M., Soulex, Y. K., Rajot, J. L., Malain, M. A., Boubkrois, S., . . . Mamadou, I. (2011). Runoff Evolution accoeding to Land use Change in a Small Sahelian Catchment. *Hydrology and Earth System Science Discussion Paper*, *8*, 1569-1607.

Doerr, S. H., Shakesby, R. A., & Walsh, R. P. D. (2000). Soil water repellency: Its causes, characteristics and hydro-geomorphological significance. *Earth-Science Reviews*, *51*(1-4), 33–65. doi:10.1016/S0012-8252(00)00011-8

Dracup, J. A., Lee, K. S., & Paulson, E. G. Jr. (1980). On the definition of Droughts. *Water Resources Research*, *16*(2), 297–302. doi:10.1029/WR016i002p00297

Edwards, W. M., Shipitalo, M. J., & Owens, L. B. (1993). Gas, water and solute transport in soils containing macropores: A review of methodology. *Geoderma*, *57*(1-2), 31–44. doi:10.1016/0016-7061(93)90146-C

Falkenmark, M., & Widstrand, C. (1992). Population and Water Resources: A delicate balance. Population Bulletin. Population Reference Bureau.

FAO. (1981). *Arid zone Hydrology for Agricultural Development*. Irrigational Drainage Paper.

Flint. (2001). Fundamental concepts of recharge in the desert South West: A regional modeling perspective. In *Ground water hydrology recharge in desert environment; South Western United States*. Academic Press.

Fuller, W. H. (1974). *Desert Soils in Desert Biology*. Academic Press.

Gleick, P. H. (1998). Water in crisis: Paths to sustainable water use. *Ecological Applications*, *8*(3), 571–579. doi:10.1890/1051-0761(1998)008[0571:WICPTS]2.0.CO;2

GOK. (2012). *National Policy for Sustainable development of Northern Kenya and other arid lands*. Nairobi, Kenya: GOK.

GOK. (2013). *The National Water Master Plan 2030, Ministry of Environment, Water and Natural Resources, Water Resources management Authority, JICA*. GOK.

Hellwig, D. H. R. (1973). Evaporation of water from sand.1: Experimental set-up and climatic influences. *Journal of Hydrology (Amsterdam)*, *18*(2), 93–108. doi:10.1016/0022-1694(73)90096-6

Hepworth, N., Agol, D., Von-Lehr, S., & O'Grady, K. (2011). *Exploring the value of water stewardship in Africa; Kenya case study*. Alliance for Water stewardship/Marks Spencer/GIZ.

Hurni, H. (1988). *Ecological issues in the creation of famines in Ethiopia*. Paper presented at the National Conference on Disaster Prevention and Preparedness Strategy for Ethiopia.

IPCC. (2007). *The physical Science Basis-summary for policy makers*. Geneva Switzerland: IPCC.

Keyantash, J. A., & Dracup, J. A. (2002). The quantification of drought: An evaluation of drought indices. *Bulletin of the American Meteorological Society*, *83*(8), 1167–1180. doi:10.1175/1520-0477-83.8.1167

Kooke, S. O., Sumba, L. A., & Kurgat, P. (2012). Rain water harvesting for drought mitigation and food security. In Water Dependencies: Systems under Stress and Societal Responses. Hydrological Society of Kenya proceedings. Kenyatta University.

Lane, J. (1977). *Transmission losses. National Engineering Handbook, Hydrology*. USDA, SCS.

Linsley, R. K., Kohler, M. A., & Franzini, J. B. (1975). *Hydrology for Engineers* (2nd ed.). New York: McGraw Hill Book Co.

Mahamane, M., Hochschild, V., Schutz, A., & Nkokuma, J. (2016). *Fragmentation rate and landscape structure of the Tillabery landscape (Sahel) Region with reference to desertification*. Academic Press.

McMahon, T. A. (1979). *Hydrological characteristics of arid zones. Proceedings the symbosium on the hydrology of areas of low precipitation*. IAHS Publication.

Miguel, V. P. (1989). *Engineering Hydrology Principles and Practices*. Prentice Hall.

Moorhead, J. E., Gowda, P. H., Singh, V. P., Porter, D. O., Marek, T. H., Howell, T. A., & Stewart, B. A. (2015). Identifying and Evaluating a suitable index for agricultural drought monitoring in the Texas high plains. *Journal of the American Water Resources Association*, 1–14.

Mundia, C. N., & Ariya, M. (2006). Dynamics of Land use/cover changes and degradation of Nairobi city, Kenya. Land degradation development. *Wiley Online Library*, *17*(1), 97–108.

Mwendwa, G. (2014). *Rainwater harvesting in Mikuyuni* (Unpublished MSc thesis). Kenyatta University.

Ntale, H. K., & Gan, T. Y. (2003). Drought indices and their application to East Africa. *International Journal of Climatology*, *23*(11), 1335–1357. doi:10.1002/joc.931

Oki, T., Caterina, V., & Kate, H. (2006). *Hydrology 2020: An integrating Science to meet world water challenges*. IAHS Publication 300.

Oki, T., & Kanae, S. (2006). Global hydrological cycles and world water resources. *Science, 313*(5790), 1068–1072.

Okoye & Kwamboka. (2016). Impact of urbanization on the morphology of Motoine/ Ngong River channel, Nairobi River Basin. *JGRP, 9*(4), 36-46.

Okpara, J. N., & Tarhule, A. (2015). Evaluation of Drought Indices in the Niger Basin, West Africa. *Journal of Geography and Earth Sciences*, *3*(2), 1–32.

Ondieki, C. M. (1994). *The Hydrology of Arid and Semi arid lands; Case of Laikipia District* (Unpublished PhD thesis). Nairobi University.

Ondieki, C. M. (2012). Water Dependencies: Systems under Stress and Societal Responses. In *Hydrological Society of Kenya (HSK) Conference Proceedings*. Kenyatta University Press.

Ondieki, C. M. (2013). Hydrology and Integrated Water Resource Management for Sustainable Watershed Management in Kenya. In H. Muga & K. Thomas (Eds.), *Cases on the Diffusion and Adoption of Sustainable Development Practices* (pp. 352–375). Hershey, PA: Information Science Reference. doi:10.4018/978-1-4666-2842-7.ch013

Ondieki, C. M. (2015). Aspects of Water Quality and Water Pollution Control in Vulnerable Environments. *Journal of Geography and Earth Sciences*, *3*(2), 83–93. doi:10.15640/jges.v3n2a5

Oyeabande, L. (1990). Aspects of Urban Hydrology and the Challenges for African Urban Environments. Urban Climatology in Africa. *African Urban Quarterly*, 39-68.

Parissopouluos, G. A., & Wheater, H. S. (1992). Effects of hysteresis on ground water recharge from ephemeral flows. *Water Resources Research*, *28*(11), 3055–3061. doi:10.1029/92WR01902

Paulo, A. A., & Pereira, L. S. (2006). Drought concepts and characterization, comparing drought indices applied at local and regional scales. *Water International*, *31*(1), 37–49. doi:10.1080/02508060608691913

Peter, F., Malchus, B., Carlton, B., Madelyn, C., & Kern, L. (2000). Land stewardship in the 21st. Century: The contributions of Watershed Management. *USDA, Forest Service Proceedings*.

Peters, J. E. (2013). Promoting Rainwater harvesting (RWH) in small Island Developing States (SIDS): A case in the Granadines. In Cases on the Diffusion and Adoption of Sustainable Development Practices (pp. 403-438). Academic Press.

Pilgrim, D. (1982). Assessment of derived rural and urban runoff coefficients. Institute of Engineers Australia. *Civil Engineering Transactions*, *24*, 235–241.

Pulwarthy, R. S., & Sivakumar, M. V. K. (2014). Information Systems in a Changing Climate: Early warning and drought risk management. *Weather and Climate Extremes*, *3*, 14–21. doi:10.1016/j.wace.2014.03.005

Quiring, S. M. (2009). Developing objective operational definitions for monitoring drought. *Journal of Applied Meteorology and Climatology*, *48*(6), 1217–1229. doi:10.1175/2009JAMC2088.1

Rose, S., & Peters, N. E. (2001). Effects of urbanization on streamflow in Atlanta area. *USA Hydrological Proceesses*, *15*(8), 1441–1457. doi:10.1002/hyp.218

Shaw, E. M. (1986). *Hydrology in Practice*. Chapman Hall.

Sungu, R. O. (2019). *An assessment of the influence of water allocation on sustainable water resources Management: A case study of the Nyando River Basin* (Unpublished PhD thesis). University of West Cape, South Africa.

Van Schaik, L. (2010). *The role of Macro pore flow from PLOT to catchment scale: A study in a semi arid area* (PhD thesis). Utrecht Netherlands Geographical Studies 390.

Wilhite, D. A., & Glantz, M. H. (1985). Understanding the Drought Phenomenon: The role of definitions. *Water International*, *10*(3), 111–120. doi:10.1080/02508068508686328

Willems, P., Olsson, J., Arnbjerg-Nielsen, K., Beecham, S., Pathirana, A., Gregersen, I. B., ... Nguyen, V.-T.-V. (2012). *Impacts of Climate Change on Rainfall Extremes and Urban Drainage Systems*. London: IWA Publishing. doi:10.2166/9781780401263

WMO. (2008). Guide to Hydrological Practice Vol. I: Hydrology – From measurement to Hydrological information (6th ed.). WMO No. 168.

WMO. (2009). Guide to Hydrological Practice Vol. II: Management of water resources and application of hydrological practices (6th ed.). WMO No. 168.

WMO. (2012). Standardized Precipitation Index. WMO 1090.

Xiao, L. (1999). *Flash floods in arid and semi arid zones*. IHP Technical documents in Hydrology. No 23. UNESCO.

Yevjevich, V. (1967). *An objective approach to defitions and investigations of continental hydrological drought. Hydrology Paper No. 23*. Fort Collins, CO: Colorado State University.

Yuan, X., Zhang, M., & Wang, L. (2016). *Understanding and seasonal forecasting of hydrological drought in the anthropocene*. Hydrology and Earth Science System Sciences; doi:10.5194/hess-2016-592

Chapter 2
Effects of Long-Term Variability on Water Resources Management Along the Boteti River in Botswana

Dominic Mazvimavi
University of the Western Cape, South Africa

Moseki R. Motsholapheko
University of Botswana, Botswana

ABSTRACT

This chapter examines how the availability and non-availability of river flows has affected benefits realized by communities residing along the Boteti River in Botswana. The Boteti River, which is about 350 km long, derives all its flows from outflows of the Okavango Delta and then discharges into the Makgadikgadi Pans. Peak outflows from the delta occur during the dry season, June to October, and during wet years such as from 1974 to 1982, water flows along the whole river from the delta to the Makgadikgadi Pans. Since 1991, outflows from the delta have only covered about 50 km, with the rest of the river being dry. There has been lack of flows on the downstream section in some years (e.g., 1929-39, 1941-47). Communities residing along the 50 km stretch that is annually flooded benefit from the river through livestock watering, flood recession crop cultivation, fishing, and harvesting of aquatic plants for food and construction. These benefits were not realized by those residing along the rest of the 300 km stretch that was not receiving flows from the delta during the 1991-2008 period.

DOI: 10.4018/978-1-7998-0163-4.ch002

The space-time variability of river flows along the Boteti River has influenced attitudes of the riparian communities towards water resources management. Communities residing along the dry 300 km stretch of the river strongly feel that initiatives aimed at ensuring that the river flows again be implemented as they encounter difficulties in obtaining water for livestock watering, and for flood recession crop production. Flood recession crop production has been shown to enhance food security of households. However, lack of flows is due to generally below average rainfall received over the years over the Okavango River basin. Thus, measures for improving water resources management along this river cannot focus on making flows available to the downstream section. The sectoral approach adopted in the implementation of government projects is also a constraint towards implementation of an integrated river basin management approach. An integrated management of resources which involves active participation of the communities has the potential to reduce current degradation of river-relate resources. Recent efforts by the Permanent Okavango River Water Commission to pilot sustainable agricultural practices has the potential to control land degradation while improving livelihoods.

1. INTRODUCTION

The high variability of river flows presents a challenge to all users of water in as far as securing a reliable supply is concerned. This is particularly true in southern Africa where most rivers run dry during the dry season (June – October), and ephemeral rivers located in the arid parts flow a few times over several years. For small-scale rural water users with not many resources at their disposal, the challenge is considerable in their quest to secure a reliable water supply for domestic use, crop production, and livestock watering. In such circumstances, conflicts over real or perceived inequitable access and use of the water frequently develop especially during the dry season or drought years between and among the small-scale users themselves and other users. When there are several water users attempting to access the same source of water for different uses, this worsens the problem (Mazvimavi and Mmopelwa, 2006). The proverbial upstream-downstream syndrome also adds to the weight of the problem. Upstream water users are often accused of abstracting an unfair proportion of the water, which results in downstream shortages. This can be aggravated in countries where water laws accord prior right of use to certain categories of water users. A river basin approach to managing water resources provides a means of resolving such problems.

There are few cases in southern Africa where an attempt was made to explore the feasibility of implementing a river basin management approach on ephemeral river basins. Most of the successes of implementing a river basin management approach are from relatively well watered regions. An attempt was made on the Kuiseb River in Namibia which has large scale commercial farmers in the upper reaches as well as communal farmers along some sections of the river, and the town of Walvis Bay is located at the mouth of this river (Botes, et al., 2003). The various groups of stakeholders were reported to have welcomed the proposal to adopt a river basin management approach. They proceeded to form a forum with the aim of eventually transforming it into a basin management committee. Even then, individual interests were evident. For example, commercial farmers and the town of Walvis Bay, with major commercial interests in improving management of the river, could justify the costs associated with the process, i.e. funding the participation of their representatives. The same however could not be said about communal farmers since the benefits that would accrue to were of a long-term nature. Besides it is doubtful whether they could fund the participation of their representatives on the forum without external support.

Manning and Seely (2005) are of the opinion that the implementation of a river basin management is possible on the Kuiseb River if the forum for integrated resources management (FIRM) approach is adopted. FIRM aims to achieve an integrated management of natural resources through the participation of resources users and other stakeholders in formulation and implementation of appropriate management systems. All the same examples from other southern African countries suggest that small-scale users face almost insurmountable problems when it comes to assuming their rightful place in such fora. For example, lack of clear and demonstrable benefits, against a backdrop of widespread poverty hindered effective participation of individual and communal small rural water users in Zimbabwe (Manzungu, 2004; Manzungu and Kujinga, 2002). This observation raises an interesting question about the feasibility of implementing a successful river basin management approach on an ephemeral river basin largely utilized by rural communal farmers. This chapter examines how variability in both space and time wields influence on use and perceptions of water resources along the intermittent-perennial Boteti River in Botswana, which is perennial on the upper part and ephemeral on the downstream part.

2. METHODS

In this study a multi-method approach was used. Secondary data were obtained from desktop searches and review of published and unpublished sources including journal articles, as well as policy documents. Some primary data were obtained from the Okavango Research Institute database, field surveillance and ground-truthing as well as interviews with key informants from the Department of Water Affairs.

3. STUDY AREA

The Boteti River is part of the endorheic Okavango Basin that occurs in Angola, Botswana and Namibia (Figure 1). The Okavango River is formed by the joining of the Cubango and Cuito Rivers along the Angolan and Namibian border, whose sources are on the Angolan Highlands close to Huambo. This river crosses the Caprivi Region in Namibia and enters the Okavango Delta in Botswana at Mohembo. The Delta is a conically shaped alluvial fan, which can be classified into the following physiographic zones; the panhandle, permanent swamps, and seasonal swamps (Mendelson and Obeid, 2004). The panhandle, which is located at the top of the Delta, is an elongated valley with a meandering channel surrounded by marshes with papyrus and reeds. Permanent swamps cover an area of about 2,000 to 3,000 km^2 and have dense papyrus, reeds and sedges. Three main channels, namely the Nqoqa, Jao, and Thaoge, distribute water within these swamps. The area covered by seasonal swamps varies from about 4,000 to 8,000 km^2 depending on annual inflows into the Delta. The Thamalakane River is the major outlet of water from the Delta. This river branches close to Maun to form two distributaries. One of the distributaries, the Nhabe River, drains into Lake Ngami, while the other is the Boteti River that flows into the Makgadikgadi Pans (Figure 2). From the 1980's to 2008, there were no flows from the Thamalakane River into the Nhabe River. All the Delta outflows collected by the Thamalakane River were flowing into the Boteti River.

The Boteti River is about 340 km long and flows in an easterly direction towards the Makgadikgadi Pans (Figure 2). A branch of this river discharges into Lake Xau, with the other continuing to join Ntwetwe Pan, which is part of the Makgadikgadi Pans. The current river, which is about 20 to 50 m wide, occupies a wide valley that is incised at some locations.

One of the first steps when planning the implementation of a river basin management approach involves the delimitation of the basin. This is problematic in the case of the Boteti River because if the topographic divide is used for delimitation, the basin will include areas not actively contributing to river flows. The basin delimitation based on the topographic divide stretches in the southerly direction for about 400 km close to Gaborone, and 140 km in the northern direction. Most of these areas do not have any relationship with regards to utilization of water along the Boteti River.

The Okavango Basin is mostly covered by Kalahari beds comprising aeolian sands and alluvial deposits whose thickness in some areas varies from 10 to 100 m. Calcrete and silcrete occur along the course of the Boteti River (Anrtzen,*et al.*, 1994). The basin is generally flat with elevations gradually decreasing towards the Makgadikgadi Pans, which occupy the position of Lake Palaeo-Makgadikgadi.

Figure 1. Location of the Boteti River within the Okavango Basin

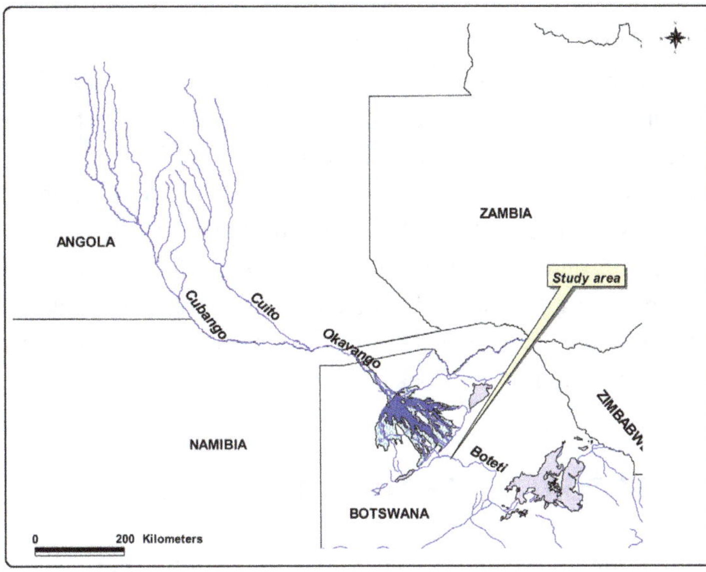

Figure 2. Location of settlements along the Thamalakane and Boteti Rivers

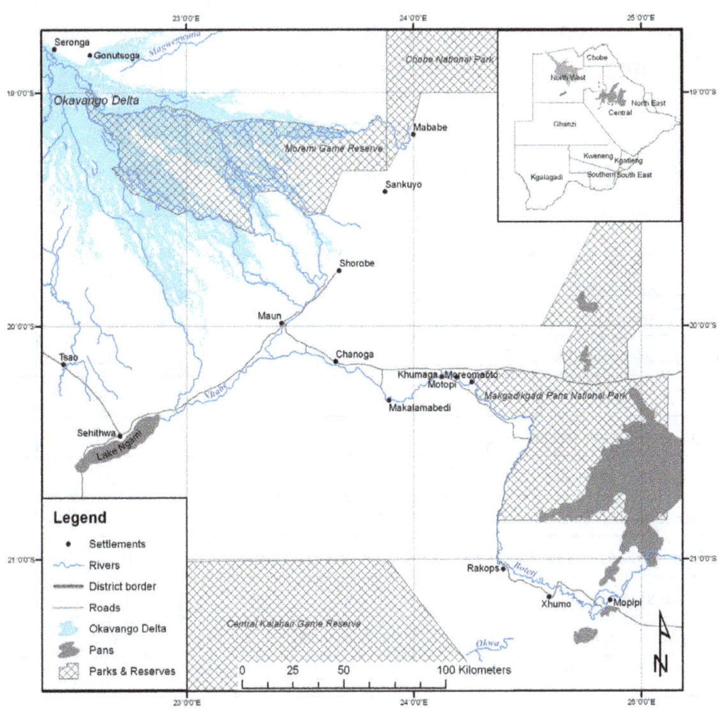

The upper part of the Boteti River falls within North West District of Botswana, up to about where it changes direction from being easterly to southerly. Thereafter it is located within the Central District of Botswana. Almost all the major settlements are located along the Thamalakane and Boteti Rivers due to the scarcity of water in this semi-arid area. The largest settlement, Maun, is located along the Thamalakane River. It is an urban village with a population of about 65693 people (Table 2). Some of the settlements located along the Boteti River are Chanoga, Makalamabedi, Motopi, Khumaga, and Rakops (Figure 2). Table 1 shows the 2001 and 2011 population figures for these settlements and some of the small settlements around them (CSO, 2003). The Boteti River forms part of the boundary between the Makgadikgadi National Park and the communal lands a situation that has occasionally led to human-wildlife conflicts due to predation of livestock.

4. WATER RESOURCES OF THE BOTETI RIVER

Rainfall varies in the Okavango Basin from 1, 300 mm/year at the source of the Kubango River in Angola to about 400 mm/year the Makgadikgadi Pans. The Delta receives an average of about 550 mm/year of rainfall on the northern part, while Maun at the extreme southern part receives 450 mm/year. Annual rainfall within the Boteti River region is highly variable over time as is illustrated in Figure 3 for Maun. The area traversed by the Boteti River is classified as being semi-arid.

Table 1. Population of major settlements along the Boteti River

Settlements	Population	
	2001	2011
Maun and associated localities	49,822	65,693
Makalamabedi and associated localities*	3,420	4065
Motopi	1,130	1436
Moreomaoto and associated localities	671	665
Kumaga and associated localities	1,252	923
Xhumo and associated localities	2,462	2304
Tsienyane/Rakops and associated localities	6,124	7240
Toromoja and associated localities	1,358	1157
Mopipi and associated localities	5,625	5247
Total	71,864	88730

*Population total of Makalamabedi in Central and North West districts
Source: CSO (2003) and Statistics Botswana (2011)

The pan evaporation rate is high and is about 2,400 mm/year at Maun. With such low rainfall, the prospects for dryland crop production are very limited. Extensive livestock production is the most feasible agricultural activity.

The Boteti River depends almost entirely for its flows on outflows from the Okavango Delta. Water in the Delta is derived from the Okavango River inflows (58%) and rainfall over the Delta (42%) (Table 2). The limited rainfall that falls along the Boteti River does not contribute any meaningful flows because of the low gradient and sandy soils that are characterized by very high infiltration rates.

At its peak the Okavango River flows enter the Delta in April. Water moves slowly due to the low gradient over the Delta. Flows take about 4 months to cover 250 km from the Mohembo at the top of the Delta to Maun (Figure 4). The outflows from the Delta into the Thamalakane and Boteti Rivers occur during the dry season. This creates a unique situation in that high flows are experienced during the dry season, which affects use and perceptions of available water resources. The occurrence of outflows during the dry season results in considerable transmission losses leading to the decrease of flows in the downstream direction unlike most rivers, which naturally increase their flows in the downstream direction. The contribution of groundwater is not well known. The few studies done on groundwater recharge rates from the Thamalakane River showed that the rates are initially very high as the outflows from the Delta advance along this river, and then gradually decrease (Sefe, 1996). High evapotranspiration rates during the dry season contribute substantially to the high

Figure 3. Variation of annual rainfall, 1922-2018, at Maun representative of the Boteti River region

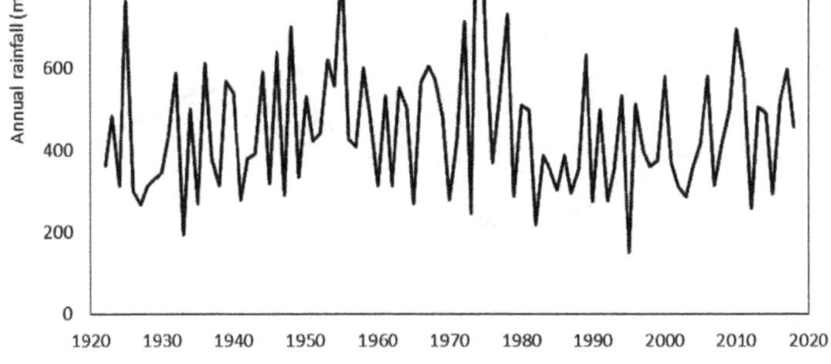

Table 2. Water balance of the Okavango Delta

		Million m³/year	%
Inputs	Okavango River inflows	8,780	58
	Delta rainfall	6,260	42
Total		**15,040**	**100**
Outputs	Outflow	260	2
	Evaporation	14,780	98
Total		**15,040**	**100**

Source: Wolski, et al. (2006)

transmission losses along the Boteti River. In addition, the low gradient results in sluggish flow which prolongs the time for seepage along the river bed. Transmission losses along the Thamalakane to the Boteti River have been estimated at about 20% of the outflows from the Delta (Sefe, 1996). Transmission losses were found to be low for small to medium flows on the Kuiseb River because the flows are confined to a part of the channel with relatively impermeable silt (Lange, 2005). The losses are substantial for high flows overflowing the river banks. Flows from the delta have very low silt loads as most of the sediment is deposited in the delta due to the very low flow velocities. The river beds of the Thamalakane and Boteti Rivers are covered mostly with sandy soil that results in high seepage losses.

Figure 4. Variation on mean monthly inflows (Mohembo) and outflows (Maun) from the Okavango Delta

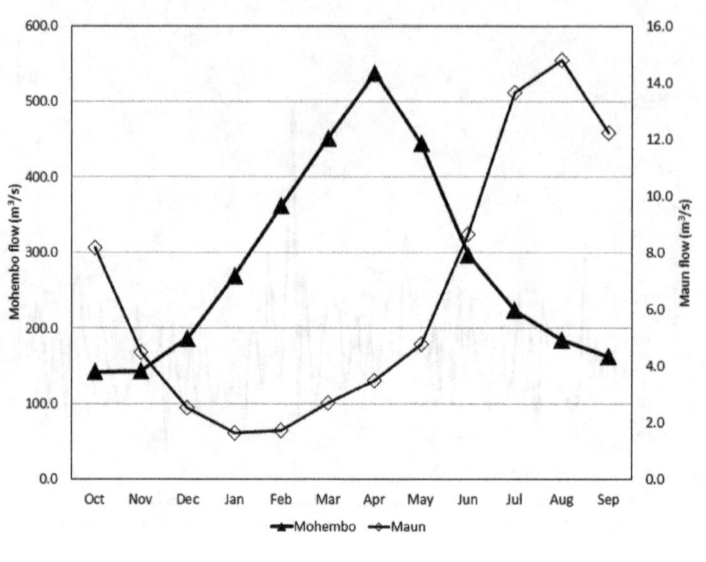

Outflows from the Delta at Maun amount to about 2% of the inputs into the Delta (Table 2). Thus evapotranspiration is the most dominant water flux over the Delta. Figure 5 compares long-term mean monthly flows along the Thamalakane River at Maun, and Boteti River at Samedupi and Rakops. These are 30 km and 240 km respectively from Maun. Rakops is located just before the Boteti River joins the Makgadikgadi Pans. The long-term annual average flow at Rakops is about 56% of the flow at Maun. The perceptions of people living along the river, particularly those in the Rakops area, are greatly influenced by water availability, which is a function of long-term average flows. The perception is that places like Rakops should at least have some flows comparable to Maun. This is not always the case as is shown in Table 3.

Annual flows on the Thamalakane River and Boteti River have considerable inter-annual variability (Figure 8), which is greater than that of the Okavango River at Mohembo. The coefficient of variation of annual flows of the Okavango River at Mohembo is 25%, while this is 88% for Thamalakane River at Maun, and 130% and 160% for Boteti River at Samedupi and Rakops respectively. During the 1991 to 2008 period outflows from the Delta went as far as some 40-50 km downstream of Maun, without reaching Rakops. Thus the first 50 km along the Boteti River can be classified as being intermittent (Figure 6), while the remaining 200 km (Figure 7) has been ephemeral in the period 1991-2008. Historical records reveal that outflows of the Okavango Delta through the Boteti River have in the past failed to reach

Figure 5. Mean monthly flows along the Boteti River showing high flows occurring during the dry season, from July to September. Rakops is downstream of Maun.

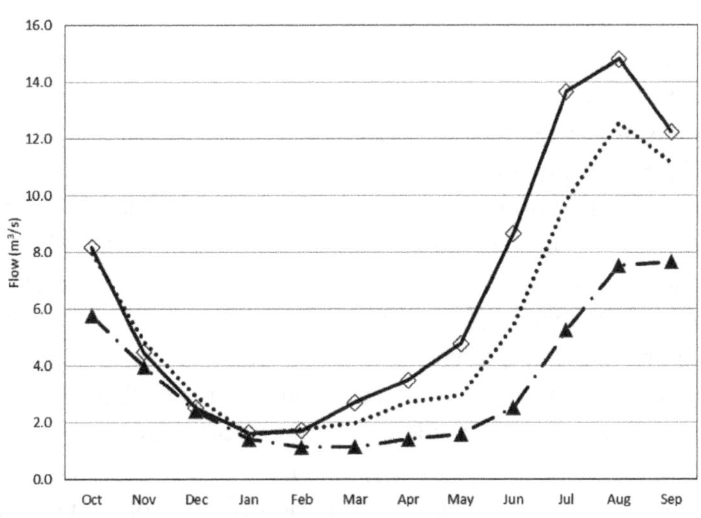

Table 3. Occurrences of flows and no flows on the Boteti River at Rakops since 1849

Period	Condition of the Boteti River at Rakops	Number of Year
2009-present	Flow	10
1991-2008	DRY	17
1989-1990	flow	9
1987-1988	DRY	2
1984-1986	flow	3
1983-	DRY	1
1974-1982	flow	9
1973-	DRY	1
1966-1972	good flow	7
1965-	poor flow	1
1961-1964	good flow	4
1960-	poor flow	1
1952-1959	good flow	8
1948-1951	low flow	4
1941-1947	DRY	7
1940-	poor flow	1
1929-1939	DRY	11
1901-1928	flow (1925: high flood)	28
1900-	DRY	1
1849-1899	flow	51

Source: Shaw (1985), and Department of Water Affairs, Botswana flow records.

Rakops for periods varying from 1 to 17 years (Table 3). Shaw (1985) confirmed that reductions of outflows from the Delta caused the drying up of Lake Ngami for some years during the 1880 – 1983 period.

Some of the residents considered the lack of flow during the 1991-2008 period on the downstream part of the Boteti River as unnatural. There have been however periods in the past when flows did not get to Rakops. Lack of flows is perceived by most residents as a consequence of increased upstream water abstraction at Maun and in Namibia. "Interference" by operators of tourism facilities within the Delta, and clearing and drenching of channels in the Delta have also contributed to the problem. Interestingly upstream abstractions amount to less than 1% of the inflows into the Delta. The lack of flows along the lower Boteti River cannot therefore be attributed to upstream abstractions. The generally low outflows during the 1991-2007 period were part of the long-term variation of flows of the Okavango River.

Mazvimavi and Wolski (2005), and McCarthy et al. (2000) attributed the seemingly decline in outflows to the presence of cyclic behaviour of Okavango River flows.

Flow duration curves constructed using 1970 to 2003 flow data show that the chance of no flow in any month is 24% for Maun, 38% at Samedupi and 57% at Rakops. The supply of water from the Boteti River for human use has a very low reliability particularly at Rakops. Alternative water sources therefore need to be developed. The most reliable source of water is groundwater. Boreholes have a very high density around most of the settlements as shown in Figure 9 for part of the North West District. Some of the boreholes have been developed for livestock watering at cattle posts and ranches.

5. WATER USES ALONG THE RIVER

The provision of water in rural areas of Botswana is carried out under the framework of the National Water Master Plan (SMEC, 1991). This aims to improve access to safe water, reduce the vulnerability of the population to water-related diseases, and create conditions for employment generation. The National Settlement Policy of Botswana provides guidelines for the type of services to be made available to settlements of different categories (DTRP, 1996). Government provides services

Figure 6. A large pool on the intermittent part of the Boteti River at Chanoga utilized for fishing and livestock watering
Source: Author, April 2005

Figure 7. The dry lower part of the Boteti River between downstream of Motopi where outflows from the Okavango Delta had not reached from 1991 to 2008
Source: Author, June 2007

Figure 8. Annual flows (Million m³/year) for Maun, Samedupi and Rakops on the Boteti River, and Okavango River at Mohembo from where the delta starts

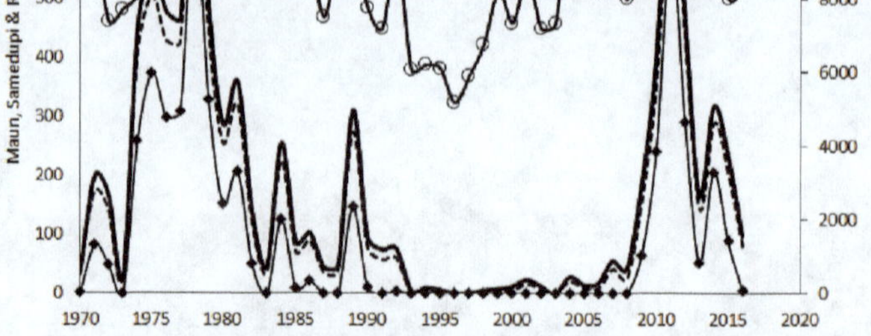

Figure 9. Location of boreholes around part of the Okavango Delta and along upper Boteti River

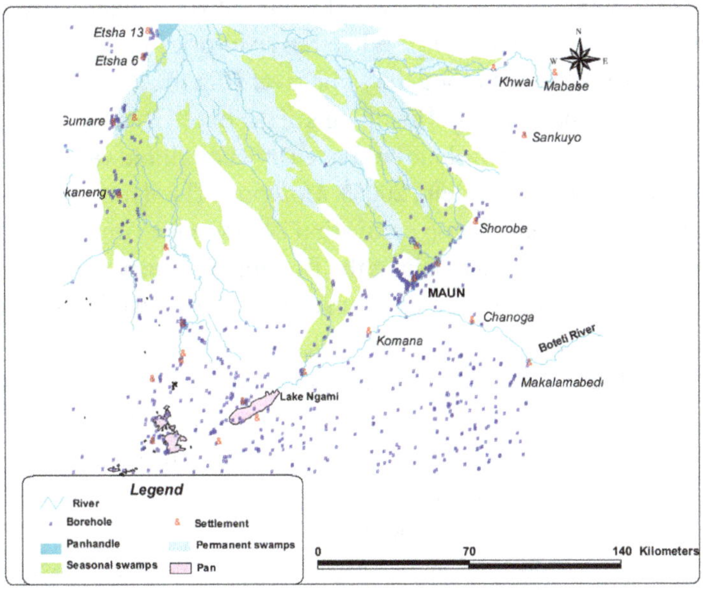

to gazetted settlements. For a settlement to be gazetted, it should normally have a minimum of 500 people, and being not less than 15 km from a parent/other gazetted settlement. The gazetted settlements fall into the following categories (which determine the institution responsible for water supply):

- Tertiary Centre (500 – 9999 people) supplied water by the respective district council;
- Secondary Centre (10000 – 19999 people) provided with water by either Department of Water Affairs or district council; and
- Primary Centre (> 19 999 people) provided by the Water Utilities Corporation, and Department of Water Affairs in some villages (DTRP, 1996).

Ungazetted settlements accounted for 16% of the national population in 2001. These occur in the form of cattle posts, temporary homesteads close to arable lands, and permanent isolated homesteads (CSO, 2003). Residents of ungazetted settlements are responsible for their own water supply.

The gazetted settlements that occur along the Boteti River are Chanoga, Makalamabedi in Ngamiland (North West District), Makalamabedi in Central Districts, Motopi, Moreomaoto, Khumaga, Xhumo, and Rakops. Residents of these settlements obtain water from stand pipes supplied with water from boreholes. There are several ungazetted settlements along the Boteti River, and for those residing along the first 50 km of this river that receives water from the Delta, 89% of the households use untreated water abstracted from the river when flow is available (Mazvimavi and Mmopelwa, 2006). When the river dries up, people obtain water from hand-dug wells located along the same river (Figure 10).

Some of the households utilize floodplains of the Boteti River for crop production on *molapo* fields. The size of the land cultivated by a household varies from 0.1 to 4.0 ha (Arntzen *et al.*, 1994; Tsholofelo, 2005). *Molapo* or flood recession cultivation is a traditional practice whereby crop cultivation is undertaken on seasonally flooded plains. When water levels recede along the floodplains, farmers take advantage of the high soil moisture to plant their crops just before the main rains. Crops produced on *molapo* fields comprise maize, sorghum, beans, water melons, and pumpkins (Rozenga, 1995; Tsholofelo, 2005). The combination of *molapo* cultivation along the floodplains when flows on the Boteti River are available with dryland cultivation, enhance the household food security (Vanderpost, 1995; Magole and Thapelo, 2005). Cashdan (1987) found that crop yields on *molapo* fields were five times higher than from fields located on sandveld in the Boteti area. Households with *molapo* fields were reported to be wealthier than those without. A similar conclusion was made by Bendsen and Meyer (2002) for other parts of Ngamiland. However, *molapo*

Figure 10. A water tanker supplied with water from a well located within the Boteti River bed. Water is pumped from the well using a motorized pump. Outflows of the Delta have not reached this location in the period 1991-2008.
Source: Author, June 2007

cultivation takes place along the river. This creates conflicts with other users of the river, due to the destruction of crops by cattle and wild animals (Arntzen *et al.*, 1994). SMEC (1987) estimated that 16 million cubic meters per year of water was required for the 1,600 ha of *molapo* along the Boteti River.

The Boteti River is important for livestock watering. About 65% of the households in both gazetted and ungazetted settlements located on the upper perennial part of this river are deriving this benefit (Tsholofelo, 2005; Mazvimavi and Mmopelwa, 2006). There are however unintended consequences. Rangeland degradation was observed along a 15 km zone around the Boteti River (Arntzen, *et al* 1994). Bare soils due to grazing and livestock movement are evident around lands that are riparian to stretches that contain pools heavily utilized for livestock watering along the upper part of the Boteti River. Wells for both livestock watering and domestic water supply were developed along the river in those areas where flow is currently unavailable during the 1991-2008 period. However, lack of river flows has during the 1991-2008 period in some of the wealthy livestock owners developing their own boreholes. These have to abide by the regulations that each borehole has to be at least 8 km from any other borehole to minimize rangeland degradation, and over-exploitation of groundwater.

Fishing for subsistence is done by about 13% of the households residing along part of the Boteti River that is annually flooded. Some residents of the nearby Maun do also undertake fishing for either subsistence or for sale. Residents of Maun who fish particularly for commercial purposes along the Boteti River, are considered by the locals as outsiders. They are seen as involved in over fishing as well as polluting water through improper handling of fish waste. Commercial fishermen are of the view that the conflicts with the locals are mainly due to most locals using inefficient fishing methods, hence the perception of over-fishing. Some trade offs have had to be done - the employment of some of the locals and the giving of commercially less preferred species to some locals has helped to reduce conflicts.

Harvesting of aquatic plants is another benefit derived from the Boteti River by residents located on the upper part that is regularly flooded. About 23% of the households in one of the settlements, Chanoga, derive this benefit from the river (Tsholofelo, 2005). The plants harvested comprise reeds (*Phragmities australis*) used for constructing houses and courtyard walls, and water lily, which is consumed as food. Erratic river flows and fires have reduced the availability of these plants.

During the 1991-2008 period, households residing along the remaining 200 km did not benefit from using the river for domestic water supply, livestock watering, *molapo* crop production, fishing, and harvesting of aquatic plants. For those households in ungazetted settlements, which are not provided with water by the district council, they obtain water for domestic purposes from either hand-dug wells along the river or ferry water using donkey carts from nearby gazetted settlements.

Hand-dug wells are also used for livestock watering. Wells located along the river obtain water from shallow aquifers within the alluvial deposits. Freshwater in these aquifers is recharged by flows of the Boteti River.

The flow of the Boteti River is therefore important for water supply when it is available, and recharging aquifers that residents depend upon when the flow is not available. This brings in an important element for implementation of a river basin management approach in that both surface water and groundwater are considered. When a river basin management approach is implemented on most rivers, the focus is often on surface water. However, both surface and groundwater have to be considered along ephemeral rivers. The same observation was made on the Kuiseb River in Namibia (Botes, et al., 2003; Seely, *et al.,* 2003).

6. CHALLENGES FOR IMPLEMENTING A RIVER BASIN MANAGEMENT APPROACH

6.1 Water Variability and Divergent Water Uses

The variability of both rainfall and river flows has resulted in numerous challenges for the residents of the drier parts of the Boteti River. The high rainfall experienced during the 1970s created conditions favourable for livestock and dryland crop production. This resulted in the development of several settlements further from the established villages but close to cultivated lands and cattle posts (Arntzen, 1994; Vanderpost, 1995). Dry conditions accompanied by droughts during the 1980's forced some of the residents to move back to established villages located along the Boteti River that had water supply systems developed by government and district councils. This resulted in the intensification of utilization of natural resources for grazing, firewood collection, and construction around these established settlements. Overgrazing, deforestation and soil erosion have become major problems around most of the settlements. This has resulted in the general decline in livelihood levels (SMEC, 1991; Arntzen *et al.,* 1994, Vanderpost, 1995). The entire length of the river supports communal livestock grazing and subsistence arable production. The only large-scale industrial activity is diamond mining in the town of Orapa about 300km downstream. The mine currently relies on groundwater from well fields developed in the Mopipi area, which has not received any water flow from the Boteti since 1991.

6.2 Molapo Fields

A potential entry point for initiating community-based river basin management activities would be to focus on improving management of *molapo* fields. This could involve determination of most suitable locations for fields along the floodplain and improving cultivation practices. Arntzen *et al* (1994) estimated that 80% to 96% of households along the mid-Boteti River had molapo fields. A study in the upper-Boteti River showed that about 40% of the households had *molapo* fields suggesting a decrease in the proportion of households deriving this benefit. However, the combination of lack of flows, and an increase in population that reduced the proportion of households with *molapo* means that the majority of the households would not have an incentive to participate in river basin management activities aimed at increasing benefits from *molapo* fields. Kgathi et al (2007) noted that access to *molapo* had also decreased in some parts of the Delta due to desiccation of channels brought about by switching of flows to other channels. The switching of flows is likely due to tectonic activity, channel blockage by papyrus, and sediment deposition (Shaw, 1985).

6.3 Domestic Water

River basin management initiatives aimed at improving water supply for domestic purposes are not likely to be very attractive. It is the policy of government to develop water supply systems for all the residents of gazetted villages. Water obtained from public stand pipes in these villages is provided at no charge, and previous attempts to charge fees for using water from these stand pipes failed (Andrews, 1994). Residents of gazetted settlements are not likely to be motivated to participate in river basin management initiatives aimed at improving domestic water supply since government and the district council are already providing water free of charge. The well intended provision of water points within gazetted villages has in few instances led to a lack of ownership by residents of these facilities. Informal discussions with residents of Makalamabedi, Motopi, Moreomaoto, Khumaga and Chanoga revealed that cases of abuse of village water supply systems, such as vandalism and night time watering of livestock are common. Livestock watering from standpipes is prohibited since the facility is prone to damage by livestock. Significantly the culprits are rarely identified.

River basin management initiatives that aim to improve access to water for domestic use are likely to be accepted by residents of ungazetted settlements. In some of these settlements along the Boteti River, 65% of the respondents were willing to contribute towards improving domestic water supply (Mazvimavi and Mmopelwa, 2006). These residents were willing to contribute cash, livestock and agricultural produce. This willingness to pay is an indicator of the problems experienced in accessing water. Such initiatives could be undertaken as part river basin management activities.

6.4 Institutional Issues

Water remains the state domain and control over water is under government departments. Some devolution has been rendered to the Water Utilities Corporation, a parastatal with the mandate to develop and maintain water supply systems to urban areas (Swatuk and Rahm, 2004). Some residents along the Boteti River have developed the attitude that water resources management is the responsibility of the government, which has the potential of hampering community participation in river basin management activities. Weaknesses of local level institutions such as village development committees do also constrain implementation of local level initiatives aimed at improving management of water resources. It is interesting to note that in the past traditional local authorities under the direction of the chief were able to mobilize communities for the development and maintenance of water supply projects. These authorities used to mobilize regiments in the construction and maintenance of water points and ensured discipline in the use of such water points (Andrews, 1994, Schapera, 1970).

A sectoral approach that is adopted by government agencies, which are directly or indirectly responsible for some aspects of water resources management, is another constraint towards the implementation of an integrated river basin management approach. The following institutions have mandates over issues that impinge on water resources management along the Boteti River; the district council, Land Board, Department of Water Affairs (DWA), Department of Animal Health and Production, Department of Crop Production and Forestry, and the Department of Wildlife and National Parks (DWNP). Each department concentrates on those issues related to its own sector. This does not always create a framework for an integrated river basin management. For instance, the Land Board discourages the allocation of land for *molapo* fields or hand dug wells along the Boteti River, because of the possible environmental damages of these activities. On the other hand, the Department of Crop Production offers technical assistance to *molapo* farmers. The DWNP has at least one borehole and water point for wildlife watering located within the riverbed.

6.5. Concluding Observations

These few examples point to an uncoordinated sectoral approach to water management which may prove to be a serious challenge to river basin management initiatives in future. Another important point to note is that the Boteti River cuts across district administrative boundaries and creating a framework for implementing an integrated river basin management approach will be problematic as this will require coordination of activities of government agencies in two districts.

However, positive efforts to reconcile policies and coordinate sectoral actions as well as community participation are already underway in the form of the Okavango Delta Management Plan project which has developed a "comprehensive, integrated management plan for the conservation and sustainable use of the Okavango Delta and surrounding areas" (Jansen, 2002). The Permanent Okavango River Basin Water Commission (OKACOM) investigated major threats to sustainable development in the whole basin (OKACOM, 2011), and formulated a Strategic Action Programme (OKACOM, 2011b). This programme aims to integrate national and basin wide initiatives. OKACOM is piloting agricultural practices such as improved production of horticultural produce irrigated using water from the Boteti River. Such efforts have the potential to improve livelihoods of the communities living along this river. The challenge however is that the benefits are likely to accrue to communities where the Boteti River regularly gets outflows from the delta.

7. CONCLUSION

The variability of river flows within and between years, and along the Boteti River poses major challenges for water resources management. The lower part of this river did have flows during the 1991-2008 period, while the upper part has received outflows from the Okavango Delta. Similar periods of no flow on the lower part of the River have been recorded since 1849. However, the 1991-2008 dry spell was the longest. Households residing along the upper part of the river, which frequently receive flows annually, derive greater benefits from using the Boteti River than those located on lower part. Flows of the Boteti River recharge aquifers that communities depend upon for water supply during periods without any flow. Abstraction of fresh groundwater from shallow aquifers located along the Boteti River has to be limited and linked to the long-term recharge rate from this river. A conjunctive management of both surface and groundwater is required along the Boteti River. The depletion of benefits from the lower Boteti River is a constraint that hinder the development of community-based river basin management initiatives.

Most of the residents along the Boteti River are communal farmers whose main priority is to sustain their households in a semi-arid environment characterized by limited livelihood options. The high variability of rainfall and river flows means that farmers focus on framing survival strategies each year. They therefore take a short-term view towards use and management of natural resources. Implementation of river basin management initiatives that increase the probability of deriving long-term benefits will remain a challenge in such an arid environment.

Government departments like in most countries have mandates over specific sectors. This sectoral approach will always be a challenge in attempting to introduce an integrated river basin management approach. The Boteti River cuts across districts, which further complicates implementation of an integrated river basin management approach. The implementation of the Okavango Delta Management Plan will provide important lessons for implementing integrated natural resources management approaches in other parts of Botswana. There is a need to increase the awareness of the community living along the Boteti River about the high natural variability of outflows from the delta, and also that benefits derived or lack of them are influenced by climatological events occurring at the Okavango River Basin scale, and not only within the delta.

REFERENCES

Andrews, J. D. (1994) Community participation in rural water village water supply in Botswana. In Proceedings: Integrated water resources management workshop. Gaborone: University of Botswana.

Arntzen, R., Chanda, R., Nkambwe, M., Ringrose, S., Sefe, F.T.K., & Vanderpost, C. (1994). *Desertification and possible solutions in the mid-Boteti River area: A Botswana case study for the Intergovernmental Convention to Combat Desertification (INCD). Findal Report Volume 1 – The Desertification process and its impact.* Ministry of Agriculture, Government of Botswana.

Bendsen, H., & Meyer, T. (2002). *The dynamics of land use systems in Ngamiland; changing livelihood options and strategies*. Harry Oppenheimer Okavango Research Centre, University of Botswana, Maun.

Botes, A., Henderson, J., Nakale, T., Nantanga, K., Schachtschneider, K., & And Seely, M. (2003). Ephemeral rivers and their development: Testing an approach to basin management committees on the Kuiseb River, Namibia. *Physics and Chemistry of the Earth Journal*, *28*(20-27), 853–859. doi:10.1016/j.pce.2003.08.028

Cashdan, E. (1987). Trade and Its Origins on the Botletli River, Botswana. *Journal of Anthropological Research*, *43*(2), 121–138. doi:10.1086/jar.43.2.3630221

Central Statistics Office. (2003). *National statistical tables, August 2001. Central statistics Office, Botswana.* Gaborone: Department of Printing and Publishing Services.

DTRP. (1996). National Settlement Policy. Department of Town and Regional Planning. Government Printer.

Jansen, R. (2002). *The Okavango Delta management plan project: application of an ecosystem-based planning approach.* Paper presented at the seventeenth Global Biodiversity Forum, Valencia, Spain.

Kgathi, D. L., Ngwenya, B. N., & Wilk, J. (2007). *Shocks and rural livelihoods in the Okavango Delta* (Vol. 289). Botswana: Development Southern Africa.

Lange, J. (2005). Dynamics of transmission losses in a large arid stream channel. *Journal of Hydrology (Amsterdam), 306*(1-4), 112–126. doi:10.1016/j.jhydrol.2004.09.016

Magole, L., & Thapelo, K. (2005). The impact of extreme flooding of the Okavango River on the livelihood of the molapo farming community of Tubu Village, Ngamiland Sub-district, Botswana. *Botswana Notes and Records, 37*, 125–137.

Manning, N., & Seely, M. (2005). Forum for Integrated Resource Management (FIRM) in Ephemeral Basins: Putting communities at the centre of the basin management process. *Physics and Chemistry of the Earth Journal, 30*(11-16), 886–893. doi:10.1016/j.pce.2005.08.035

Manzungu, E. (2004). *Water for all: Improving water resources governance in southern Africa.* The Gatekeeper Series No. 113 on the Natural Resources Group of the International Institute for Environment and Development.

Manzungu, E., & Kujinga, K. (2002). The theory and governance of water resources in Zimbabwe. *Zambezia, 29*(2), 195–216.

Mazvimavi, D., & Mmopelwa, G. (2006). Access to water in gazetted and ungazetted rural settlements in Ngamiland, Botswana. *Physics and Chemistry of the Earth Journal, 31*(15-16), 713–722. doi:10.1016/j.pce.2006.08.036

Mazvimavi, D., & Wolski, P. (2006). Long-term variations of annual flows of the Okavango and Zambezi Rivers. *Physics and Chemistry of the Earth Journal, 31*(15-16), 944–951. doi:10.1016/j.pce.2006.08.016

McCarthy, T. S., Cooper, G. R. J., Tyson, P. D., & Ellery, W. N. (2000). Seasonal flooding in the Okavango Delta, Botswana – recent history and future prospects. *South African Journal of Science, 96*, 25–33.

Mendelsohn, J., & Obeid, S. (2004). *Okavango River: The flow of lifeline.* Cape Town: Struik Publishers.

Nadia Manning, N., & Seely, M. (2005). Forum for Integrated Resource Management (FIRM) in Ephemeral Basins: Putting communities at the centre of the basin management process. *Physics and Chemistry of the Earth Journal*, *30*(11-16), 886–893. doi:10.1016/j.pce.2005.08.035

OKACOM. (2011a). *Cubango-Okavango River Basin Transboundary Diagnostic Analysis*. The Permanent Okavango River Basin Water Commission.

OKACOM. (2011b). *Strategic action programme (SAP) for the sustainable development and management of the Cubango-Okavango basin*. The Permanent Okavango River Basin Water Commission.

Rozenga, W. (1995). *Ecologically sustainable land use in Lower Boteti, Central District, Botswana*. Utrecht, The Netherlands: Department of Geography of Developing Countries. Utrecht University.

Schapera, I. (1970). *Tribal innovators: Tswana and social change 1795-1940*. London: Athlone Press.

Seely, M., Henderson, J., Heyns, P., Jacobson, P., Nakale, T., Nantanga, K., & Schachtscschneider, K. (2003). Ephemeral and endoreic river systems: Relevance and management challenges. In Transboundary rivers, sovereignty and development: Hydropolitical drivers in the Okavango River basin. African Water Issues Research Institute and Green Cross International.

Sefe, F. T. K. (1996). Maun groundwater development project: Phase 1 – exploration and resource assessment. Hydrological analyses. Eastend Investment Report to the Department of Water Affairs, Government of Botswana.

Shaw, P. (1985). The Desiccation of Lake Ngami: An Historical Perspective. *The Geographical Journal*, *151*(3), 318–326. doi:10.2307/633012

SMEC. (1987). Southern Okavango Integrated Water Development Phase 1: Final Report Technical Study, Volume III, Water resource and development. Report by Snowy Mountains Engineering Corporation to the Department of Water Affairs, Ministry of Mineral Resources and Water Affairs, Government of Botswana, Gaborone.

SMEC. (1991). *Botswana national water master plan study: final report. Prepared by Snowy Mountains Engineering Corporation Ltd, WLPU Consultants and Swedish Geological International AB*. Gaborone: Ministry of Mineral Resources and Water Affairs.

Swatuk, L. (2005). Political challenges to implementing IWRM in sounthern Africa. *Physics and Chemistry of the Earth*, *30*(11-16), 872–880. doi:10.1016/j.pce.2005.08.033

Swatuk, L., & Rahm, D. (2004). Integrating policy, disintegrating practice: Water resources management in Botswana. *Physics and Chemistry of the Earth*, *29*(15-18), 1357–1364. doi:10.1016/j.pce.2004.09.011

Tsholofelo, M. (2005). *The contribution of the Boteti river to the livelihoods of the Chanoga Community. Winter School report*. Harry Oppenheimer Okavango Research Centre.

Wolski, P., Savenije, H. H. G., Murray-Hudson, M., & Gumbricht, T. (2006). Modelling of the flooding in the Okavango Delta, Botswana, using a hybrid reservoir-GIS model. *Journal of Hydrology (Amsterdam)*, *331*(1-2), 58–72. doi:10.1016/j.jhydrol.2006.04.040

Chapter 3
The Effects of Streamflow Variability on Salt Flux in a Semi-Arid Tropical River Basin in Eastern Kenya

Johnson Utu Kitheka
South Eastern Kenya University, Kenya

ABSTRACT

This chapter presents the results of a study on the influence of streamflow variability on salt fluxes in a semi-arid Tiva River Basin located in Eastern Kenya. Measurements of salinity, total dissolved solids (TDS), conductivity, and river discharges were undertaken in sampling stations located within the main branch of the river as well as in the tributaries, namely Kauwi, Kalundu, Nzeeu, and Mwitasyano Rivers in the Upper Region of the Tiva River Basin. The mean and maximum river discharges for the sub-basins of the river ranged from 11 to 33 m^3s^{-1} and from 118 to 210 m^3s^{-1}, respectively. The peak river discharge at the main Tiva branch was 270 m^3s^{-1}. The study shows that there is a significant relationship between the variability of streamflow and the variability of salt fluxes in the river as exemplified by variation of salinity, conductivity, and TDS in the river.

DOI: 10.4018/978-1-7998-0163-4.ch003

BACKGROUND

The relationships between streamflow and salinity were best represented by power functions rather than linear regression functions. The relationships were negative such that the levels of river water salinity, conductivity and TDS decreased with an increase in river discharge. Salinity, TDS and conductivity were also inversely correlated to turbidity. An increase in turbidity corresponded to a decrease in conductivity and hence salinity and TDS. The low conductivity at high streamflow conditions were attributed to the dilution effect of increased volume and also by the presence of high proportion of non-conductance organic and inorganic materials such sand and clay. The highest TDS, conductivity and salinity values were measured during the low flow (baseflow) conditions and the concentrations were lowest during high streamflow conditions. It is postulated that the high concentrations were a result of high evapotranspiration and seepage of subterranean water from bank storage and groundwater aquifers, and that low concentrations were a result of dilution and flushing effect of high streamflow. Inter-subbasin variations in the levels of salinity were attributed to differences in landuses, lengths and sizes of the sub-basins. The effects of upstream irrigation were evident in one of the main sub-basins-Mwitasyano river where the highest salinity levels (max: 3.4%; mean 1.1%) were measured. The total salt flux from Upper Tiva river basin was estimated to be 100,344 tonnes.yr^{-1} with basin salt production rate of 27.87 tons.m^{-2}.yr^{-1}. This salt flux was attributed to the nature of the basement complex metamorphic rocks (e.g. Kankur limestone) through which the river drains. The influence of irrigation upstream was noted to be important in the Mwitasyano sub-basin that contributed 61% of the total salt load. The use of Tiva river water for irrigation during dry season was noted to cause soil salinisation which impairs crop production in irrigation plots along the river. The study emphasizes the need for water resources and agricultural development programmes in the semi arid Tiva River Basin to promote sustainable irrigation and landuse practices. It is suggested that construction of water reservoirs in the Tiva basin would help in controlling salinity levels in the river.

INTRODUCTION

Seasonal rivers in arid and semi arid lands (ASALs) of Africa are important sources of water to communities and their livestock. These rivers are sources of water for various uses such as livestock watering, domestic uses and irrigation. Therefore the variations of water quality of seasonal rivers in ASALs have the potential of impacting the various uses of water in regions that are otherwise characterized by scarcity of water. However, despite the important role played by seasonal rivers in such lands, few studies have been undertaken to unravel their hydrologic and water quality characteristics (Kitheka, 2013 & 2014). There is thus scanty of data and information on river discharges and material fluxes particularly the fluxes of nutrient, sediment and dissolved salt in seasonal rivers of Sahelian Africa. The contribution of African seasonal rivers in terms of global nutrient and material fluxes is therefore little understood, since most of the hydrological studies have focused on large perennial river systems (Ohowa et al., 1997; Busulwa and Bailey, 2004; Waziri and Ogugbuaja, 2010; Waziri and Ogugbuaja, 2012; Elmoustafa, 2013; Kitheka, 2013 & 2014). Seasonal rivers in semi arid lands are unique in that they usually flow for relatively short period during rainy season. Their water quality features can therefore change rapidly. Data and information on their hydrology are required to advice on the water resources and agricultural development programmes. This is becoming critical as a result of climate change and also due to current programmes that are aimed at opening up arid and semi arid lands for development. Seasonal rivers could be the only reliable source of water that can be used to develop ASALs. This is because in these regions, other potential sources of water such as groundwater aquifers are also characterized by poor water quality conditions.

Most of the studies on seasonal rivers in Kenya have focused on water pollution, soil erosion and nutrient fluxes (see Kithia, 1997; Ohowa et al., 1997). Some studies on Kenya have shown that some seasonal rivers at the coastal basin are characterized by high nutrients loads in rainy season as compared to dry season (Ohowa et al., 1997). Recent studies on Kenya's largest river-Tana have demonstrated the effects of modification of streamflow and sediment load as a result of damming of the river in the Upper Tana Basin (Kitheka, 2013 & 2014).. However, despite the impact of damming, Tana river still receives substantial sediment load from seasonal streams such Tiva river that drains into the lower Tana Basin (Figure 1). To our knowledge, no studies have been undertaken on salt fluxes in rivers of Kenya. However, past studies on salinity conducted elsewhere have shown that the main driving forces acting in shaping seasonal variation of salinity include low natural drainage density of the catchment, which limits the salt loads induced by the natural runoff processes, and the runoff in the catchment area which promote dissolved salt dilution during the high-flow period (cf. Kaabata *et al*., 2012).

Tiva river is one of the seasonal rivers that forms part of the larger Tana River Basin-the largest river system in Kenya (Figure 1). Tiva river basin receives highly erratic and seasonal rainfall which is one of the major factors limiting community access to water, agricultural production and food security in Kitui and Tana River Counties where it traverses (see also Lasage et al., 2008). The seasonal river and its tributaries has been source of water for majority of the people and livestock in Kitui County for many generations. In order to address the chronic shortage of water to rural communities in the county, an effort is now being focused on the exploitation of water resources of Tiva river through development of sand dams (Beimers et al., 2001a-b; Burger et al., 2003; Bossenbroek and Timmermans, 2003: Munyao et al., 2004; Puttemans, 2004; Borst and De Haas, 2006; Lasage et al., 2008). However, such development programmes needs to be based on the assessment of suitability of water for domestic, livestock and agricultural uses. Such an approach can safeguard public health, protect fragile semi arid lands soil (e.g. from salinisation, sodification, etc) and enhance water security for the local communities. In the past, there have been little concern on the water quality of seasonal rivers targeted for development in ASALS. This lack of concern was partly due to the assumption that seasonal rivers in marginalized ASALs are not subject of major anthropogenic influences that can cause significant pollution. This study shows that while anthropogenic drivers of pollution may not be critical in ASALs, natural factors associated with hydrogeochemical and catchment characteristics are important determinant of water quality of seasonal rivers such as Tiva. However, it is important to note that anthropogenic influences are also progressively becoming important in ASALs of Kenya and cases of water pollution due to municipal wastewater discharges have been reported in Kitui (Abila et al., 2012; Karanja et al., 2015).

This research was partly motivated by the need for baseline studies to establish the levels of basic water quality parameters (e.g. salinity, conductivity, TDS) in seasonal rivers draining ASALs of Africa. Such studies will provide data and information against which future changes due to development activities can be determined. This is important in view of rapid population growth and expansion in ASALs and recent campaigns to open up ASALs for development in Africa. Establishment of the levels of key physico-chemical parameters and their associated natural and anthropogenic drivers is equally important in that these parameters are important determinant of water security, public health and agricultural crop production. Consumption of water of unquestionable quality is associated with various waterborne and water related diseases. Also, use of water with high levels of salinity in small-scale irrigation leads to deterioration of soil quality and subsequently leads to reduced crop production in ASALs (cf. Puttemans, 2004). High salinity in rivers can also cause major damages such as corrosion and plugging of pipes and water fixtures in housing and industry (U.S. Department of the Interior, 2011). Such damages can impose high operation and maintenance costs to rural water supply schemes in ASALs that are often run by communities with limited resources (see also Lasage et al., 2008). In some areas within the Upper Tiva river basin, it has been observed that communities using highly saline river water during dry seasons have been affected negatively as evidenced by high proportion of people with dental florosis which is associated with the consumption of water with high concentration of fluoride. Fluoride is one of the dissolved salts in the river water. Recently, there has been reports of loss of endangered species of wildlife such as Rhino in the lower parts of the basin following consumption of highly saline water in the Tsavo East National Park (Save the Rhino, 2018; The Guardian, 2018).

This study was focused on the determination of the patterns of salt fluxes in the Tiva river basin. This was achieved through measurement of salinity, total dissolved solids (TDS) and conductivity in the Tiva river sub-basin. These physico-chemical parameters have been used describe the overall water salinity in rivers. Although conductivity is of little human concern, high conductivity is usually an indication of existence of a source of dissolved ions and is therefore a good indicator of other water quality problems in rivers (Bhattet al., 1999; Vaishal and Punita, 2013). There is also a relationship between the concentration of dissolved solids in water and salinity. The high amount of dissolved solids means high concentration of ions in water and this causes high salinity (Bhattet al., 1999). Some studies have noted that high levels of conductivity and cations are the product of decomposition and mineralization of organic materials (cf. Abida, 2008). Conductivity therefore provides a clear view of the total ionic strength and extent of salinity and helps in assessing the level of TDS in a water body (APHA, 1992; Massdam and Smith, 1994). TDS is an indicator of the amount of dissolved mineral salts in water and is usually composed mainly of carbonates, bicarbonates, chlorides, phosphates and nitrates of calcium, magnesium, sodium, potassium and manganese, organic matter, salt and other particles (Mahananda, 2010). At high flows, the TDS concentrations is diluted by surface runoff and for most rivers there is an inverse correlation between discharge rate and TDS concentration (Charkhabi and Sakizadeh, 2006; Srivastava et al., 2011). Some but not the entire dissolved solids act as conductors and contribute to the conductivity level in water. River water with high TDS are unpalatable and potentially unhealthy (Vaishal and Punita, 2013). The WHO drinking water standard for TDS is $500mgl^{-1}$ and water with greater concentration is unpalatable. Excessive TDS in water imparts a bad taste in water due to mineralization of various salts. A TDS concentration $>2{,}000mgl^{-1}$ has been reported to produce a laxative effect as a result of the presence of magnesium sulphate and some sodium sulphate (Kumaraswamy, 1991, Dembere 1998). Sodium affect the cardiac part and women suffer toxemia during pregnancy. The maximum permissible limit of TDS in drinking water is $1{,}000mgl^{-1}$ (WHO). For irrigation, maximum permissible limit of TDS is $500mgl^{-1}$ and above this limit, salinity has detrimental effect on crop production. Salinity is closely related to TDS and conductivity. An increase in TDS and conductivity therefore leads to a corresponding increase in salinity. Past studies elsewhere have shown that the concentration of salinity usually decrease in rainy season as a result of dilution of dissolved ions by the surface runoff (cf. Srivastava et al., 2011).

This study sought to establish the salinity fluxes in the semi arid Tiva River Basin in Eastern Kenya. The specific objectives of the study were; (i) provide baseline data and information on the levels of salinity, TDS and conductivity in a typical seasonal river; (ii) determine the influence of streamflow on the variability of salinity, conductivity and TDS and (iii) establish the magnitude of salt flux in a typical seasonal river.

DESCRIPTION OF THE STUDY AREA: TIVA RIVER BASIN

Geographical Characteristics

Tiva River Basin is found in lower Eastern Kenya and is one of the sub-basins of the Tana Basin-the largest river system in Kenya draining into the Indian Ocean (Figure 1). The total surface area of the basin determined using My Google Map Length and Area Computation Tool is 18,150 km^2 and length of the river from its headwaters in Kitui County to the lower Tana basin near Garsen is 380km. The upper Tiva basin covers a surface area of 3,600 km^2. The width of the main river varies from about 40m in the upper basin to 100m in the middle reaches and <50m in the lower zone. The topography of the basin can be divided into an upland and lowland zone. The upland zone includes the Yatta Plateau and the Kitui Mountain Ranges in the north-west of the basin. Elevations in the upland area vary between 600 and 1800 m above sea level. The lowland area is found to the east of the basin and this area covering the majority of the sub-basin is a gently eastward-sloping peneplain with elevation varying from 400 to 600 m above sea level. Few isolated inselbergs are found in this zone (Borst and de Haas, 2006).

The climate in the sub-basin is generally hot and dry with erratic and unreliable rainfall typical of arid and semi-arid climatic zones (Lasage et al., 2008). There are two rainy seasons, one from March to May (long rains) and one from October to December (short rains) (Lasage et al., 2008). The rest of the year is hot and dry. The total annual rainfall average is between 750 and 1150 mm with 40/60 percent reliability. Air temperature ranges between 16°C and 34°C with mean maxima of 28°C and minima of 22°C. The average temperature is 24°C (Horst and de Hass, 2006). The prevailing winds are the north and south easterly monsoon winds. Potential evaporation is high, 1500-1600 mm.yr^{-1}, largely exceeding the mean annual rainfall of about 1000mm.yr^{-1} in the headwaters of the basin (Lasage et al., 2008). Runoff generation is high due to low infiltration capacity of soils (see also Ngigi et al., 2005).

Vegetation Cover

Vegetation in the Tiva basin is characterised by scrublands and wooded bushland. In the upper region of the basin, the Kitui hilltops are usually characterised by upland dry forest ecosystems dominated by *Drypetes*, *Combretum*, *Vepris* and *Croton* species (Lind & Morrison, 1974). The lower zones of the Kitui hills are covered with Acacia-*Commiphora* and *Strychnos-Combretum* or *Commiphora* bushland vegetation type (Lind & Morrison, 1974). At higher altitudes close to 1200 m, the vegetation is dominated by *Croton megalocarpus*, *Rawsonia lucida*, *Manilkara discolor* and *Drypetes* species, resulting in semi-evergreen dry upland forests. In the lowland zone, below 500m asl, the vegetation is mainly *Acacia -Commiphora* complex.

Geology and Topography

The Tiva sub-basin is within the Mozambique belt and is generally occupied by the basement complex system consisting mainly of metamorphic rocks (Nyamai et al., 2003). The rocks are dominated by massive metamorphic granulites which are highly impermeable. These are usually composed quartz and feldspars, red garnets and pyroxenes or less. The rock outcrops are composed of metamorphic granulites composed of quartz and feldspars, mafic pyroxenes, biotite and red almandine garnets (cf. Nyamai et al., 2003). The river channel is usually sandy with well sorted and mature sand without or with minimal silt/clay content and with water being retained in their sub-surface pores. Most of the river courses are composed of metamorphic granulites and gneisses. Mafic granulites rich in garnets and black heavy minerals, iron ore and pyroxenes have weathered into deep red soils rich in black heavy mineral sands. In some places are found well foliated biotite gneiss which allows some infiltration of rainwater into the ground. Kunkar limestone and white patches of sodium chloride are common along the river banks making the water slightly brackish, particularly during periods of low flow.

The lower part of the sub-basin is found in the lower Tana Basin in lowlying area with elevation less than 500m asl. This region is covered with the Quaternary sediment deposits. The river forms a large inland delta (36km long and 4km wide A= 180km^2 Ndarapo Swamp) in its lower course in Tana River County and subsequently flows to Tana through Laga Kokani where it joins the river in the area north of Garsen (Figure 1). The middle zone of the river before it forms the inland delta appears to be covered by degraded sodic lumic vertisols. High sodicity of soils in the lowlying basin perhaps accounts for whitish patches around the delta that can be discerned from the satellite image (Figure 2). Part of the Tsavo East National Park is located on the southern parts of the Tiva river basin.

Sub-Basins and Drainage

The main tributaries of the Tiva river occurs in its upper basin. These are Kauwi, Kalundu, Nzeeu, Mikuyuni and Mwitasyano, among others that flow only during rainy seasons. The seasonal rivers are generally characterised by low flows in dry season and high flows during rainy seasons, i.e. April-May and November-December. Most of the ephemeral streams generally become dry within one month after the rainy season (cf. Borst & De Haas, 2006). The basins are also highly degraded due to various anthropogenic activities such as cultivation, settlements and livestock grazing. The upper parts of the Tiva river basin in which this study was undertaken is one of the relatively 'high potential' areas within Kitui County and therefore in the recent years, there has been a significant increase in human and livestock population. There has been an increase in small scale irrigation on the banks of the river. As it will be demonstrated in this study, small-scale irrigation projects along the river are becoming unsustainable due to soil salinisation. This occurs in dry seasons when irrigation of crops is most critical.

Socio-Economic Characteristics

The population within the basin is estimated to be 1,060,000. The population density is relatively low being about 33 people per km. Rainfed agriculture is an extremely challenging activity given the sporadic and low rainfall in the basin which is often poorly distributed on spatial-temporary basis. The region particularly Kitui County has enormous potential for development of mining industries in view of its vast mineral resources. Plans are underway to commence mining of coal, iron ore and marble

Figure 1. Map of Tana Basin showing Tiva river sub-basin to south and flowing into the Tana river as Laga Kokani in the area north of Garsen Town. The elevations above sea level are expressed in feet.

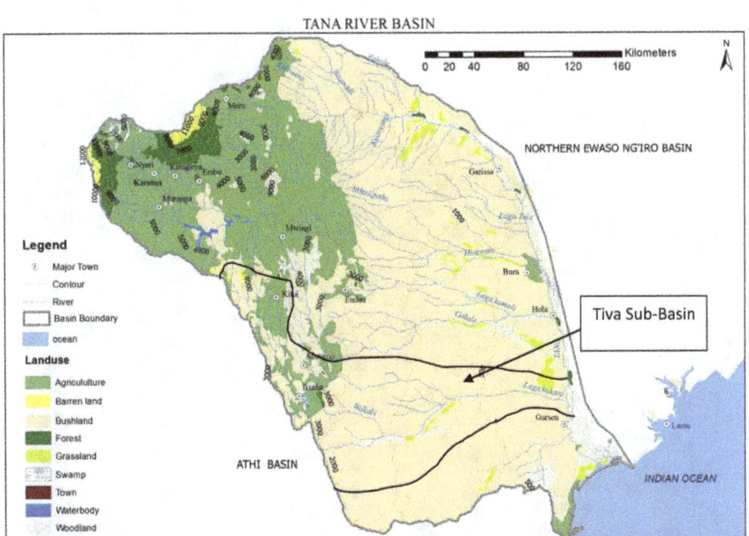

(limestone). The development of mining industry is expected to affect water quality in the Tiva sub-basin. Livestock grazing by Orma community is the predominant socio-economic activity in the dry lower zones of the Tiva sub-basin that extends into the Tana river County. However, in the upper zones, major landuse activities are mixed farming characterised by cultivation of crops and livestock grazing. Most of the farms are small-scale mixed farms <3ha, with the exception of the upper parts of the basin (Mwitasyano and Kauwi sub-basins) where large-scale cultivation of crops is common landuse activity in addition to livestock grazing.

METHODOLOGY

Streamflow Measurements

This study was undertaken within the main Tiva river and also in seasonal rivers draining into the river. These include Mwitasyano, Kauwi, Nzeeu and Kalundu. Sampling stations were established within each of the sub-basins of the Tiva river

Figure 2. Satellite image of southern region of Kenya including Kitui region and Tiva Sub-basin (Source: USGS and NASA). The border to the west is the Yatta Plateau that forms the divide with the Athi river.

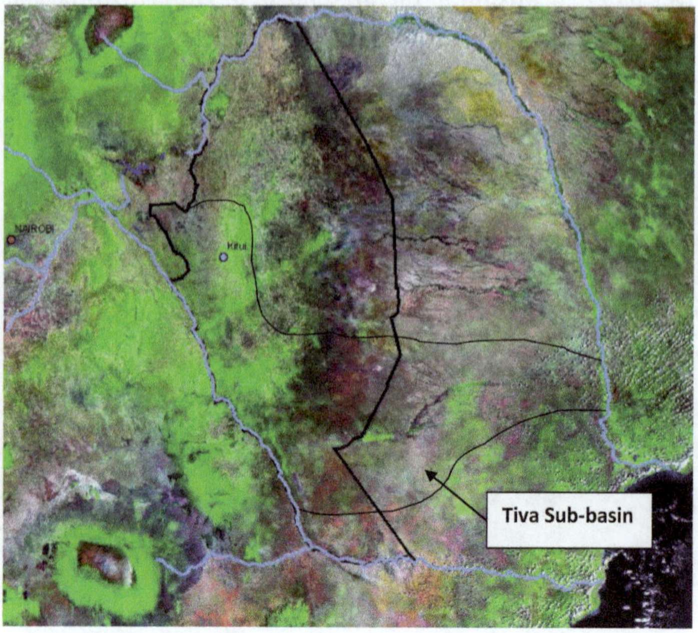

system. In these stations, measurements of river discharges and physico-chemical parameters were undertaken in the period between March 2013 and December 2015.

River discharges were measured on weekly and monthly basis at sampling stations using standard hydrological procedures as described in Linsley *et al.* (1998). River discharges were determined using the cross-sectional-area velocity approach (see also Chapman, 1996 and Bartram and Ballance, 1996). Manning roughness coefficient was computed according to equation 1.

$$Q = \frac{1}{n} A R^{2/3} S^{1/2} \tag{1}$$

where: Q = maximum discharge in riverbed section (m^3/s); n = Manning roughness coefficient of riverbed; A = wetted cross-sectional area (m^2); P = wetted perimeter (m); R = hydraulic radius (m) = A/P, and, S = slope of riverbed (m/m).

River discharge data were correlated with the data on salinity, conductivity and TDS in each of the sampling stations. The data was also used to compute monthly and annual salt flux rates. The stream morphology was determined through measurement of the gradient, width and depth of the river channel. The sediment characteristics were based on the field observations and particle size analysis of sediment samples.

Measurement of Physic-Chemical Parameters

The main water quality parameters that were measured in the field are turbidity, temperature, total dissolved solids concentration, conductivity and salinity. Turbidity was measured using Hanna Instruments HI93703 Microprocessor turbidity meter capable of measuring turbidity in the range of 0 to 1000FTU. TDS, salinity, temperature and conductivity were measured using Martini Instruments Mi306 EC/TDS/NaCl/Temperature meter.

The study was undertaken in both dry and wet seasons. During rainy season, the frequency of sampling was daily in order to capture highly variable streamflow. During dry seasons, measurements of baseflow and sampling of physico-chemical parameters was undertaken once every week and occasionally once every month later in dry season. Because low flows were more frequent as compared to high streamflows, our measurements tended to have a greater concentration of low flow measurements as opposed to high flow ones.

Analysis of Hydrologic Data

The data analysis involved regression and correlation analysis that were performed to establish the relationship between streamflow and salinity, conductivity and TDS and other control variables (e.g turbidity). Correlation coefficient (*r*) and Coefficient of Determination (R^2) were generated using Microsoft Excel Statistical Analyses Tools ($p = 0.05$). The best regression curves for the relationships were fitted using either linear, logarithmic or power functions depending on the strength of the relationship as represented by R^2 coefficient. Also, we determined standard deviations, mean values and maximum and minimum values of the measured parameters.

Determination of Salt Fluxes

The computation of salt fluxes in the main branch of the Tiva river as well as in the main sub-basins were based on the results of river discharge and TDS measurements that were undertaken simultaneously in established monitoring stations. The mean monthly TDS concentrations $\bar{C}$ was computed according to the following equation 2

$$\bar{C} = \sum_{i=1}^{n} \frac{C_i}{n} \tag{2}$$

In the above equation, $\bar{C}$ is the mean monthly concentration and C_i is the monthly TDS concentration. The mean river discharge $\bar{Q}$ was computed using the following equation 3

$$\bar{Q} = \sum_{i=1}^{n} \frac{Q_i}{n} \tag{3}$$

In the above equation, $\bar{Q}$ is the mean monthly river discharge and Q_i is the monthly river discharge.

The mean monthly salt fluxes were computed using the following equation 4

$$S_f = \sum_{i=1}^{n} \overline{\bar{Q}C} \tag{4}$$

In the above equation, S_f is the mean monthly salt flux, $\bar{C}$ is the mean TDS concentration and $\bar{Q}$ is the mean river discharge. The TDS concentrations (C) in the river occurring during the time between two samplings were extrapolated according to constant concentration and constant flux approaches (Chapman, 1996). The monthly salt fluxes (Sf) were extrapolated to obtain the annual salt fluxes for the main river as well as for the tributaries draining from the sub –basins.

Monthly measurements of river discharges, TDS, salinity and conductivity that were undertaken in the main branch of the river in the period between March 2014 and December 2014, were used to determine salt flux in the upper Tiva river basin. However, in the tributaries, only spot measurements were undertaken in dry and wet seasons to obtain insight on the magnitude of variability of streamflow and salt fluxes from the Kalundu, Mwitasyano, Nzeeu and Kauwi river sub-basins.

RESULTS

Variation of Streamflow

Tiva River Basin drains mainly from Kitui Mountain Ranges that are part of the Mozambique Basin Basement Complex composed of a variety of weathered metamorphic rocks (cf. Nyamai et al., 2003). The mountains are essentially anticlines rising to an altitude of 1800m a.s.l in Kitui Central in the upper parts of the basin. The major tributaries draining into the main Tiva river from the mountains are Kalundu, Nzeeu and Kauwi rivers. Mwitasyano river - a major tributary further to the north also receives flow contribution from the volcanic Yatta Plateau and Yatta Canal as irrigation return flow.

The flow of Tiva river is influenced by rainfall distribution in the upper parts of the basin in Kitui County. There are normally two periods when there is flow in the river, namely the long and short rainy periods. Rainfall is normally of relatively greater magnitude and is well distributed in most parts of the basin in the period between October and December. During the south–east monsoon in the period between March and May, rainfall amounts are relatively lower and rainfall is generally poorly distributed in the basin. The significant river discharges occur in the October-December period (short rains) and in the March-May period (long rains). The dry periods when there is no streamflow in the river are mainly January-March and June-September periods.

There is a significant variability in the river discharge rates of the sub-basins that form the larger Tiva basin as a result of influences of catchment sizes, morphology and landuse characteristics. Based on data collected in the period between March 2014 and December 2015, the mean discharges for the main tributaries of Tiva river range from 11 to 33 m^3s^{-1}. The maximum discharges for the main tributaries ranged between 118 and 210 m^3s^{-1}. The peak river discharge at the main Tiva River branch was 270 m^3s^{-1}. It must however be noted that streamflows < 0.4 m^3s^{-1} are more frequent in the main branch of the river. The flow hydrograph generated for one of the sub-basins showed that streamflow is characterized by relatively steep rising and falling limbs, short duration peaks and long recession period (Figure 2). The relatively long recession period indicates return of water into the main channel from bank storage and groundwater aquifers bordering the river. The low streamflows for most of the sub-basin streams were more frequent as they occurred >75% of the time as compared to high flows that occurred in < 15% of the time. The features are consistent with the characteristics typical seasonal rivers in the semi-arid regions. For all sub-basins, the high streamflows were characterized by relatively high turbidity that reached 1000 FTU. This was attributed to high rates of soil erosion in the basin and consequently high sediment load (see also Kithia, 1997; Kitheka, 2014).

Variation of Salinity as a Function of Streamflow

The salinity in the sub-basins of Tiva river shows significant variability and streamflow seemed to play an important role in the variability. During dry season in Mwitasyano river with streamflow < 0.1m^3s^{-1}, conductivity and TDS were 1,159 $\mu S.cm^{-1}$ and 2,324 mgl^{-1}, respectively, which are way above the WHO Drinking

Figure 3. Kalundu river streamflow hydrograph based on measurements conducted on 28-29th March 2013. The long recession period is evident in the stream.

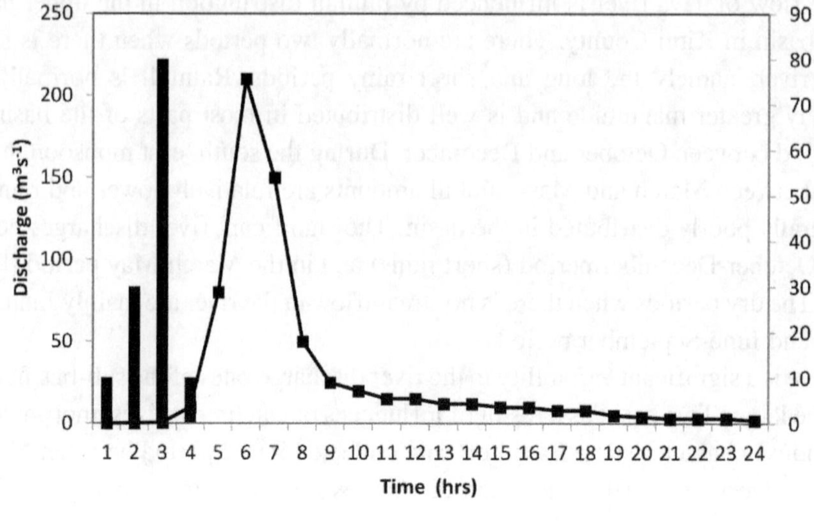

Water Standards (WHO, 1993, 2006 and 2008). However, during wet season with streamflow > 2 m^3s^{-1}, the conductivity and TDS ranged 217 - 491 $\mu S.cm^{-1}$ and108-246 mgl^{-1}, respectively. The wet season salinity ranged 0.4-0.8% and turbidity ranged 137-1000 FTU.

The variability in salinity was related to streamflow and therefore there was a rapid change in salinity flowing a decrease in river discharge. For instance, on 1st April 2013 with the streamflow rate of 118 m^3s^{-1}, the conductivity, TDS and salinity were 217.7 $\mu S.cm^{-1}$, 108.1 mgl^{-1} and 0.4%, respectively. However, the cessation of river discharge in the following day (1st May 2013) led to an increase of conductivity, TDS and salinity to 1,786 μScm^{-1}, 894 mgl^{-1} and 3.45%, respectively. The same patterns were observed in the case of Kauwi river. In this river, on 28th March 2013, we observed that when streamflow was 168 m^3s^{-1}, the levels of conductivity, TDS and salinity were generally low being 82$\mu S.cm^{-1}$, 40mgl^{-1} and 0.1%, respectively. However, following cessation of streamflow on 1st May 2013, the conductivity, TDS and salinity increased considerably to 400 $\mu S.cm^{-1}$, 200 mgl^{-1} and 0.8%, respectively

In Nzeeu sub-basin, on 28th March 2013 we observed that when the river discharge was 141m^3s^{-1}, the levels of conductivity, TDS and salinity were generally low being 63 $\mu S.cm^{-1}$, 31.4 mgl^{-1} and 0.1%, respectively. However, following cessation of streamflow on 1st May 2013, the levels of conductivity, TDS and salinity increased to 198 $\mu S.cm^{-1,}$ 98 mgl^{-1} and 0.4%, respectively. There are also day to day variations in the levels of salinity that are a result of day to day variations in streamflow which in turn is a response to the day to day variations of rainfall. For instance, on 29th March 2013 when the streamflow dropped to 5.95 m^3s^{-1}, the levels of conductivity, TDS and salinity increased to 159 $\mu S.cm^{-1}$, 79.5 mgl^{-1} and 0.3%, respectively.

In Kalundu sub-basin, the high flows (210 m^3s^{-1}) experienced on 28th March 2013 led to relatively low levels of conductivity, TDS and salinity that were 81 $\mu S.cm^{-1}$, 40 mgl^{-1}, 0.1%, respectively. However, 31st March 2013 following reduction in streamflow to 1.8 m^3s^{-1}, levels of conductivity, TDS and salinity increased to 156 μScm^{-1}, 78 mgl^{-1}, 0.3%, respectively. On 1st May 2013, when the stream was low (0.2 m^3s^{-1}), the levels of conductivity, TDS and salinity were generally high being 246 $\mu S.cm^{-1}$, 123 mgl^{-1}, 0.4%, respectively.

In the main Tiva river branch which receives streamflow from Nzeeu, Kalundu and Kauwi rivers, we observed similar patterns. For instance on 28th March 2013, the high streamflow rate of 182 m^3s^{-1} resulted in relatively low levels of conductivity, TDS and salinity that were of the order 112 $\mu S.cm^{-1}$, 80 mgl^{-1}, 0.2%, respectively. Following reduction of streamflow to < 0.01 m^3s^{-1} on 1st May 2013, the levels of conductivity, TDS and salinity increased to 455 $\mu S.cm^{-1}$, 227 mgl^{-1}, 0.9%, respectively. In December 2015 during low streamflow discharge of 0.4 m^3s^{-1}, the conductivity, TDS and salinity were relatively high being 534 $\mu S.cm^{-1}$, 263 mgl^{-1} and 1.0%, respectively. Thus, it can be concluded that the high streamflows were generally characterized by relatively low conductivity, TDS and salinity. There is in generally a rapid increase in salinity following cessation of high streamflow in the Tiva river which probably indicates the influence of water entrapment within the river channel including returnflow from bank storage and groundwater aquifers.

Variations of Salinity, TDS, and Conductivity in Nzeeu and Kalundu Sub-Basins

Table 1 shows the mean, maximum and minimum vales of various physico-chemical parameters associated with salinity that were measured in Nzeeu and Kalundu sub-basin rivers. The mean salinity and TDS for Nzeeu sub-basin were 0.3% and 76.3 mgl^{-1}, respectively. The mean salinity and TDS for Kalundu river were 0.3% and 86.7 mgl^{-1}, respectively. The lowest salinity and TDS were recorded in periods of maximum river discharges. On the other hand, the maximum salinity and TDS were recorded in periods of low river discharges and during periods of baseflow consisting partly of water trapped within the river channel in addition to effluent flows that originates from bank storage and groundwater aquifers.

The mean salinity and TDS including the streamflow characteristics were generally similar for the two sub-basins that originates from a forested Kitui hills at an elevation ranging 1400-1800m asl. This is despite the fact that Kalundu river recorded significant higher maximum river discharge as compared to Nzeeu river. This is attributed to the differences in landuse characteristics in the two sub-basins. Kalundu sub-basin is more degraded and is characterized by more extensive small-scale cultivation as compared to Nzeeu sub-basin.

Variations of Salinity, TDS, and Conductivity in the Tiva Sub-Basins

Table 2 shows the data on the various physico-chemical parameters measured at the Tiva sub-basins at the upper and lower sections of the river (Kabati and Kwa Vonza sampling stations). The mean salinity for the main branch of Tiva river was 0.3% while TDS ranged between 88 and 92 mgl^{-1}. The maximum salinity and TDS were 0.8% and 225 mgl^{-1}, respectively. These occurred during dry season when river discharges were <0.01 m^3s^{-1}. The lowest salinity, conductivity and TDS were measured during periods of high river discharge ranging between 137 and 216 m^3s^{-1} with the turbidity averaging 799 FTU.

Table 1. The statistical values for various physico-chemical parameters measured at Nzeeu and Kalundu rivers

	Turb (FTU)		Sal (%)		Cond (µS.cm^{-1})		TDS (mg.l^{-1})		Q_f (m^{3}s^{-1})	
	Nzeeu	Kalundu	Nzeeu	Kalundu	Nzeeu	Kalundu	Nzeeu	Kalundu	Nzeeu	Kalundu
Mean	452.4	383.8	0.3	0.3	152.8	173.9	76.3	86.7	11.3	13.1
Max	1000	1000	0.4	0.5	234.4	262.9	117.1	131.5	141.8	210.6
Min	53	47	0.1	0.1	59	81	29.5	40	0.1	0.2
STD	340.7	314.8	0.1	0.1	51.7	57.4	25.8	28.5	28.6	43.3

Table 2. The statistical values for various physic-chemical parameters at Tiva near Kabati and Tiva at Kwa Vonza

	Turb (FTU)		Sal (%)		Cond (µS.cm^{-1})		TDS (mg.l^{-1})		Q_f (m^{3}s^{-1})	
	Tiva (Kabati)	Tiva (Upper)	Tiva (Kabati)	Tiva (Upper)	Tiva (Kabati)	Tiva (Upper)	Tiva (Kabati)	Tiva (Upper)	Tiva (Kabati)	Tiva (Upper)
Mean	798.75	680.3	0.3	0.3	175.4	183.8	88.1	92.4	32.6	54.6
Max	1000	1000	0.8	0.9	441	455	223	227	136.5	216
Min	84	13.7	0.1	0.1	81	84	40.5	41.4	0	0
STD	362.9	308.3	0.2	0.2	121.7	117.8	61.6	58.3	56.1	77.4

Variations of Salinity, TDS, and Conductivity in Kauwi and Mwitasyano Sub-Basins

Table 3 shows the key statistical parameters for the Kauwi and Mwitasyano rivers draining into the Tiva river from the northern parts of the basin in an area with extensive large-scale cultivation and to some extent irrigation. The mean salinity and TDS for the two rivers were not significantly different. However, the maximum salinity and TDS were significantly difference. Mwitasyano river had the highest salinity and TDS of 3.4% and 894 mgl^{-1}, respectively, as compared to the values for Kauwi that were 0.8% and 200 mgl^{-1}, respectively. These differences could be attributed to the differences in the maximum river discharges experienced in the two rivers. Kauwi river experienced much higher river discharges as compared to Mwitasyano river during the period of measurements. However, Kauwi river tended to experience the lowest concentrations of salinity and TDS during high flow conditions. On the other hand, Mwitasyano river experienced the highest concentrations during low flow conditions.

Table 3. The Statistical Values for Kauwi River (Kabati) and Mwitasyano River

	Turb (FTU)		Sal (%)		Cond (μS.cm^{-1})		TDS (mg.l^{-1})		Q_r (m^3s^{-1})	
	Kauwi	Mwita Syano	Kauwi	Mwita Syano	Kauwi	Mwita Syano	Kauwi	Mwita Syano	Kauwi	Mwita Syano
Mean	674.1	467.2	0.33	1.1	164.2	727.7	83.4	364.0	30.3	32.60
Max	1000	1000	0.8	3.4	400	1786	200	894	168	118.13
Min	6.5	0.8	0.1	0.4	82	217.7	40	108.1	0.01	0.01
STD	423.7	476.1	0.23	1.4	112.4	714.9	55.4	358.1	59.9	57.18

Relationship Between Streamflow and Salinity

There is a relationship between variations of streamflow and salinity in various streams draining the sub-basins of the Tiva river. Figures 4, 8 and 12 shows the relationship between streamflow and salinity in the Nzeeu, Kalundu and Tiva rivers. The relationships are inverse so that an increase in streamflow results into a decrease in salinity, and the vice versa is true. Thus high streamflows are characterized by relatively low salinities as compared to low flows that are characterized by relatively high salinity. For Nzeeu, Kalundu, Tiva and Mwitasyano sub-basins, the relationships were relatively strong with correlation coefficients r being 0.55, 0.83, 0.88 and 0.96, respectively. The relationships were best represented by the regression power functions as opposed to linear regression functions. However, the extent to which streamflow can be used to predict variations of salinity in Nzeeu river was limited by the low coefficient of determination R^2 of 0.30 indicating that variations in streamflow explained salinity variations by 30% . The R^2 values were however much higher for Kalundu, Tiva and Mwitasyano sub-basins; 0.69, 0.78 and 0.93, respectively. This means the power functions derived for the relationship between streamflow and salinity in Kalundu, Tiva and Mwitasyano sub-basins can be used to predict salinity quite effectively. The streamflow directly influences salinity such that level of salinity reduces as the streamflow increases. For the most part, higher streamflows dilute salinity, a finding that has also been reported in the case of Colorado river in the US (U.S. Department of the Interior, 2011).

Relationship Between Streamflow and TDS

There is also a strong relationship between streamflow and TDS concentration in the sub-basin rivers that drains to the larger Tiva river. Figures 5, 10, 14 shows the results of the relationship between streamflow and TDS concentration in Nzeeu, Kalundu and Tiva sub-basins. The relationships were also relatively strong with the *r* values of 0.44, 0.79 and 0.86 and 0.99 for Nzeeu, Kalundu, Tiva and Mwitasyano rivers, respectively. The relationships are inverse so that an increase in streamflow leads to a decrease in TDS concentration. The high streamflows were characterized by relatively low TDS concentrations as compared to low streamflows that were characterized by high TDS concentrations. The relationships were also represented best by regression power functions. The extent to which streamflow can be used to predict TDS concentrations using the regression power functions was found to be good in the case of Kalundu river (R^2=0.62), Tiva (R^2=0.75) and Mwitasyano river (R^2=0.99). For Nzeeu river, the variability of streamflow explained variability in TDS concentrations by only 19% (R^2 = 0.19) with the implication that the derived regression power function cannot be used to predict variability of salinity in the river. The high TDS concentrations during the low streamflow conditions were attributed to accumulation of mineral salts as a result of seepage of water from bank storage and groundwater aquifer as well as evaporation of water in the sandy river channel.

Relationship Between Streamflow and Conductivity

There is a relationship between streamflow variability and the variations of water conductivity in all sub-basins of the larger Tiva Basin. Figures 6, 9 and 13 shows the plots of the relationship between streamflow and conductivity in Nzeeu, Kalundu, and Tiva sub-basins. The relationships were inverse with relatively strong *r* values of 0.79, 0.90 and 0.97 for Kalundu, Tiva and Mwitasyano rivers respectively. The *r* value for Nzeeu was however low (r = 0.44). The relationships were also inverse indicating that conductivity declines as streamflow increases. The highest conductivity values were measured during periods of low streamflow conditions with water entrapped within the sandy river channel. The relationships were also best represented by regression power functions, with the extent to which streamflow can be used to predict conductivity varying from one sub-basin to the other. The R^2 values were also relatively high for Kalundu, Tiva and Mwitasyano rivers (R^2values of 0.62, 0.80, 0.93, respectively) meaning that the regression power functions derived for these sub-basins can be used predict variations of conductivity. The R^2value for Nzeeu river was relatively low (R^2= 0.20) thus limiting the extent to which the derived equation can be used to predict variations of conductivity in the river.

Relationship Between Salinity and Turbidity

There is a relationship between turbidity and salinity in all the sub-basins of the larger Tiva basin. Figures 4, 5 and 8 show the plots of the relationship between turbidity and TDS in Nzeeu, Kalundu and Tiva sub-basins. The relationships were inverse so that an increase in turbidity was associated with a decrease in TDS and subsequently salinity and conductivity. Thus, periods of high streamflows with highly turbid water were characterized by relatively low TDS concentrations and conductivity. Periods of low turbidity were associated with low streamflow conditions. However, TDS, salinity and conductivity tended to be high during the low flow periods. The relationships were best represented by linear regression equations with the exception of Kalundu river where a logarithmic regression function described the relationship more satisfactorily. The *r* values for Nzeeu, Kalundu, Tiva and Mwitasyano rivers were 0.88, 0.87, 0.87 and 0.93, respectively. The equations that were derived from the relationships had relatively high R^2 values of 0.77, 0.84, 0.75 and 0.93 for Nzeeu, Kalundu, Tiva and Mwitasyano rivers, respectively.

The variations of streamflow have an important influence on the variability of turbidity. During high flow conditions, the streams receive highly turbid runoff derived from degraded land. This explains the high turbidity (700-1000 FTU) that was measured in streams during periods of high flows. Turbidity during low flow conditions were of the order 0.7 FTU indicating presence of relatively clear water. Observations also showed that during periods of high river discharges, the sediment load consisted mainly of suspended sediment load with high proportion of clay, sand and silt (Kitheka, 2013 and 2014). There is thus high concentration of materials that do not conduct electricity more effectively, Thus, in addition to the dilution effect of increased freshwater volume, the low conductivity during high flows can also be attributed to the presence of high proportion of non-conductance inorganic materials such as clay, sand and silt (see also Bhattet al., 1999; Vaishal and Punita, 2013).

Figure 4. Relationship between streamflow and salinity in Nzeeu river

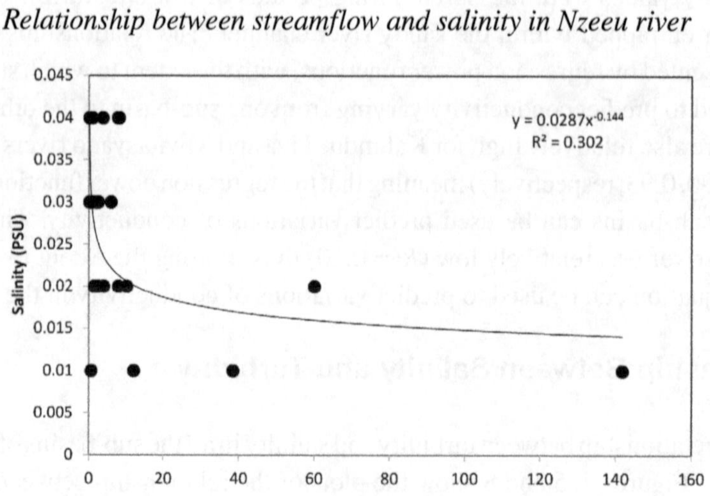

Figure 5. Relationship between stream flow and TDS in Nzeeu river

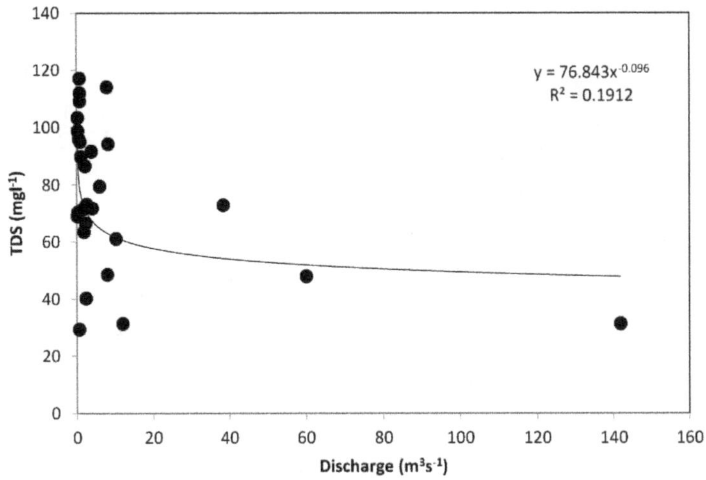

Figure 6. Relationship between stream flow and conductivity in Nzeeu river

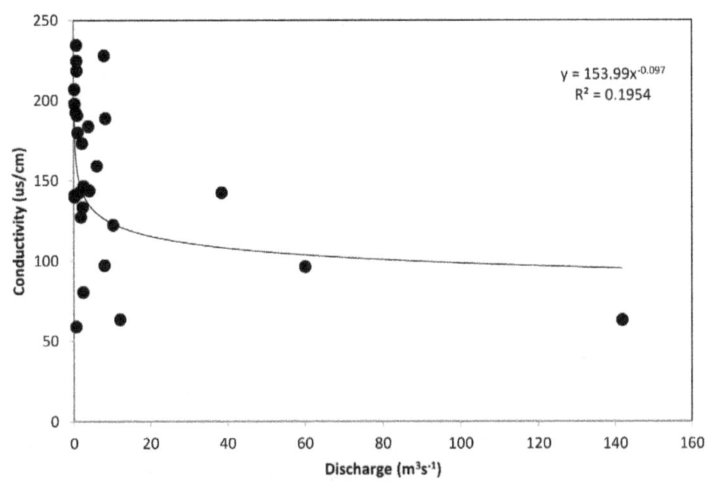

Relationship Between Streamflow and Salinity-Combined Data for All Sub-Basins

Data for all tributaries draining into the Tiva river were combined and correlation and regression analyses conducted to establish the extent of the relationship between streamflow and physico-chemical parameters. The results of regression and correlation analyses showed weak relationships. For instance, the relationships between streamflow and conductivity yielded a R^2 value of 0.34 while that

Figure 7. Relationship between TDS and turbidity in Nzeeu river

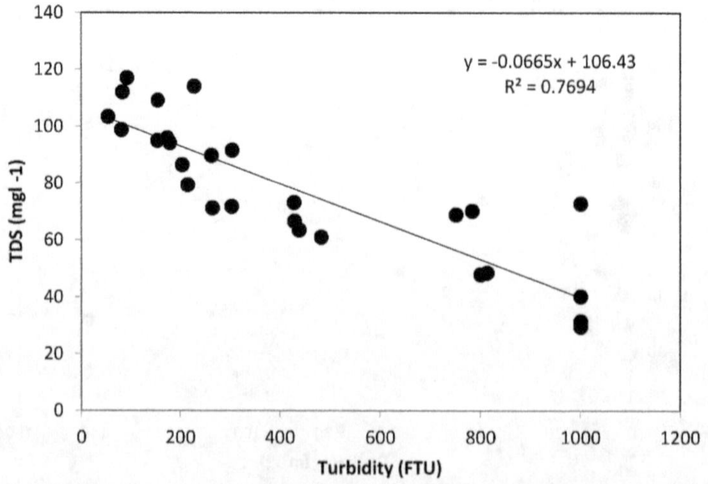

Figure 8. Relationship between stream flow and salinity in Kalundu river

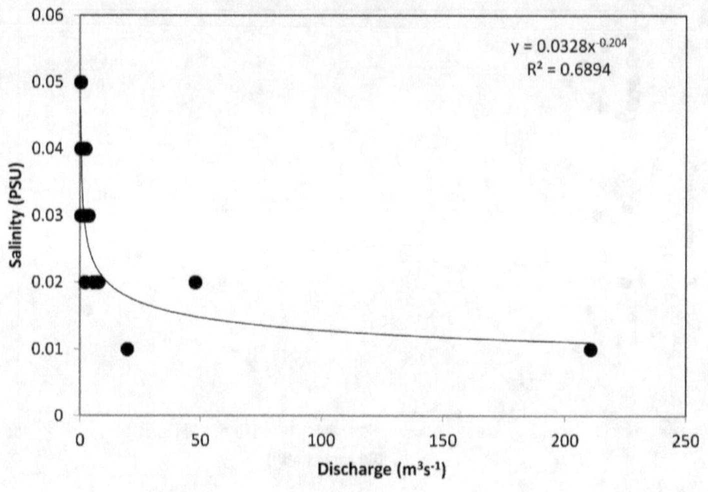

between streamflow and TDS yielded a R^2value of 0.28. The relationship between streamflow and salinity yielded a R^2 value of 0.44. Although the relationships were best represented by power equations of the form $y = bX^{-e}$, the weak relationships indicate that these power function equations cannot accurately predict the related variables effectively. The weak relationships were attributed to the fact that each river subbasin presents unique combination of hydrologic, geologic, and landuse characteristics that contributes to the measured levels of salinity in their outlets.

Figure 9. Relationship between stream flow and conductivity in Kalundu river

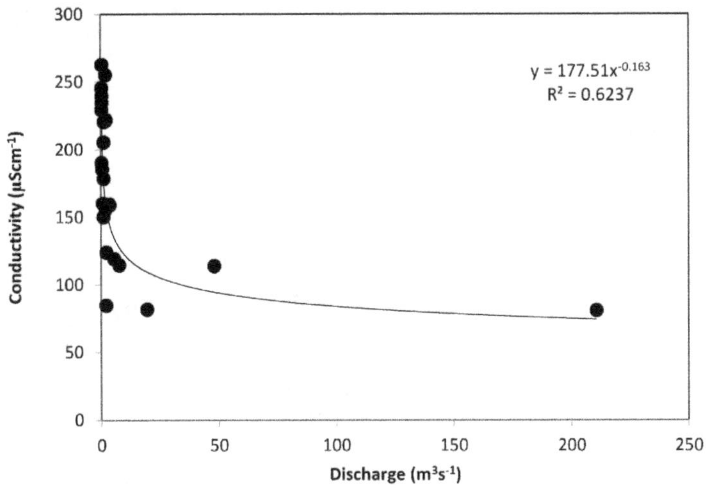

Figure 10. Relationship between stream flow and TDS in Kalundu river

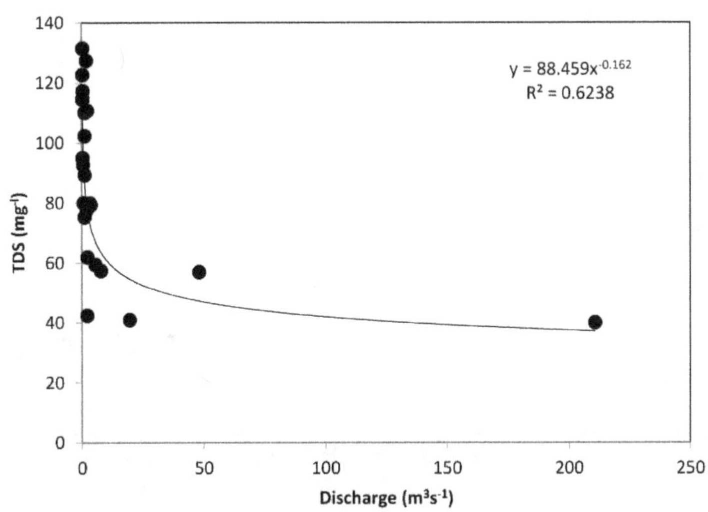

Salt Fluxes and Discharge of Salt Downstream

The results showed significant fluxes of salt in the Tiva river. The salt is derived from the sub-basins that drains into the main river. Tables 4, 5, 6 and 7 shows the computed mean, maximum and minimum salt fluxes in the major rivers in the Upper Tiva Basin. Table 4 shows the total salt fluxes computed as a product of the mean streamflow rate (Q) and TDS concentrations and expressing the products in terms

Figure 11. Relationship between TDS and Turbidity in Kalundu river

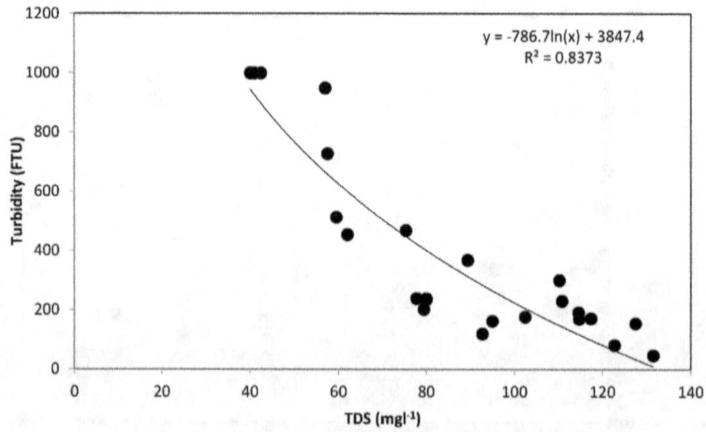

Figure 12. Relationship between streamflow and salinity in Tiva river

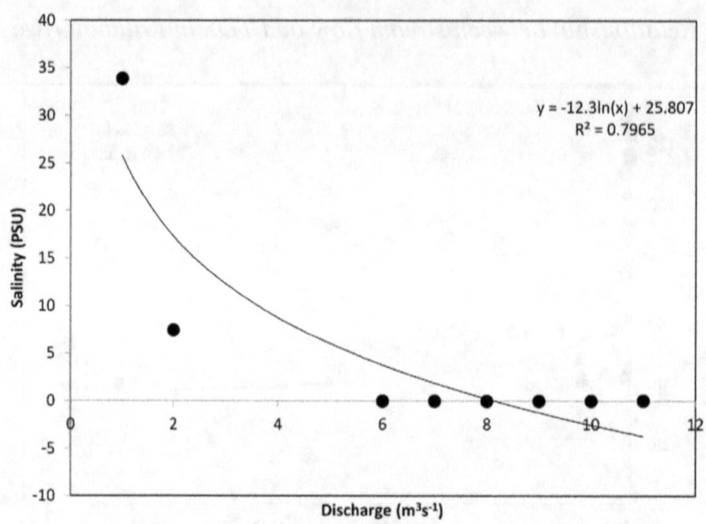

of salt flux per day and subsequently salt flux per year. The computations are based on an estimated period of 60 days when the river experienced streamflow that can be considered average. The assumption is that the seasonal rivers experience average streamflow rates for a period of 30 days in each of the short and long rainy seasons (see also Lasage et al., 2008).

The results showed that the total salt flux in the upper Tiva basin is 100,344 tons.y^{-1} (Table 7). The basin salt production rate varies from 3.74 to 61.53 tons.m^{-2}.yr^{-1}. Off all the sub-basins considered in this study, Mwitasyano had the highest

Figure 13. Relationship between stream flow and conductivity in Tiva river

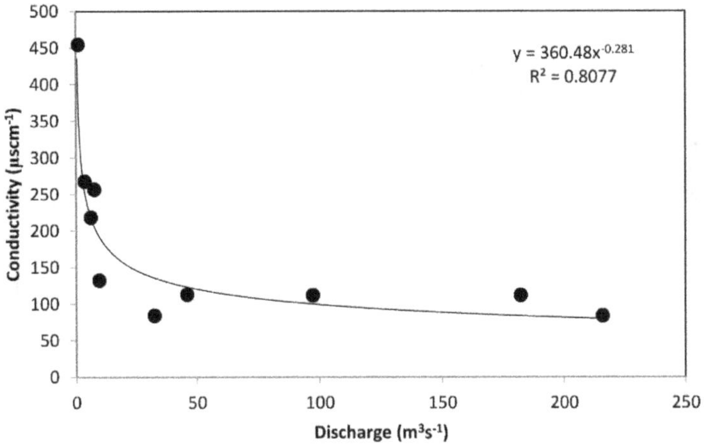

Figure 14. Relationship between stream flow and TDS in Tiva river

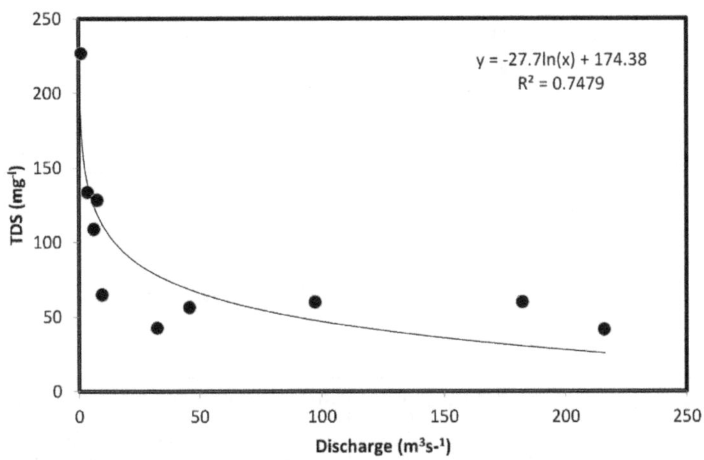

salt flux that was attributed to the high salinity measured in the river. Thus, the majority of salt load emanates from Mwitasyano sub-basin that contributes 61% of the total salt flux in the Tiva Basin. The other sub-basins contributes about 40% of the total salt load. There is a possibility of additional salt input into the Tiva river as flows downstream since a number of other small tributaries joins the river south of Tiva Market.

Most of the salt load ends up in the lower Tana Basin in an area where the river forms a large inland delta (Ndarapo Swamp). The deposition of salt in the lower basin explains the whitish patches that are visible in the LANDSAT Satellite image

Figure 15. Relationship between TDS and Turbidity in Tiva river

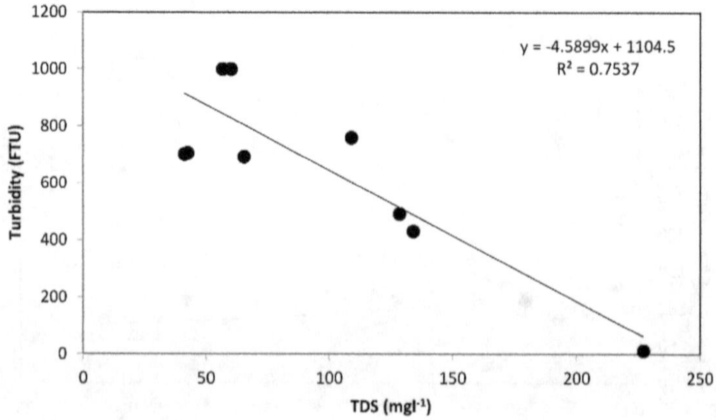

Table 4. The computed values for salt flux in the Nzeeu and Kalundu Rivers

	TDS (mgl^{-1})		River Discharge (m^{3}s^{-1})		Salt flux (tons.day^{-1})	
	Nzeeu	**Kalundu**	**Nzeeu**	**Kalundu**	**Nzeeu**	**Kalundu**
Mean	76.33	86.69	11.35	13.10	74.86	98.11
Max	117.1	131.5	141.75	210.6	1,434.15	2,392.75
Min	29.5	40	0.108	0.18	0.28	0.63
STD	25.84	28.46	28.61	43.27	63.87	106.38

Table 5. The computed values for salt flux in the Tiva River (Kabati) and Tiva (Kwa Vonza)

	TDS (mgl^{-1})		River Discharge (m^{3}s^{-1})		Salt Flux (tons.day^{-1})	
	Tiva (Kabati)	**Tiva (Vonza)**	**Tiva (Kabati)**	**Tiva (Vonza)**	**Tiva (Kabati)**	**Tiva (Vonza)**
Mean	88.13	92.42	32.59	54.59	248.13	255.60
Max	223	227	136.5	216	2,629.97	944.39
Min	40.5	41.4	0	0	-	0
STD	61.58	58.31	56.09	77.35	298.45	332.08

(Figure 3). It is not certain whether all salts transported by Tiva river enters into Tana river through Laga Kokani or are retained within Ndarapo swamp. This is an area requiring further investigation since the river seems to lose substantial part of its volume once it leaves the swamp. Tiva basin salt flux is can be considered low

Table 6. The computed values for salt flux in the Kauwi near Kabati and Mwitasyano

	TDS ($mg.l^{-1}$)		River Discharge (m^3s^{-1})		Salt Flux ($tons.day^{-1}$)	
	Kauwi	Mwitasyano	Kauwi	Mwitasyano	Kauwi	Mwitasyano
Mean	83.44	364.03	30.29	32.60	218.36	1,025.47
Max	200	894	168	118.13	2,903.04	9,124.16
Min	40	108.1	0.02	0.01	0.05	0.08
STD	55.36	358.07	59.901	57.18	286.51	1,768.95

Table 7. Total salt flux in the main Upper Tiva River at Tiva Market

	Sub-basin River	Sub-basin Area (Km^2)	Mean Annual Salt Flux ($tons.yr^{-1}$)	Mean Annual Salt Flux per unit area ($tons.m^{-2}.yr^{-1}$)	Contribution to the total salt flux (%)
1.	Nzeeu	1,200	4,491	3.74	4.48
2.	Kalundu	800	5,887	7.36	5.87
3.	Mwitasyano	1,000	61,528	61.53	61.32
4.	Tiva (Lower)	400	15,336	38.34	15.28
5.	Kauwi	200	13,102	65.51	13.06
	Total	**3600**	**100,344**	**27.87**	**100**

when compared to the salt flux in the Colorado River System in United States where between 1940 and 1980, an average of approximately 9.4 million tons of salt per year were carried down the river every year. Since 1981, on average, approximately 8.8 million tons of salts have been measured in the Colorado river each year (U.S. Department of the Interior, 2011).

DISCUSSION

Influence of Streamflow on Salinity

The variability of salinity, total dissolved solid and conductivity in the seasonal Tiva River Basin is influenced by processes associated with streamflow as well as flow of subterranean water into the river channel. The variability in salinity is related to streamflow and therefore there is a rapid change in salinity following a change in river discharge. There are also day to day variations in the levels of salinity such that the decrease of streamflow subsequently leads to an increase in salinity. There is usually a rapid increase in salinity during low flow conditions and this occurs within a period of < 1 month following cessation of rainfall in the basin. The rapid increase in TDS and salinity were attributed to the bank storage return flow and the subterranean groundwater flow to the river channel during low flow conditions. Water flowing into the river channel from groundwater aquifers and bank storage has high TDS concentration (cf. Kumaraswamy, 1991; Dembere 1998; Jain, 1998). The entrapment of water in deep metamorphic geological substratum leads to high residence time of subterranean groundwater. This allows more dissolution of rock minerals, thus increasing the level of TDS, salinity and conductivity. The mineralization of ground water due to entrapment and groundwater recharge of solubilised minerals from rocks and soils thus explains the high TDS and hence high conductivity. Higher mineralization imparts bad taste on potable water (Kumaraswamy, 1991; Dembere 1998; Jain, 1998).

The conductivity levels in Nzeeu river ranged 59 to 234 μS.cm^{-1} with STD of ±51.4 μS.cm^{-1} while that in Kalundu river ranged 81 to 262 μS.cm^{-1} with STD of 57.4 μS.cm^{-1}. The conductivity in the lower Tiva (kwa Vonza) ranged 81 to 445 μS.cm^{-1} with STD of 121.7 μS.cm^{-1}. The conductivity in Nzauwi river ranged 82 to 400 μS.cm^{-1} with STD of 112.4 μS.cm^{-1}. The level of conductivity in Mwitasyano river ranged 217.7 to 1786 μS.cm^{-1} with STD of 714.9 μS.cm^{-1}. The conductivity levels were generally much higher than ranges of 83 to 204 with STD of 45μS.cm^{-1}reported by Waziri and Ogugbuaja (2012) in Yobe river in Nigeria. A study in polluted Thome river in Nairobi reported conductivity ranging 160-496 μS.cm^{-1} (Karanja et al., 2015) which is higher than that recorded in all Tiva sub-basins, with the exception of Mwitasyano sub-basin where relatively much higher values were measured. In most cases, the maximum conductivity and TDS were higher than the maximum permissible drinking water conductivity of 150 μS.cm^{-1} and 500 mgl^{-1}, respectively (WHO, 1993). The mean conductivity in Mwitasyano of 727.7μS.cm^{-1} (Salinity = 1.1%; TDS = 364 mgl^{-1}) was generally much higher as compared to other sub-basins of the Tiva river. As will be discussed in the later sections, landuse seems to be playing an important in determining salinity levels in the Mwitasyano sub-basin.

The highest TDS, conductivity and salinity value were measured during the low flow conditions and lowest during high flow conditions. The low salinity levels during turbid high streamflow conditions can be attributed to increased ionic compounds in the water. This results in decrease in TDS, salinity and conductivity with an increase in turbidity as has been reported elsewhere by Waziri and Ogugbuaja (2012). The highest values measured during the low flow conditions were also attributed to the high evaporation rates that ranges between 1500 and 1600 mm.yr^{-1} (Lasage et al., 2008). However, in the case of Mwitasyano river, the effects of high evaporation rates on salinity are aggravated by intensive cultivation/irrigation activities upstream.

In this area, land approximating 30km^2 has been placed under intensive cultivation including irrigation using water from the Yatta canal. The irrigation return flows are usually channeled into Mwitasyano river. Irrigation increases the salt concentration of the source water by consuming water (evapotranspiration) and by dissolving salts in the underlying soil and geologic formations (U.S. Department of the Interior, 2011). In the case of other sub-basins of Tiva river, irrigation activities did not seem to an important factor in the variability of TDS, conductivity and salinity.

The relatively higher salinity levels in Mwitasyano river as compared to other sub-basins could also be explained by the nature of the drainage basin. The sub-basins such as Nzeeu and Kalundu have steep gradients and are composed of 1st and 2nd order streams with relatively short slope lengths and hence experiences rapid flow response to rainfall. In such streams, surface runoff generated by rainfall storms reaches the main river channel quickly and hence there is little time to dissolve soil/rock minerals. This explains the low salt production rates in the Nzeeu and Kalundu sub-basin (3.74 and 7.36 tons.m^{-2}.yr^{-1}, respectively). However, in the sub-basins such as Mwitasyano which is 3rd order stream receiving flow contributions from 1st and 2nd order streams with relatively gentle slope and longer slope lengths, the surface runoff generated by rainfall storms is able to infiltrate and percolate into the soil slowly enabling greater dissolution of soil/rock minerals. The subterranean water loaded with dissolved salts subsequently flows into the main river channel during periods when streamflow is subsiding. The maximum flow of the subterranean water into the river is attained when the hydraulic gradient is favourable during periods when streamflow has subsided.

During periods of low streamflow (and hence baseflow), there is a tendency for TDS concentrations to exceed the maximum permissible limit in drinking water of 1,000 mg.l^{-1} (WHO, 1993). This imparts unpalatable taste to water and makes water unsuitable for irrigation due to heavy mineralization. However, the maximum TDS concentrations in the river are still lower than 2,000 mg.l^{-1} above which production of laxative effect as a result of the presence of magnesium and sodium sulphate is expected (Kumaraswamy, 1991, Dembere 1998). The use of water for irrigation crop production is problematic since the TDS levels in the Mwitasyano river exceeds the maximum permissible limit of 500 mg.l^{-1}. It is thus expected that high TDS concentrations and hence high salinity levels in Mwitasyano river would affect irrigation crop production in dry season when irrigation of crops is most critical. This is attributed to the fact that the use of high salinity water for irrigation causes salinisation of the soils (Radiro, et al., 2003; Sijali et al, 2003).

Measurements undertaken during periods of high streamflow of 270m^{3}s^{-1} yielded TDS concentrations of 200 mgl^{-1} in the main Tiva river branch. During periods of low streamflow of 0.4m^{3}s^{-1}, TDS concentration increased to 263 mgl^{-1}. Thus the

TDS concentrations in the main branch of the river seem to be within the maximum permissible limit for drinking water and irrigation.

Non-Linear Relationships Between Streamflow and Salinity

This study established non-linear relationships between streamflow and salinity, conductivity and TDS in all the seasonal rivers draining into the Tiva river that were subject of this study. Most of the relationships were best described using a regression power function as opposed to linear regression function (cf. Williams, 1966). The non-linear relationship established in this study can be attributed to the variable nature of the conductive mobility of the ionic species in the water. The high turbidity of Tiva river and its sub-basin streams during high flow conditions is due to the high suspended sediment load (>80%). The suspended sediment load is usually composed of non-conducting clays, silt and sand materials. The relatively high levels of these materials, in addition to dilution, leads to relatively lower conductivity of water during periods of high streamflow (cf. Waziri and Ogugbuaja, 2012). Past studies conducted elsewhere have shown that complex particulate matter consisting of both inorganic and organic chemicals, as well as microbial entities affects TDS and conductivity (Kaufman et al, 2005; Centeno, et al., 2006; Waziri & Ogugbuaja, 2010). There are also seasonal signals in the variability (Koning and Ross, 1999) that are related to the seasonal variations of rainfall and subsequently streamflows.

Salt Fluxes Into the Lower Basin

The results of computation river salt loads showed that Tiva river transports a significant volume of salt into its lower basin (Tables 4 -7). The river on average basis discharges 100,344 tonnes of salt per year. The average total salt production per unit basin area was computed as 27.87 tons.m^{-2}.yr^{-1}. Measurements undertaken during periods of high streamflow of 270m^{3}s^{-1} yielded a TDS concentration of 200 mgl^{-1} in the main Tiva river branch. Using this data, the maximum salt flux at the main Tiva branch is estimated to be 4,665 tons.day^{-1} which translates to 279,900 tons.yr^{-1}. Thus, it can be stated that salt flux in the upper region of the Tiva river ranges between 100,000 and 280,000 tons.yr^{-1}. The results show that most of the salt emanates from Mwitasyano sub-basin where salt production rate per unit area was 61.53 tons.m^{-2}.yr^{-1}. Salt production in river basins is mainly through weathering of metamorphic rocks. The discharge and deposition of salt into lower basin perhaps explains the whitish patches that are visible in satellite image as the river forms a large inland delta before it reconstitutes and flows to the Tana river as Laga Kokani in the area north of Garsen Town. It is unlikely that all salts transported by Tiva river enters into Tana river due to the possibility of high deposition in the inland delta/swamp with complex drainage network (Figure 3). This is however an area requiring further research investigations. It is however important to note that salt flux of between 100,000 and 270,000 tons per year is much lower than that for Colorado River System in United States where an average of approximately 8.8 million tons of salts have been measured in the river each year (U.S. Department of the Interior, 2011). The relatively low salt fluxes can be attributed to relatively lower stream discharge rates and lack of extensive irrigation activities in the basin with the exception of Mwitasyano sub-basin where irrigation is practiced to a certain extent.

Implications on Water Resources and Agricultural Development

This study has demonstrated the dynamics of salinity production and transport in the upper Tiva river basin, as determined by the measurements of salinity, TDS concentrations and conductivity. While, the salinity, TDS concentrations and conductivity levels in the river are below the maximum permissible levels for drinking water and irrigation, there is a possibility of salinity increasing in future. This increase is possible due to climate change as reduced rainfall in the basin will consequently lead to the reduction of infiltration and percolation of water to the groundwater aquifers. This will reduce the extent to which the high salinity groundwater can be diluted by rainwater percolating through the geological substratum. The increased evapotranspiration due to increasing temperatures will also have the effect of concentrating salts in the shallow subterranean water systems. These salts will subsequently reach the groundwater system via infiltrating and deep percolation water. In addition to climate change, anthropogenic activities have also the potential of increasing salinity in water. This is possible due to increased irrigation activities in the basin. In the recent past, there has been an increase in the number of small scale irrigation projects in the basin, particularly along the banks of the Tiva river. This increase in irrigation activities is driven by the increasing unreliability of rainfed crop production which is characterized by high failure rates. Irrigation is seen by policy makers and the communities in the basin as the most appropriate approach of adapting to climate change. Unfortunately, small-scale irrigation projects are becoming unsustainable due to salinisation of the soils. This is already evident in the basin as exemplified by a high number of failed small-scale irrigation projects along the Tiva river channel. There is also a possibility of increasing river water salinity due to rapidly increasing pumping of groundwater in the basin. In the recent past, both the national and county governments have intensified the exploitation of groundwater through drilling of boreholes. Numerous deep and shallow boreholes are being drilled throughout the region in order to address the water scarcity problem. These boreholes normally pump a lot of groundwater particularly in dry seasons when other sources of water dries up. The use of highly saline groundwater for irrigation will also lead to the intensification of soil salinisation problem in the basin. In addition, during rainy season, salts in highly salinised soils will be eventually flow into the river channel and therefore intensify salinity problems in the river. High levels of salinity will make water unpalatable and reduce the extent to which river water can be used for domestic purposes. This will require the use of expensive and unsustainable water treatment technologies (e.g reverse osmosis) whose costs are beyond the reach of most people in the basin. In the recent past, there has been evidence that both livestock and wildlife are also affected following long consumption of highly saline water. Recently, the country lost nine endangered species of rhino that were translocated to Tsavo East National Park which is found in the lower parts of the Tiva River basin. The investigations done by wildlife officials revealed that the rhinos perished following consumption of highly saline water (Save the Rhino, 2018 and the Guardian, 2018). It is also important to note that use of highly saline water from the river can also cause corrosion and plugging of water supply pipes and fixtures in housing and industry (see also U.S. Department of the Interior, 2011). This has the potential of making rural water supply projects expensive and unsustainable due to high operation and maintenance costs.

While natural causes of salinity are difficult to control, anthropogenic factors that can lead to escalation of the salinity problem in the basin can be managed through targeted interventions (see U.S. Department of the Interior Report, 2011). There is a need for water resources and agricultural development programmes in the semi-arid Tiva River Basin to promote sustainable irrigation practices and control land degradation, particularly extensive clearance of vegetation to open land for agriculture and settlement. Retention of sufficient vegetation cover would increase infiltration and percolation of water during rainy season and therefore dilute saline groundwater that eventually flows into the rivers. Construction of large water storage dams in the Tiva river is also recommended. This will modify the flow of the river downstream, and hence significantly reduce and alter the salinity variability in the downstream areas. On long term basis, reservoirs can greatly reduce the levels of salinity in rivers through dilution of salinities in the river water (see also U.S. Department of the Interior, 2011). This will be beneficial to the local communities and their livestock and will also greatly support wildlife conservation in the basin and nearby areas. There is also a need to control exploitation of groundwater system through enforcement and or restriction of drilling for groundwater. Uncontrolled drilling of groundwater has the potential of leading to over exploitation of groundwater aquifers and this will lead to increased intrusion of saltwater into the aquifers. This water will eventually reach the river systems through groundwater flow. In addition, proper planning for human settlements is also required to limit the number of too many and widely scattered rural settlements and trading centres that exerts a lot of pressure of scarce water resources in the semi arid Tiva river basin.

CONCLUSION

This study focused on establishing the extent to which salinity in the seasonal Tiva river basin is influenced by streamflow. The study has established that salinity in the basin is subject of complex interrelationships between river hydrology, geology and to a certain extent landuse. The salinity levels are generally low with the exception of Mwitasyano tributary that receives irrigation return flow. There is a significant seasonal variation of salinity including the related parameters-TDS concentrations and conductivity. Low levels of salinity, TDS and conductivity were experienced during periods of high streamflow when the water was highly turbid. The low conductivity experienced during this period was attributed to the dilution effect of high streamflow and also due to the presence of high quantity of non-conductance inorganic materials such as sand and clay. The low streamflows were characterized by high levels of TDS and hence salinity and conductivity. This was attributed to the accumulation of dissolved solids associated with the bank storage return flow and seepage of subterranean groundwater into the river channel. High evaporation rates and lack of sufficient flow to flush out dissolved solids further intensifies the accumulation of salts during dry season. The salinity increases rapidly after cessation of streamflow in rainy season. The study showed that Tiva river transports between 100,000 and 270,000 tonnes of salt to the lower basin every year. While most of the salinity can be attributed to the nature of metamorphic rocks, irrigation seems to be contributing to high dry season salinity associated with low flow in Mwitasyano sub-basin. The study calls for water resources and agricultural development programmes in the basin to promote sustainable irrigation practices and control land degradation, particularly extensive clearance of vegetation to open land for agriculture and settlement. The construction of water storage reservoirs can help in controlling increases in salinity in the Tiva river. The study also calls for controlled exploitation of groundwater aquifers as well as proper planning for human settlements in the basin.

REFERENCES

Abida & Harikrishna. (2008). Study on the Quality of Water in Some Streams of Cauvery River. *Journal of Chemistry*, *5*(2), 377–384.

Abila, R., Muthangya, M., Mutuku, E., Mutati, K., Munguti, M., & Musyoka, C. M. (2012). Physico - chemical and bacteriological quality assessment of shallow wells in Kitui town, Kenya. *Jnl. Env. Sci. Wat. Res*, *1*(2), 27–33.

APHA (American Public Health Association). (1992). *Standard methods for the examination of water and wastewater* (18th ed.). Washington, DC: American Public Health Association.

Batram, J. B., & Balance, R. (1996): Water quality monitoring: a practical guide to the design and implementation of freshwater quality studies and monitoring programmes. Published on behalf of UNESCO, WHO and UNEP by E$FN SPON.

Borst, L., & de Haas, S. A. (2006). *Hydrology of Sand Storage Dams - A case study in the Kiindu catchment, Kitui District, Kenya*. Free University of Amsterdam.

Beimers, P. B., van Eick, A. J., Lam, K. S., & Roos, B. (2001a). Improved design sand-storage dams, Kitui District, Kenya. Project report, Delft University of Technology.

Beimers, P. B., van Eick, A. J., Lam, K. S., & Roos, B (2001b). Building sand-storage dams, SASOL Foundation Kitui District, Kenya. Practical work report, Delft University of Technology.

Bhatt, L. R., Lacoul, P., Lekhak, H. D., & Jha, P. K. (1999). Physicochemical characteristics and phytoplankton of Taudha Lake Kathmandu. *Pollution Research*, *18*(14), 353–358.

Bossenbroek, J., & Timmermans, T. (2003). *Setting up a measuring program at Kisayani, to measure the affected area by sand storage dams, Traineeship Report*. Delft University of Technology.

Burger, A. S., Malda, W., & Winsemius, H. C. (2003). *Research to Sand-storage dams in Kitui district*. Delft University of Technology.

Busulwa, H. S., & Bailey, R. G. (2004). Aspects of the physico-chemical environment of the Rwenzori Rivers, Uganda. *African Journal of Ecology*, *42*(1), 87–92. doi:10.1111/j.1365-2028.2004.00467.x

Centeno, J. A., Cook, A., & Weinstein, P. (2006). Environmental Toxicology and exposure to natural dust: The role of trace elements. *Chinese Journal of Geochemistry*, *25*(1), 222–223. doi:10.1007/BF02840174

Chapman, D. (1996). Water quality assessments; a guide to the use of biota, sediments and water in environmental monitoring (2nd ed.). Published on behalf of UNESCO, WHO and UNEP by E$FN SPON.

Charkhabi, A. H., & Sakizadeh, M. (2006). Assessment of spatial variation of water quality parameters in the most polluted branch of the Anzali Wetland, Northern Iran. *Polish Journal of Environmental Studies*, *15*(3), 395–403.

Dhembere, A. J., Pandhe, G. M., & Singh, C. R. (1998). Groundwater characteristics and their significance with special reference to public health in Pravara area. *Maharastra. Poll. Res.*, *17*(1), 87–90.

Elmoustafa, A.M. (2013). Sustainable water management for seasonal rivers deltas, case study: Coporolo river, Angola. *IJRRAS, 15*(3).

Jain, C. K. (2004). Groundwater quality of District Dehradun, Uttranchal. *Ind. J. Env. & Ecoplan.*, *8*(2), 475–484.

Karanja, A., Kotut, K., & Gitonga, N. M. (2015). Assessment of Physico-chemical Properties of Kasarani Section of River Thome, Nairobi. *Journal of Geography. Environment and Earth Science International*, *3*(4), 1–5.

Kaufman, Y.J., Koren, I., Remer, L.A., Tanre, D., Ginoux, P., & Fan, S. (2005). Dust transport and deposition observed from the Terra-Modis spacecraft over the Atlantic Ocean. *Jour. Geographical Research, 110*(D10S12), 1-16.

Kawabata, Y., Kawa, M., Yamada, M., Nwona-Agyeman, S.O, Aparin,V., Ollibekov, B.J., … Kataya, M. A. (2012). Seasonal Changes in Water Quality of Rivers and Ground Water in Karakalpakstan, Uzbekistan. *Journal of Arid Land Studies*, 171-174.

Kitheka, J. U. (2013). The hydrologic alteration of the Tana river and the impacts in the Lower Tana Basin. *Proceedings of the National Conference on the Tana River organized by the University of Nairobi and National Environment Management Authority (NEMA).*

Kitheka, J. U. (2014). Assessment of modification of the Tana River runoff due to developments in the Upper Tana Basin. *Proceedings of the 2nd Hydrological Society of Kenya (HSK) Workshop: "Hydrology in Water Cooperation and Security for Sustainable Economic Development". Sub-Theme: Hydrology and Sustainable Development*, 178-191.

Kithia, S. M. (1997). Landuse changes and their effects on sediment transport and soil erosion within the Athi drainage basin, Kenya. In *Human Impact on Erosion and Sedimentation Proceedings of Rabat Symposium S6*. IAHS Publ.no. 245.

Koning, N., & Ross, J. C. (1999). The continued influence of organic pollution on the water quality of the turbid Modder River. *Water S.A.*, *25*(3), 285–292.

Kumarswami, N. (1991). An approach towards assessment of dug well water quality by physico-chemical characteristics-a case study. *Pollution Research, 10*(1), 13–20.

Lasage, R., Aerts, J., Mutiso, G. C. M., & de Vries, A. (2008). Potential for community based adaptations to droughts: Sand dams in Kitui, Kenya. *Physics and Chemistry of the Earth, 33*(1-2), 67–73. doi:10.1016/j.pce.2007.04.009

Lind, E. W., & Morrison, M. E. S. (1974). *East African Vegetation*. London: Longman.

Linsley, R. K., Kohler, M. A., & Paulhus, J. L. H. (1988). *Hydrology for engineers*. McGraw-Hill Book Company.

Ghazanfar, B., & Moat. (2003). Flora of tropical East Africa: quantitative analyses of the flora and its composition. *Proceedings of the XVIII session of AETFAT*.

Mahananda, M. R. (2010). Physico-Chemical analysis of surface water and ground water of Bargarh District, Orissa, India. *International Journal of Research and Review in Applied Sciences*, *2*(3), 284–295.

Massdam, R., & Smith, D. G. (1994). New Zealand's national river water quality network. 2. Relationships between physico -chemical data and environmental factors. *New Zealand Journal of Marine and Freshwater Research*, *28*(1), 37–54. doi:10.1080/00288330.1994.9516595

Munyao, J. N., Munyoki, J. M., Kitema, M. I., Kithuku, D. N., Munguti, J. M., & Mutiso, S. (2004). *Kitui sand dams: Construction and operation*. Nairobi, Kenya: SASOL Foundation.

Neessen, D. (2004). Regional water balance modelling of a semi-arid catchment in South Kitui District, Kenya. Katholieke Universiteit Leuven Faculteit Landbouwkundige en Toegepaste Biologische Wetenschappen.

Ngigi, S. N., Savenije, H. H. G., Thome, J. N., Rockström, J., & de Vries, F. W. T. P. (2005, April). Agro-hydrological evaluation of on-farm rainwater storage systems to supplement irrigation in Laikipia District, Kenya. *Agricultural Water Management*, *73*(1), 21–41. doi:10.1016/j.agwat.2004.09.021

Nyamai, C. M., Mathu, E. M., Opiyo-Akech, N., & Wallbrecher, E. (2003). A re-appraisal of the geology, geochemistry, structures and tectonics of the Mozambique belt in Kenya, east of the Rift System. African Journal of Science and Technology (AJST). *Science and Engineering Series*, *4*(2), 51–71.

Ohowa, B. O., Mwashote, B. M., & Shimbira, W. S. (1997). Dissolved Inorganic Nutrient Fluxes from Two Seasonal Rivers into Gazi Bay, Kenya. *Estuarine, Coastal and Shelf Science*, *45*(2), 189–195. doi:10.1006/ecss.1996.0187

Puttemans, S. (2004). *Potential for small scale irrigation from groundwater dams in South Kitui*. Katholieke Universiteit Leuven Faculteit Landbouwkundige en Toegepaste Biologische Wetenschappen.

Radiro, M. O., Sijali, I. V., & Maingi, P. M. (2003). Integrated soil management for sustainable use of salt affected soils of Taveta Division. Salt affected soils. *Proceedings of a workshop. Taveta. 22-27 August 2003. IDRP Workshop Report, Nairobi, Kenya. KARI Irrigation and Drainage Research Programme*, 7-14.

Save the Rhino. (2018). *All Rhinos translocated to Tsavo East National Park have died*. Retrieved from https://www.savetherhino.org/africa/kenya/all-rhinos-translocated-to-tsavo-east-national-park-have-died/

Sigleo, A., & Frick, W. (2015). Seasonal Variations in River Flow and Nutrient Concentrations in a Northwestern USA Watershed. Environmental Protection Agency, Western Ecology Division, Newport, and U.S. Environmental Protection Agency, Ecosystems Research Division.

Sijali, I. V., Radiro, M. O., & Maingi, P. M. (2003). *Salt affected soils: Proceedings of a workshop*. IDRP Workshop Report. Nairobi, Kenya: KARI Irrigation and Drainage Research Programme.

The Guardian. (2018). *Eight of 14 Rhinos die after move to Kenya National Park*. Retrieved from ttps://www.theguardian.com/world/2018/jul/13/kenya-rhinos-die-after-being-moved-between-national-parks

U.S. Department of the Interior. (2011). *Quality of water Colorado River Basin Progress Report No. 23*. Bureau of Reclamation.

Vaishali, P., & Punita, P. (2013). Assessment of seasonal variation in water quality of River Mini, at Sindhrot, Vadodara. *International Journal of Environmental Sciences, 3*(5).

Wass, P. (Ed.). (1995). *Kenya's Indigenous Forests: Status, Management and Conservation. IUCN*. Cambridge, UK: Gland.

Waziri, M., & Ogugbuaja, V.O. (2010). Inter relationships between physico-chemical water pollution indicators: A case study of River Yobe-Nigeria. *AJSIR, 1*(1), 76-80.

Waziri, M., & Ogugbuaja, O. V. (2012). Prediction of some water quality indices in River Yobe-Nigeria, through annual projections. *Frontiers in Science, 2*(4), 58–61. doi:10.5923/j.fs.20120204.02

Williams, W. D. (1966). Conductivity and the concentration of total dissolved solids in Australian lakes. *Australian Journal of Marine and Freshwater Research, 17*(2), 169–176. doi:10.1071/MF9660169

WHO. (1993). *Guidelines for drinking water quality* (2nd ed.). Geneva, Switzerland: WHO.

WHO. (2006). *WHO Guidelines for Drinking Water Quality* (3rd ed.; Vol. 1). WHO.

WHO. (2008). *Guidelines for Drinking-water Quality* (3rd ed.). Geneva, Switzerland: World Health Organization.

Chapter 4
Water Allocation Assessment and Hydrological Simulation on Mukurumudzi River Basin in Kenya

Jacob Mutua Katuva
University of Nairobi, Kenya

Christian Thine Omuto
University of Nairobi, Kenya

John P. O. Obiero
University of Nairobi, Kenya

ABSTRACT

Lack of water allocation plans in semi-arid river basins increases the risks of hydrological droughts. The Mukurumudzi River Basin was experiencing extremely low flows which led to a water allocation assessment to quantify the available resource amidst competing demands from domestic, commercial, and environmental needs. An abstraction survey was conducted along the entire length of the river followed by a low flow study and water balance modelling using rainfall-runoff model. Observed flows at source and mouth were 0.028 m^3/sand 0.043 m^3/s, respectively. Simulated flows estimated an environmental flow (Q95) deficit of0.023 m^3/s and 0.010 m^3/s in July and August, respectively. Normal flow (Q80) of 0.190 m^3/s and flood flow (Q50) of 0.520 m^3/s were estimated. Compensating the environmental flow deficit in July and August was recommended. A more detailed hydrogeological study should be conducted to elucidate the potential of the groundwater resource in meeting current demands.

DOI: 10.4018/978-1-7998-0163-4.ch004

INTRODUCTION

The Water Resources Management Authority (WRMA) in collaboration with the Water Resources Users Association (WRUA) of the Mukurumudzi River Basin in Southern Coast of Kenya developed a Sub-Catchment Management Plan in 2012 which highlighted an existing conflict between domestic, commercial irrigation and industrial water demands in the catchment (WRMA & Mukurumudzi WRUA, 2012). The flow at the river gauging stations on Mukurumudzi River (3KD06) on 13/03/2013 was estimated at approximately 0.006 m^3/s (6 l/s) ($\cong$520 m^3/d), which was 53% lower than the recorded lowest daily flow for the month of March (11.28 l/s, $\cong$974 m^3/d). This meant that there might be over abstraction of water from the river and this was likely to increase due to unknown abstractors in the catchment, unknown levels of abstraction, over allocation of the water resource and unauthorized abstractions. The low flows were attributed to lack of a water allocation framework which is a tool used to mitigate conflicts over water resources and to equitably allocate the finite and scarce water resources. The environmental flow requirements are usually compromised when there is over- abstraction and diminishing water resource leading to environmental degradation and conflicts among water users (Linsleyet. Al. 1982; Linsleyet. Al. 1989; Maidment 1993). Previous studies have established that changes in the hydrology of Mkumudzi Basin have occurred in the recent past on a decadal scale (Kibson Consult, 2010; Streamtec Pty, 2011). It appears inevitable that such changes will continue to occur and there is a need to understand them before any realistic attempt is made to determine the impacts of Kwale Mineral Sands Project and the Sugarcane irrigation scheme on the hydrology of Mukurumudzi River Basin.

The above issues formed the basis of this study which aimed at developing a water allocation framework that could be used to equitably distribute the available water resource for the different water uses in the Mukurumudzi basin (GoK, 2007). This was done in order to mitigate conflicts among water users and also ensure environmental flows are protected. The specific objectives were (i) quantify and classify the water abstractors and level of abstractions in the Mukurumudzi river basin through an abstraction survey, (ii) assess the surface water resource availability through simulation modelling based on analysis of existing hydrological data and (iii) develop a water allocation frame work for Mukurumudzi river basin.

STUDY AREA

The Mukurumudzi River basin (MRB) covers an area of 207 km^2 and is located in the southern part of the Kenya coast approximately 50 km south of Mombasa. The basin has a bi-modal rainfall pattern with half of the mean total annual rainfall falling in April, May and June. The annual average precipitation ranges between 1,000 mm and 1,200 mm. The Mukurumudzi river starts 30 km inland from the Indian Ocean. The river is about 40 km long, originating from the Shimba Hills and draining into the Indian Ocean. The catchment has one River Gauging Station coded 3KD06 (sited at location L3 in Figure 1) that covers 69 km^2. The catchment also has two major industries, a mining company and a sugarcane irrigation farm. The mining is conducted by the Kwale Mineral Sands Project (KMSP) while the irrigation of 5,500 ha of land is conducted by the Kwale International Sugar Company Limited (KISCOL). Both the mining and irrigation companies have so far developed 3 dams (Mukurumudzi dam, Lower and Upper Koromojo dam) with a total storage of 14 Million m^3 on the Mukurumudzi River basin to meet their water demands.

METHODOLOGY

Abstraction Survey

A low flow gauging exercise was conducted in collaboration with WRMA and the Mukurumudzi WRUA. Eight sites were selected (L01 to L08) during a reconnaissance

Figure 1. Study Area

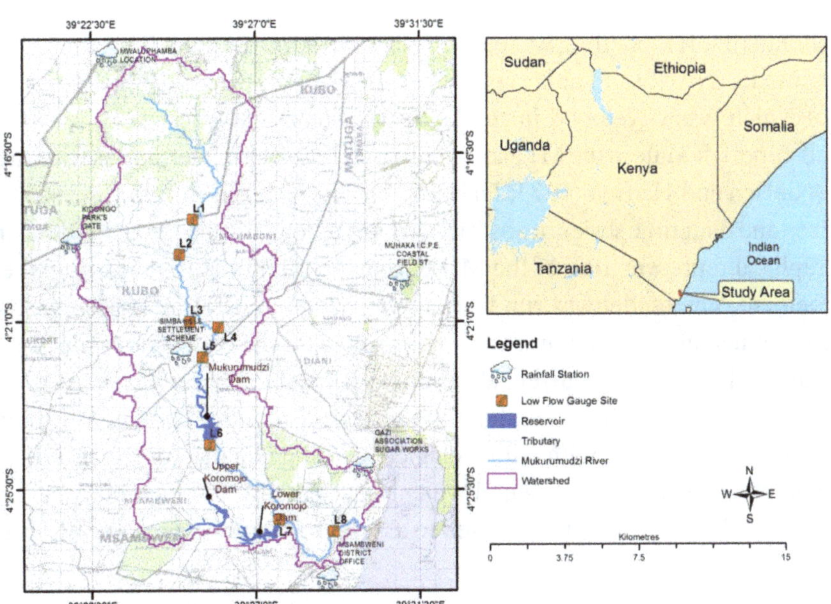

visit on the Mukurumudzi river. Prior to the field work, consultative meetings were held with the WRUA, the County Commissioner, the Kenya Forestry Service and the Red Cross. The flow measurements were made from upstream to downstream using acurrent meter hired from University of Nairobi, Environmental and Bio-systems Department. The mid-section method for velocity determination was used to compute the instantaneous river discharge. This was followed by abstraction survey during the dry season to capture the abstractions when water demand was typically at its highest and resource availability at its lowest. The survey started from the edge of the forest where the river leaves the Shimba Hills National Reserve. It was known from the outset that there was only one abstraction point inside the Reserve, which was accessed after obtaining approval from the Kenya Wildlife Service to enter the Reserve. The abstraction survey team comprised of two WRMA professionals, two members of the WRUA, and a Community Liaison Officer from the Kwale Minerals Sands Project. Since the river was narrow it was easy to survey both banks at the same time starting from the source and walking all the way to the mouth of the river at Gazi Bay. The survey was conducted using questionnaires developed in consultation with WRMA. Every abstractor was interviewed in order to obtain information regarding inter alia, the location of abstraction (GPS), volume abstracted, the purpose of abstraction, storage, permit status of the abstraction point. This information was used in the analysis of the abstractors to quantify the abstraction amounts, types and permit status as well as the water balance.

Rainfall Modelling

The rainfall runoff modelling was carried out with the NAM model, a Danish rainfall-runoff model. NAM is an abbreviation for the Danish word Nedbør-Afstrømnings Model which was developed in the 1970s (Nash & Sutcliffe 1970). NAM is the rainfall-runoff module in the M11 software package and also interphases with ArcGIS 10.0 (Nielsen and Hansen, 1973; DHI, 2011). The water users and reservoirs were digitized and their time series of abstraction rates uploaded onto them. Figure 2 shows the graphical representation of the MIKEbasin NAM model framework in the GIS environment. The model was run without the abstractions' time series to establish the environmental water requirements at the mouth of the river. The model was then run again with all existing current abstractions to establish the observed flows' time series at the mouth of the river, this time series was used to compute the monthly Q95 flow values (a flow with a 95% chance of being equalled or exceeded) which were compared to the Environmental Water Requirements (EWR) at the mouth of the river (Pyrce, 2004). This was also used to assess the water balance and allocations.

Figure 2. Graphical representation of Mike Basin Model Interphase

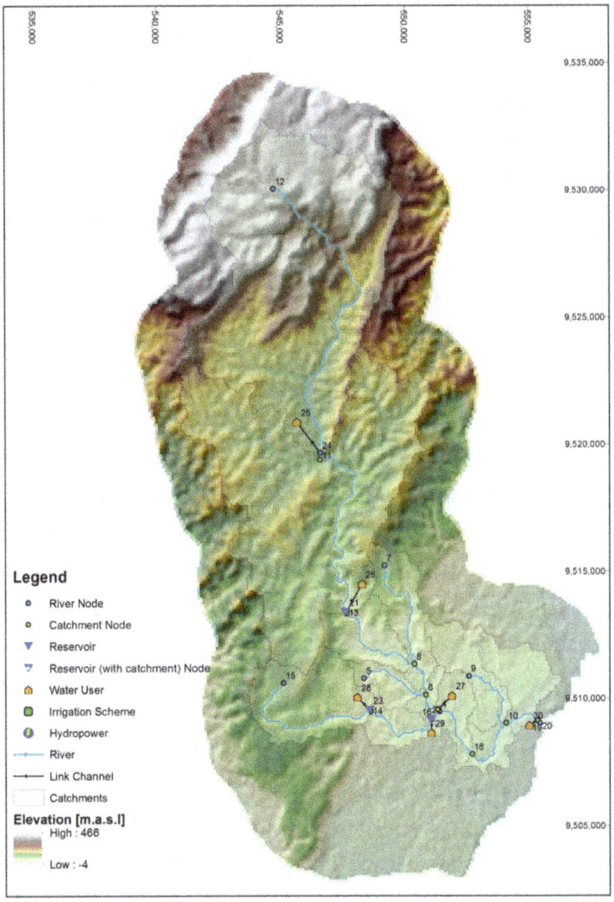

RESULTS AND DISCUSSION

Assessment of Low Flows

Figure 3 shows the flow profile from upstream towards the mouth of Mukurumudzi river. The river flow generally increased downstream due to inputs from tributary streams.

Abstraction Survey

Abstractors were broadly categorized as individuals, groups or institutions. Figures 4 and 5 provide details on the category of abstractors by number and volume of total abstractions respectively. Individual abstractors (45%) account for only 9% of the total

Figure 3. Daily low flow profile from forest edge to estuary

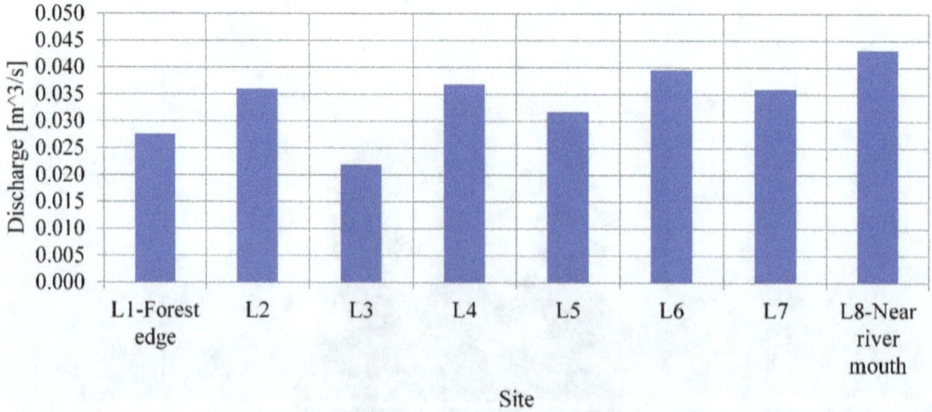

abstraction volume and the other 91% of the total abstraction volume is as a result of 22% of the abstractors. This 22% comprises community managed organisation water projects (CBO). The irrigation and mining company were not abstracting water at the time of the abstraction survey. The main water lifting technologies used for abstractions were categorized as portable pumps, weir and furrow, weir and fixed pump, and dam (Figures 6 and 7). The most common method of abstraction was the use of portable pumps (44%) which accounted for only 9% of the total abstraction volume. Conversely, dams and weirs with either fixed pumps or furrow account for 56% of the abstractors but weirs with fixed pump accounted for 91% of the total abstracted volume; this was because dams and weirs with furrow were not abstracting any water at the time of abstraction survey. Figure 8 shows that CBOs accounted for 91% of the abstracted volume. In terms of defining a strategy for enforcement, these results show that WRMA should focus on companies and CBOs.

Data Reliability

A double mass curve was plotted for Shimba Hills Settlement Scheme rainfall versus Msambweni District office rainfall and Kwale Agricultural Department rainfall versus Shimba Hills Settlement Scheme rainfall (Figures 10 and 11).

The relationship between the Msambweni District Office and Shimba Hills Settlement Scheme rainfall was fairly close although there were some slight deviations. There was a similar pattern for the Kwale Agricultural Department and Shimba Hills Settlement Scheme. The river water level data obtained from WRMA, was converted to flows using a rating equation deduced from the relationship between gauge height and the corresponding observed river discharge. The rating equation used was:

Figure 4. Types of Abstractor

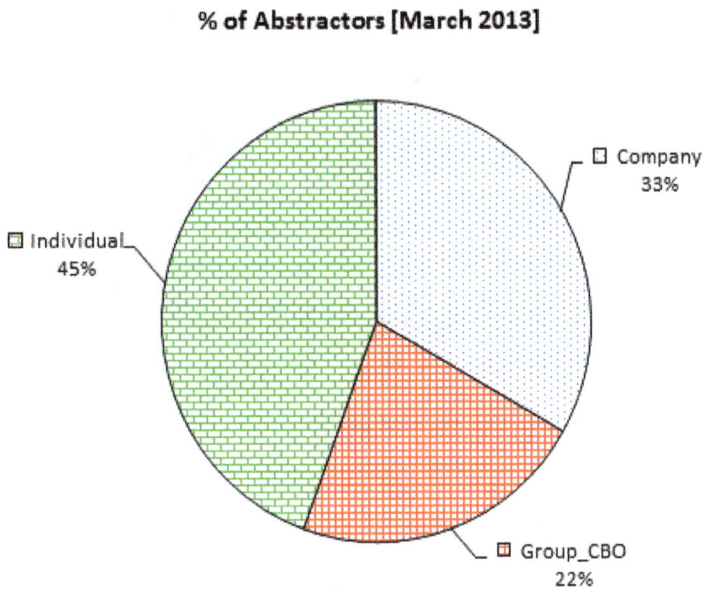

Figure 5. Type of abstractor by % of total volume of abstraction

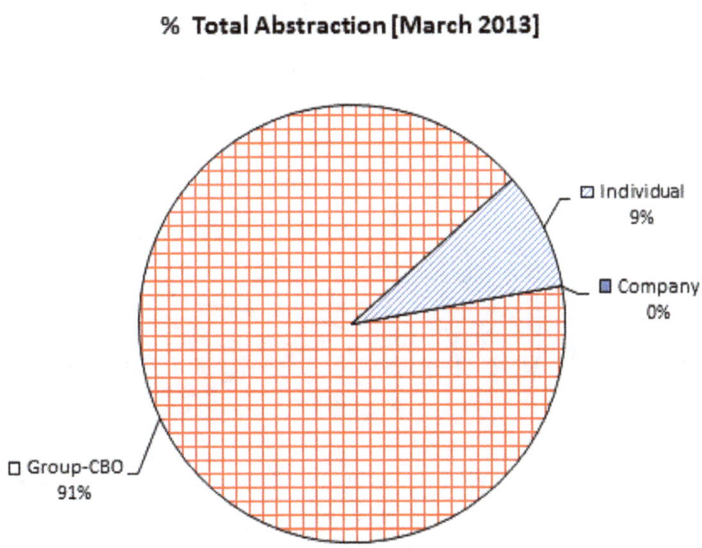

Figure 6. Method of abstraction

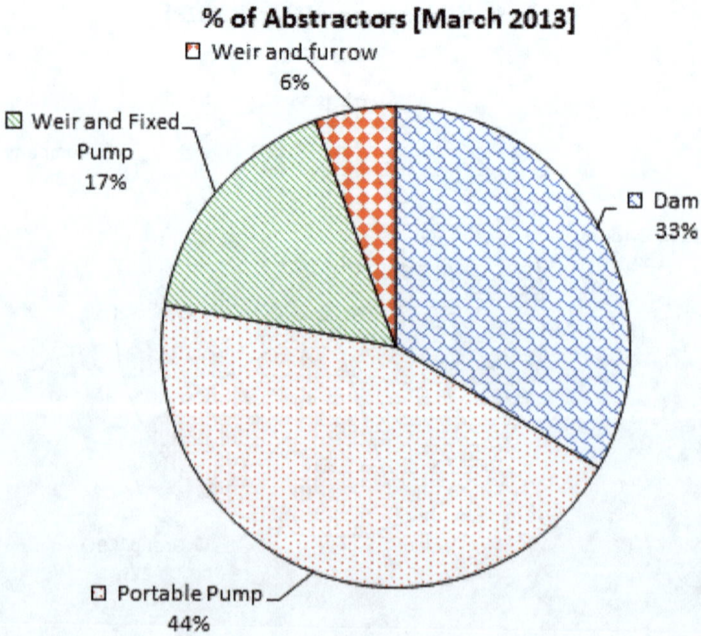

Figure 7. Method of abstraction by % of total volume of abstraction

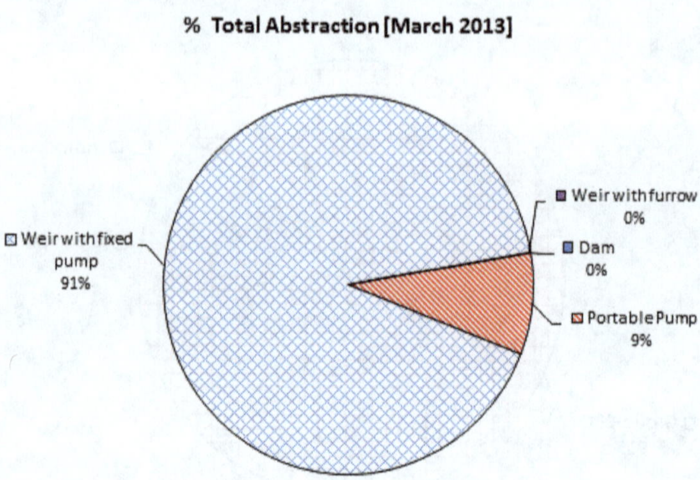

Figure 8. Type of abstractor and method of abstraction

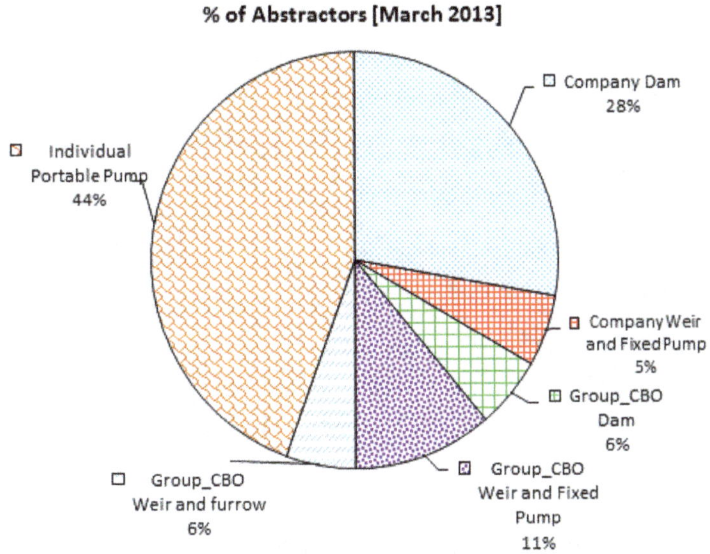

Figure 9. Type of abstractor and method of abstraction by % of total abstraction

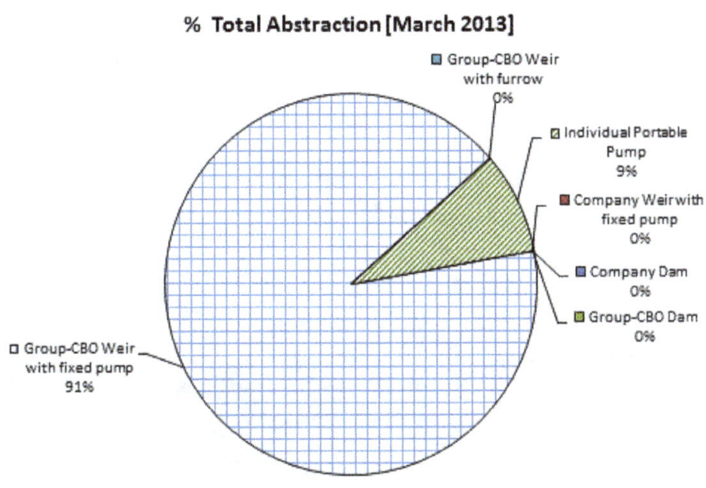

$$Q = 3.951\ H^{1.670}$$

In the above equation, Q is the riverdischarge in m^3/s and H is the water level in m. The gauge height/discharge data records for a maximum gauge height of 0.64 m while the water level data had maximum value of 2.5m. The water level above 0.64 m lacked a degree of confidence in discharge determination. A double mass

Figure 10. Double Mass curve for rainfall at Msambwenivs.rainfall at Shimba Hills

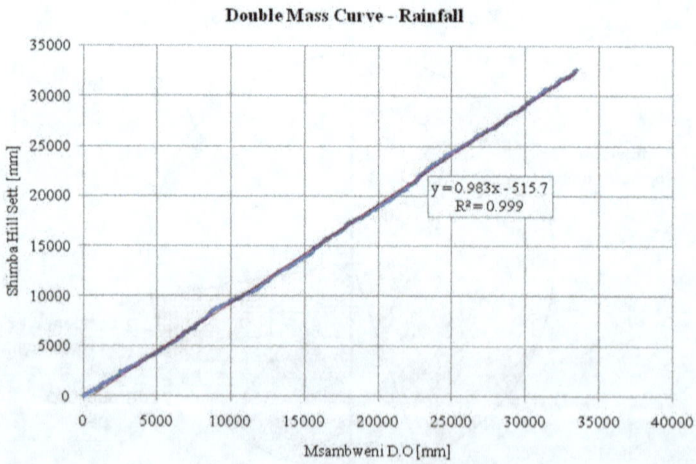

Figure 11. Double Mass Curve for rainfall at Shimba Hills vs. Rainfall at Kwale Agricultural Department

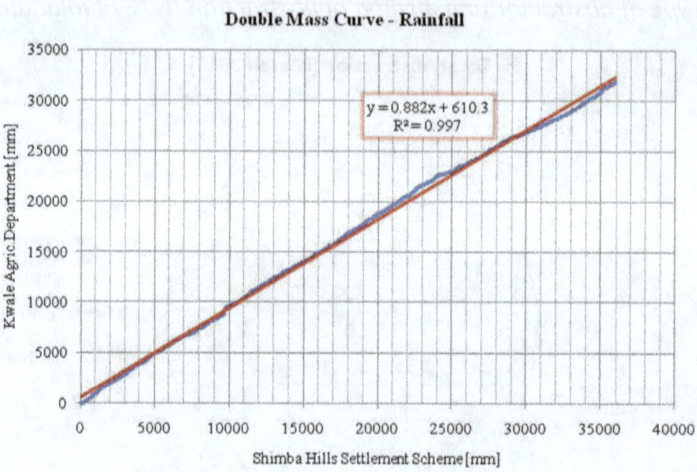

curve between river discharge and rainfall at 3KD06 was developed and analysed to establish the consistency of the discharge data. The Shimba Hills Settlement Scheme rainfall was used to check the consistency of the discharge at 3KD06. The double mass curve showed a change in regime in the discharge dataset from 21st August 1973 to 16th November 1980. The period prior to this year had a different regime as shown in Figure 12. This could be attributed to change in land cover, the possible change of flow control characteristics for 3KD06, or the blockage of gauging weir was blocked by debris, among other factors. The gap was eliminated and the double mass curve re-plotted which then yielded a better relationship with a correlation coefficient r of 0.9956 (Figure 13).

NAM Model Results

The discharge data cleaned through the DMC analysis was used in the NAM model. A section of the discharge data that had continuous data was compared to the rainfall and evapo-transpiration time series to ensure there were no gaps. These time series were then used in the rainfall-runoff (RR) model to calibrate the nine parameters and generate the runoff time series for the whole period. The model was calibrated for the period of 1/2/1959 to 28/10/1988. The data for this period was good and it represented a climatic period. Figure 14 show the rainfall- runoff simulation results and water balance respectively. The coefficient of determination was 0.368 while the coefficient of correlation was 0.63 which represented a good fit (ASCE, 1993).

Flow Duration Curves Analysis

The Flow Duration Curve (FDC) for the simulated river discharge and observed discharge is shown in Figure 15 and depicts a good relationship at low flows. The simulated discharge was used to fill the gaps of the observed flow time series since the water balance and the flow duration curve for observed and simulated discharge

Figure 12. Double Mass Curve: Simba Hills Rainfall vs 3KD06 Discharge using all raw data

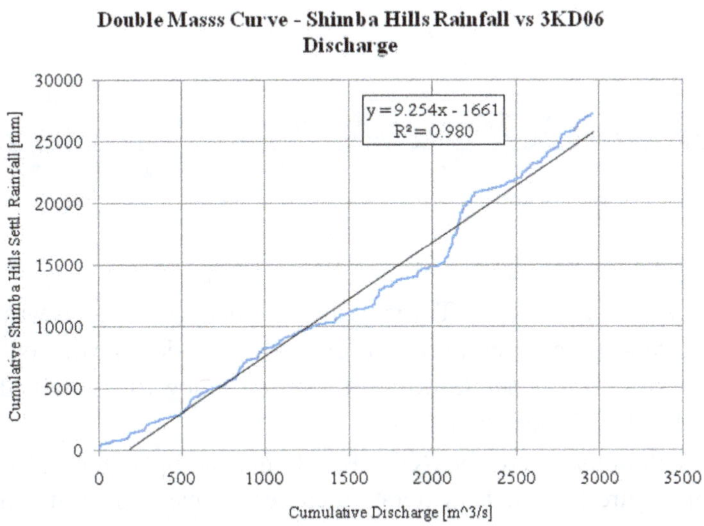

Figure 13. Double Mass Curve: Simba Hills Rainfall vs 3KD06 Discharge after data improving quality

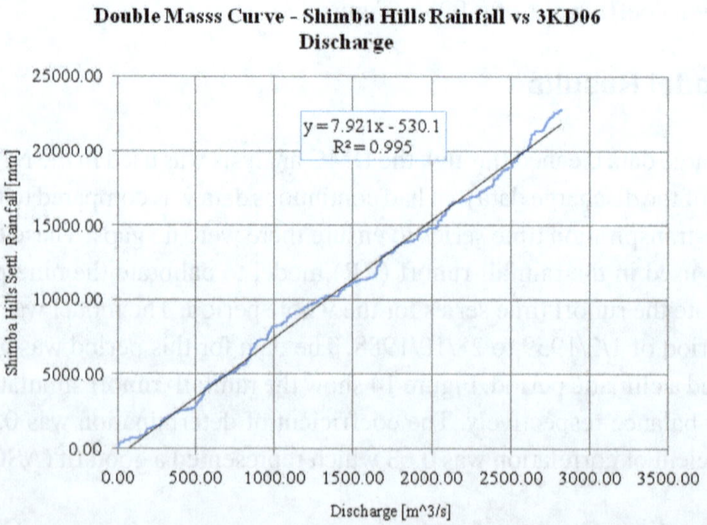

Figure 14. Simulated River Discharge vs Observed River Discharge at RGS

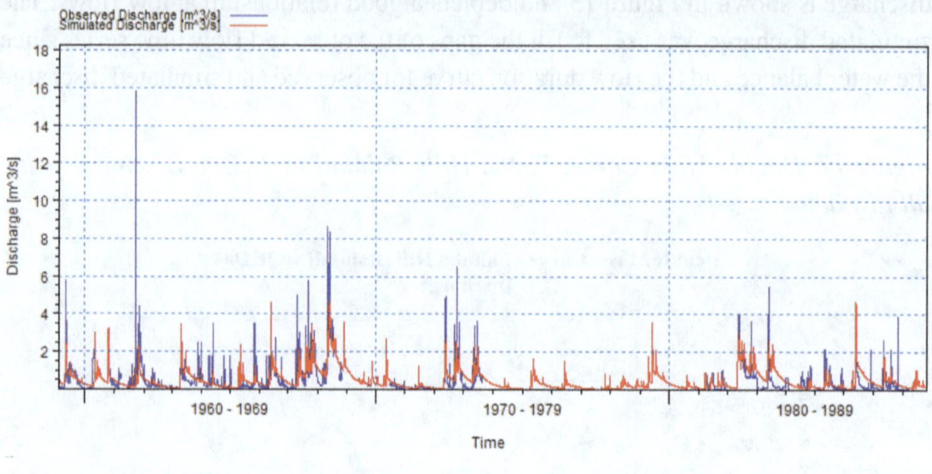

matched very well at low flows. The gap filled discharge time series was converted into specific run-off and used in the MIKE BASIN model to carry out the water balance analysis so as to generate the EWR. Figure 16 shows the gap filled discharge time series data for 3KD06 RGS.

The flow duration curve for RGS 3KD06 and the operational dam sites are presented in Figure 17. The flow duration curves are based on all the naturalized

Figure 15. Flow Duration Curve for Simulated River Discharge and Observed Discharge at RGS

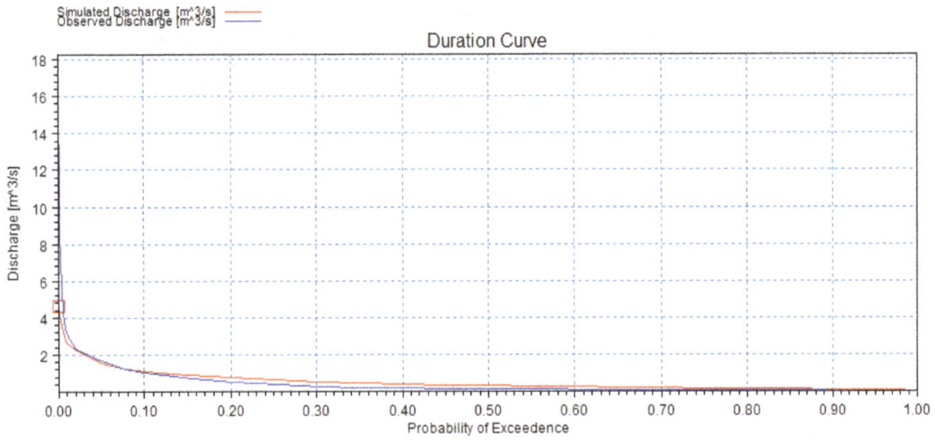

Figure 16. Gap filled discharge for river gauge station 3KD06

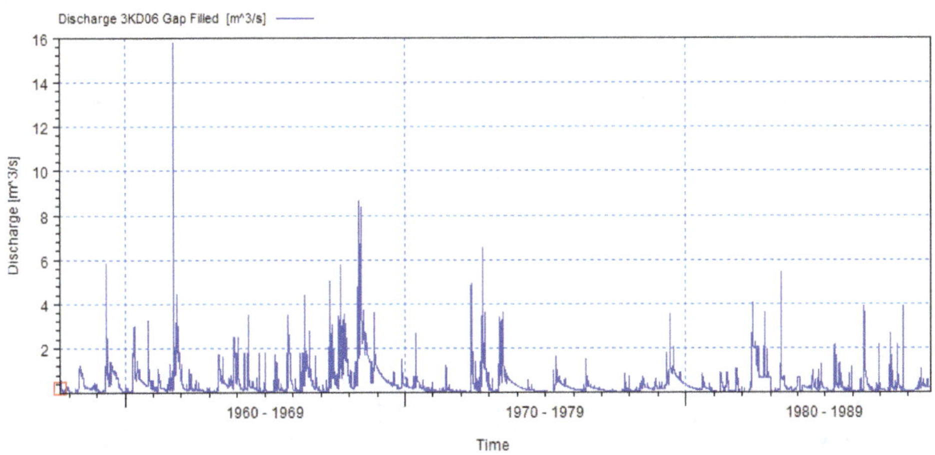

daily discharge values for the period 2/09/1957 to 30/09/1988. The Q95 and Q 80 represent the thresholds for the Reserve and Normal Flow as defined by the Water Resource Management Rules (2007).

The FDC was also analysed on a monthly basis paying special focus on the EWR (Q95) at the outlet of the river. The monthly EWR (Q95) values give a reflection of the seasonal variations of the discharge in the river system, hence a better indicator of the EWR compared to a single Q95 value. Figure 19 shows the monthly Q95 flow values for 3KD06 and the dams in the catchment. The Monthly EWR for the mouth of the river was also analysed and the results are shown in figure 20.

Figure 17. Flow Duration Curve (FDC) of the Mukurumudzi River at RGS 3KD06 and the Dams

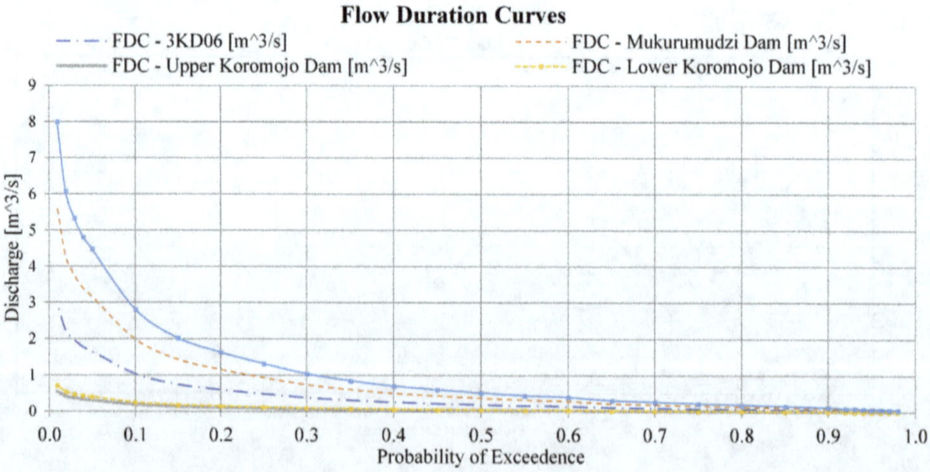

Figure 18. Flow Duration Curves (FDC) for RGS 3KD06 including the Q75 and Q98 curves

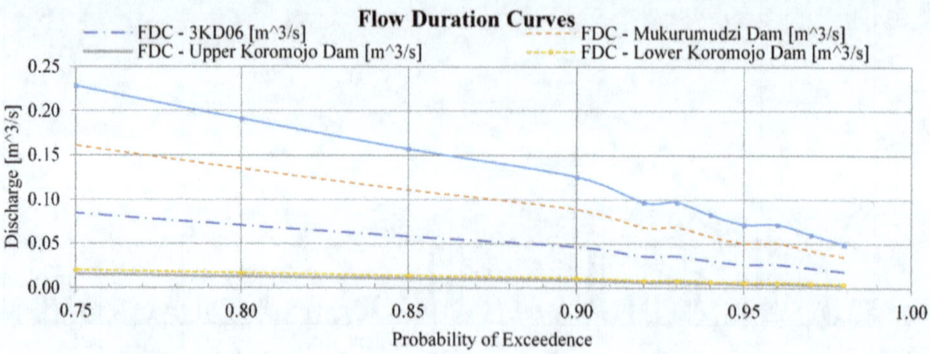

Water Balance and Allocation

Apart from the abstraction points that existed at the time of the study, there were plans to construct additional dams in the catchment (Table 1). The MIKE Basin Model was used to simulate the water balance when all the abstractions are in place. The monthly Q95 values were analyzed from the time series generated at the mouth of the river Mukurumudzi with all abstractions in place. These values were compared

Figure 19. Monthly EWR for 3KD06 and the Dams

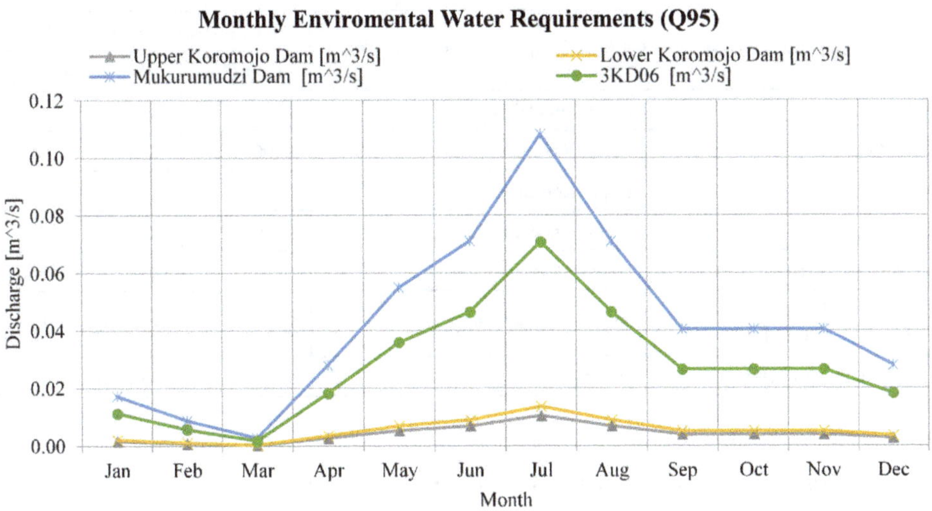

Figure 20. Monthly Environmental Water Requirements (EWR) at Mouth of the Mukurumudzi River.

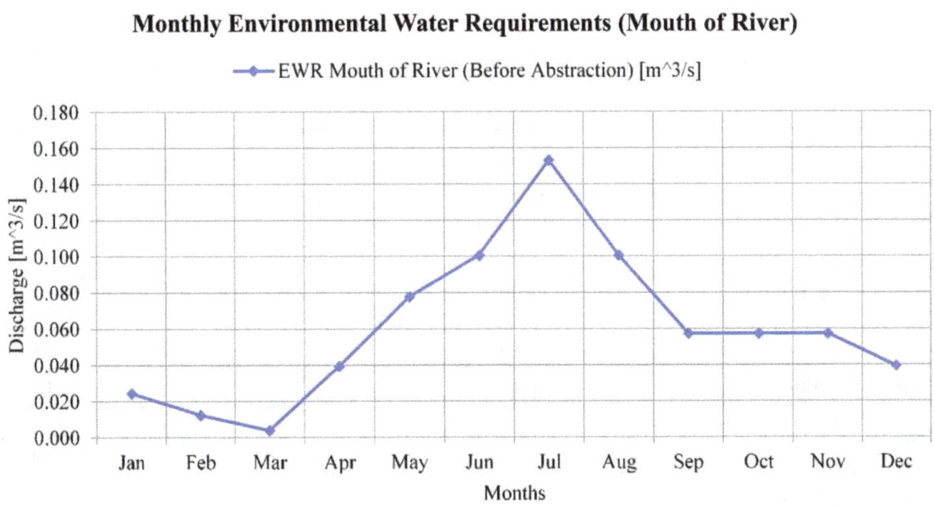

to the expected naturalized monthly EWR at the mouth of the river. Table 2 shows the comparison and the deficit expected to occur through the year. The values are with assumption that the reservoirs in the catchment are releasing their respective EWR which will otherwise change if there is no provision for releasing the EWR at the dam sites. A graphical representation of Table 2 is shown in Figure 21.

Table 1. Water Apportionment in Mkrumudzi River Basin as at March 2013

Name of Applicant	Infrastucture at Abstraction Point	Authorised Amount(m^3/d)	
		Flood Flow	**Normal Flow**
KICOL	Mukurumudzi Weir	30,000	-
KISCOL	Lower Koromojo Dam	30,000	-
KISCOL	Upper Koromojo Dam	30,000	-
KMSP	Mukurumudzi dam	22,000	2,000
Majimboni Water Project	Weir with fixed pump		125 [measured]
MajimboniMuungano Self Help Water Project	Weir with fixed pump		337 [measured]
Individual Abstractors	Portable Pump		44 [measured]

Table 2. Simulated EWR at Mouth of Mukurumudzi River

Month	EWR Mouth of River (Before Abstraction) [m/s]	Simulated EWR Mouth of River (After Abstraction) [m/s]	Deficit
Jan	0.024	0.030	No Deficit
Feb	0.012	0.017	No Deficit
Mar	0.004	0.010	No Deficit
Apr	0.040	0.042	No Deficit
May	0.078	0.084	No Deficit
Jun	0.101	0.106	No Deficit
Jul	0.153	0.130	-0.023
Aug	0.101	0.090	-0.010
Sep	0.057	0.067	No Deficit
Oct	0.057	0.065	No Deficit
Nov	0.057	0.063	No Deficit
Dec	0.040	0.045	No Deficit

*There will be a deficit of 23 l/s and 10 l/s during the months of July and August respectively.

Figure 21. Simulated and Expected EWR at Mouth of the Mukurumudzi River

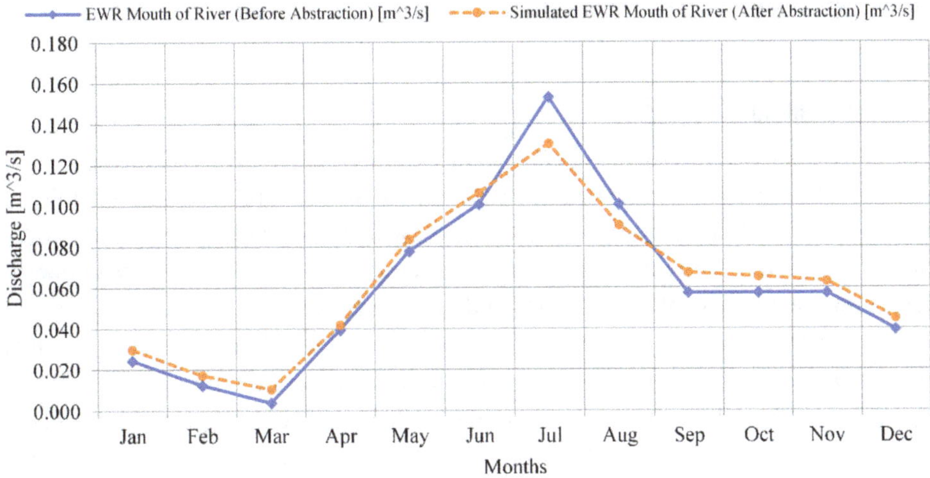

CONCLUSION

This study established that there is a marginal increase in the flow of River Mukurumudzi downstream along the profile. This was attributed to groundwater discharge during the dry season. The catchment receives an average rainfall of 1300 mm per year and has a daily evapo-transpiration rate of 4.4 mm/day. The catchment had a total of 18 abstractors as at April 2013. The total authorised normal flow abstractions with valid documentation was 2,000 m^3/d while the total authorised flood flow abstractions with valid documentation was 104,000 m^3/d. The abstractors abstracting the highest amount of water were the community based organisations as by that time the companies were not abstracting any water as they were still putting up infrastructures for water abstraction. The environmental water requirements (EWR) of the river were satisfied in all the months except the months of July and August where there was a deficit of 23l/s and 10 l/s in the months of July and August respectively. The EWR at the mouth of the river will be affected when more dams and other abstraction points are constructed in the catchment.

The study also found that as of March 2013, 91% of the river water abstraction volume was abstracted by community based organisations water supply groups. These are groups that should be involved in the WRUA and should be aware of requirements for compliance to the WRMA Rules (2007). In a catchment where water use conflicts can arise due to the high demands and limited resources, specifically during periods of low flows, WRMA and the WRUA should be proactive in engaging the abstractors in dialogue and compliance and also be vigilant to deal with unauthorized abstractions in the basin. They should also communicate information regarding flow conditions, water allocations and authorised abstractions, in order to mitigate against water use conflicts.

RECOMMENDATIONS

The Water Resources Management Authority (WRMA) should ensure that the Water Resource Users Association (WRUA) members and the wider community are aware of the natural discharge data, and make reference to this when there are concerns regarding the impacts of abstractions and in-stream works.

All the abstractors in the catchment should comply with releasing the monthly EWR (Q95) at their abstraction points including dams. In the months of July and August, they should release an extra cumulative 23 l/s and 10 l/s respectively (176,774 m3 every year) to satisfy the EWR for the survival of the mangrove forest at the mouth of the river. This flow will not be necessary if river flows equal or exceed the EWR in those months.

This study was carried out during dry season when the flows were generally low. Thus high flow gauging is required in order to improve the understanding of the river flow dynamics during the wet season and how this can affect the dams' water levels. The dam's storage dynamics (operational rules) should be studied to quantify the sustainability of the volumes stored for the abstraction purposes. More research should focus on hydrological and hydrogeological assessments to elucidate the potential of the underlying aquifer in the study area in providing water resource for domestic, irrigation and industrial uses for growth and development.

ACKNOWLEDGMENT

The authors gratefully acknowledge Financial support from Rural Focus Ltd, Eng. MichaelKariuki Thomas and Kwale Mineral Sands Project (Base Titanium Ltd). A special acknowledgement is given to Geoffrey Mwania, DamarisNyambura, and the Water Resources Users Association for their support in data collection. The authors would also like to thank the Water Resources Management Authority (www.wrma.org.ke), Kenya Meteorological Department (www.kmd.ac.ke), Kwale Minerals Sands Project (www.basetitanium.com) and Kwale International Sugar Company Ltd (www.kwale-group.com) for sharing data.

REFERENCES

Albert, M. J. (2004). *Hydraulic Analysis and Double mass curves of the middle Rio Grande from Cochiti to San Marcial, New Mexico* (Masters Thesis). Colorado State University. Fort Collins, CO.

ASCE. (1993). Criteria for Evaluation of Watershed Models. Journal of Irrigation and Drainage Engineering. *ASCE, 119*(3), 429 - 442.

ASCE. (1993). Criterion for Evaluation of Watershed Models. *Journal of Irrigation and Drainage Engineering, 119*(3), 429 - 442.

DFID. (2003). *Handbook for the Assessment of Catchment Water Demand and Use.* HR Wellingford U.K.

DHI. (2003). *Mike Basin: Rainfall–Runoff Modelling Reference Manual.* Copenhagen, Denmark: Danish Hydraulic Institute.

DHI. (2006). *Mike Basin Users Manual.* Copenhagen, Denmark: DHI Water and Environment.

DHI Water, . (2008). *Environment, Inc. and Lane Council of Governments June.* McKenzie River Mike Basin Model.

GoK (Government of Kenya). (2007). Water Resources Management Rules, 2007. LN 171, Government Press.

Ireson, A., Makropoulos, C., & Maksimovic, C. (2004). *Water Resources Modelling under Data Scarcity: Coupling MIKE BASIN and ASM Groundwater Model.* Academic Press.

KibsonConsult. (2010*). Appendix 8-3 Hydrological Assessment Report on Mukurumudzi River.* Author.

Linsley, K. R., Kohler, A. M., & Paullus, H. L. (1982). *Hydrology for Engineers (3rd ed.).* New York: McGraw Hill Company.

Linsley, R. K., & Franzini, J. (1989). *Water Resources Engineering, 3rd.* New York: McGraw-Hill Inc.

Liya, M. (2010). *Mango.* Modeling the Effect of Land Use and Climate Change Scenarios on the Water Flux of the Upper Mara River Flow.

Maidment, R. D. (Ed.). (1993). Handbook of Hydrology. McGraw Hill.

Nash, J. E., & Sutcliffe, V. J. (1970). River flow forecasting through conceptual models: Part 1. A discussion of principles. *Journal of Hydrology, 10*(3), 282 - 290.

Notter, B., MacMillan, L., Viviroli, D., Weingartner, R., & Liniger, H. (2007). Impacts of Environmental change on water resources in the Mt. Kenya region. *Journal of Hydrology, 343*, 266 - 278.

Pyrce, R. (2004). Hydrological Low Flow Indices and Their Uses. Watershed Science Centre, Trent University.

Streamtec Pty Ltd 2011, Environmental Water Requirements of the Mukurumudzi River, Kwale, Kenya (Streamtec Report 8/2011). (2011). Ukunda, Kenya: Prep. by Davies PM, Davies LM and Craig IR for Base Titanium Ltd.

Streamtec Report 10/2011. (2011). *Measuring Flows in the Mukurumudzi River, Kwale, Kenya*. Author.

Tolson, A. B., & Shoemaker, A. C. (2007). Cannonsville Reservoir watershed SWAT2000 model development, Callibration and Validation. *Journal of Hydrology, 337*, 68 - 86.

WRMA & Mukurumudzi WRUA. (2012). *Sub-Catchment Management Plan for Mukurumudzi River*. Author.

Chapter 5 Effects of Irrigation Management Practices on Water Allocation Among Farmers in Kiladeda Sub-Catchment, Tanzania

Mwadini Khatib
Ministry of Lands, Water, Energy, and Environment, Tanzania

Joy Obando
Kenyatta University, Kenya

Shadrack Murimi
Kenyatta University, Kenya

ABSTRACT

Kiladeda River in Pangani Basin, Tanzania plays a vital role of providing water for agricultural activities of the sub-catchment. However, it is experiencing a problem of inequitable distribution of irrigation water among farmers. Cross-sectional data was collected from farmers both in upstream and downstream using questionnaires, while river discharge measurements were conducted in referenced spot gauging stations. WEAP model was used to analyze water demand and allocation among farmers. Furrow irrigation (94%) and plastic buckets (6%) were the main irrigation water management practices. The model results revealed a water shortage of 46.4% of the total irrigation water required. The annual irrigation water demand and unmet demand were 13.93mm^3 and 7.47mm^3, respectively, and are both expected to increase twice in 2020. This high water demand for irrigation could be the main cause of excessive water abstraction. A partnering approach is recommended to improve irrigation water management, reviewing of laws, regulations, and water rights.

DOI: 10.4018/978-1-7998-0163-4.ch005

INTRODUCTION

The global water demand for irrigation farming has already been exceeded due to an uneven distribution of water resources and population densities (Qadirn et al., 2010). Among the uses of water in several sectors, agricultural water has been considered as a powerful factor for providing food security and providing countless opportunity for crop diversification to feed the growing population in the world (Hussain et al., 2011). Nonetheless, it is clear that worldwide irrigation farming is the most important off stream water uses and placing an ever increasing pressure on the available water resources (Booker, et al., 2012). It consumes more than 70% of the total water withdrawn for human activities (WWAP, 2012) and recent studies show that global agricultural water demand will increase by 5% from 2,743 km^3 in 2008 to 3,858 km^3 in 2050 (FAO, 2011; WWAP, 2012). Similarly, irrigated agriculture contributes significantly to the most Sub Saharan African countries' economies, employments, and food supplies (FAO, 2011; IFAD, 2012). It contributes an average of 30% of GDP and 67% of employment in the region, thus, it is essential for accelerating agricultural growth and stabilizing the livelihoods of the region. However, the proportion of water going to irrigated agriculture is often much higher. It accounts for more than 90% of the total water withdrawal (FAO, 2011). Still, water availability for irrigation in these countries is already limited, and water is distributed inequitably among farmers in the upstream and downstream of the rivers. Irrigated agriculture is the backbone of most of the East African countries. For example, it contributes 25% of the GDP and 51% of total employment in Kenya (Ntale and Litondo, 2013). It also contributes 40% of GDP and 80% of employment in Tanzania (URT, 2009). However, the demand for irrigation water is increasing alongside rapidly increasing populations. For instance, agriculture consumes 92.9% of the total water withdrawn in Tanzania and 88.7% of this water is consumed by irrigation farming. Water availability for irrigation is a growing challenge in most countries in Africa (UN, 2013). This situation has been attributed to insufficient and inappropriate irrigation infrastructure leading to low water use efficiency and water losses of about 85% of water (MOW, 2010). In Pangani Basin, Tanzania, irrigation farming is commonly practised by smallholder farmers, consuming 400 to 480Mm^3 of water annually which is more than 80% of total water abstracted from the rivers (IUCN, 2009). It is believed that farmers utilise most of the available water but with very low efficiencies leading to water stress (Komakechet al., 2012). Nonetheless, despite the vast amounts of fresh water available, farmers are still faced with inequitable water distribution for irrigation leading to water shortages in the sub catchment. The increase of irrigation water demand in the sub catchment is associated with the rapid population growth, social and economic growth and agriculture expansion. In addition, lack of knowledge on the irrigation water management practices is the main problem which hampers farmers from taking part in the management practices. Thus, the need to understand irrigation management practices is imperative for a better water resource management which lies under principles of integrated water resources management (IWRM). Therefore, this paper presents results of irrigation water management practices adopted by farmers and their effects on irrigation water demand and allocation in Kiladeda Sub-catchment to enhance the efficient irrigation water use.

Description of the Study Area

Kiladeda sub-catchment has an area of 20 km^2 located within the North Western part of Pangani Basin, Tanzania. It originates from the Shiri springs located 7km from Moshi town. The river basin is bounded by latitudes 3°00' S and 3°30' S and longitudes 36° 30' E and 37°15' E (Figure 1). The river is a tributary of Weruweru sub catchment (250.646 km^2) in Pangani Basin. The river is fed by underground springs and catchment runoff during the wet season. The average flow rate is approximately 400 litres per second. Stream discharge generally increases during March and peaks during April gradually decreasing in May. Discharge fluctuates widely during the wet season and begins to decline in June and stabilize as base flow from July onward throughout the dry season. The river has relatively low discharges in the dry season and extremely high flow during rainy season (Mromba, 2012). Kiladeda sub catchment is characterized by steep slope valleys to gentle slopes, lowlands and in a few areas rolling topography. The altitude ranges between 840 m and 1,060 m above sea level. The area experiences bimodal rainfall with two distinct rainy seasons divided into short rains lasting from October to November while long rains is from mid-March to mid-May or June, as large variations are sometimes experienced causing rainfall unreliability. The sub catchment experiences lowest temperature of 14^0C to 18^0C which occurs during July to August and the highest temperature of 32^0C to 35^0C occur during January to February as described on temperature trend. A variety of food crops are grown including maize, beans, bananas, vegetables and different types of fruits. On average the sub-catchment receives less than 1000 mm of rainfall per year.

Kiladeda sub-catchment cuts across four villages in these districts namely; Shiri Njoro and Shiri Mgungani (Hai district), Chekereni (Moshi rural) and Shiri Matunda (Moshi Urban) with total population of population of 16,586 people and 3,560 households (URT, 2012). The sub catchment serves about 1,112 farmers. The geology of the sub catchment is classified as Neogene volcanics, sediments and Precambrian basement complex. However, the large part of the sub catchment falls under the Neogene volcanics of Mount Kilimanjaro. Large parts of the Kiladeda

Figure 1. Kiladeda sub-catchment, Tanzania

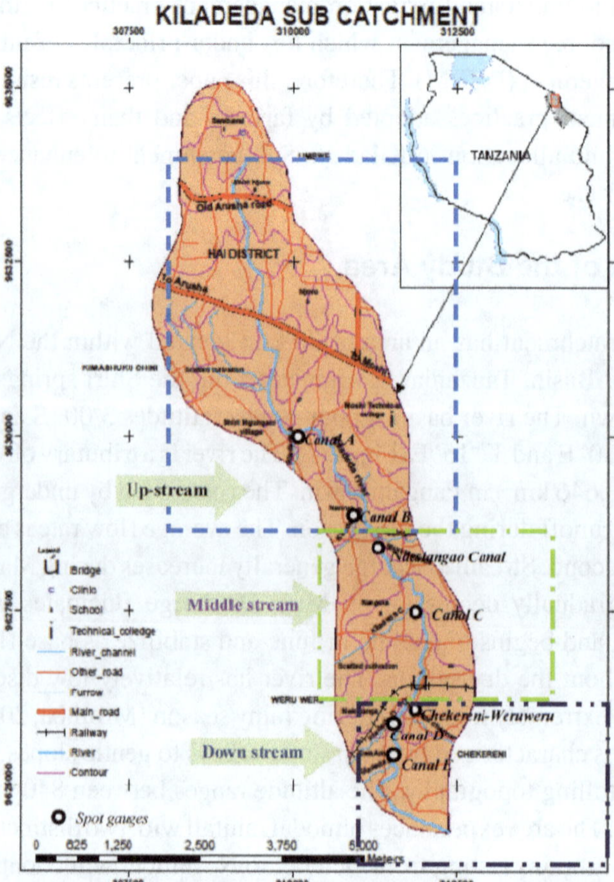

sub catchment are occupied by forests and built up land, with approximately 21.12% and 57.72% of the total area respectively (Chiwa, 2012).

METHODOLOGY

Data Collection

River discharge measurements were conducted in referenced spot gauging stations in Kiladeda sub catchment. Measurements were also conducted in irrigation furrows. GPS coordinates were recorded from each measurement point and transferred to the sub-catchment map in Quantum GIS. Measurements were taken in two months during Dry and Wet season in a daily basis. The OTT C2 current meter was used to measure flow velocity in the river and furrow sections. Discharge measurement was done by using Wading method. River cross section was divided into numerous sub sections perpendicular to the stream flow. The area of each sub-section was determined by direct measurement of width and depth. The width was measured by using a tape measure while the depth was measured using a wading rod which marks every section depth. Water velocity in each sub-section was measured using current meter. The velocity was determined by placing a current meter in the river and counting the number of revolutions in a measured amount of time (40 seconds in this case) in each sub-section. Data for all measurements were recorded in a discharge rating table sheet. Discharge of each sub-section was calculated by multiplying cross-sectional area and average velocity. The total river discharge was obtained by the sum of the discharges in all the sub-sections. In addition, the study employed questionnaires that consist of open and closed ended questions that were administered to 100 farmers. Since the study seeks to compare difference between strata, therefore, 50 farmers from each upstream and downstream were chosen based on equal proportional theory of (Kothari, 2013). Then the farmers were selected in randomly sampling method each zones of the river. The questionnaire explored irrigation techniques adopted, irrigation water allocation, benefits and constraints of adopted irrigation technique as well as the laws and regulations on irrigation water use.

DATA ANALYSIS

Water Evaluation and Planning (WEAP) Model

The study employed Water Evaluation and Planning (WEAP) model to investigate water demand and allocation in three zones of Kiladeda sub-catchment. WEAP model is an integrated water management tool for evaluation of water use and allocation in sub-catchments (Chiwa 2012). It governs the allocation of available water to meet the different water needs for farmers. The model was selected in comparison to other models such as SWAT and Mike Basin because of its ability of using a wide range of secondary data of hydrological, demographic and meteorological/climate data to simulate water demand and allocation (Mounir et al., 2011, SEI, 2011, Yates et al., 2011). It accommodates extensive primary and secondary spatial data sets and allows the simulation of various water allocation and demand scenarios. This study analysed two scenarios which were reference scenario and irrigation expansion scenario.

Calculation of Irrigation Water Requirement

Irrigation water requirement for WEAP model was calculated by using evapotranspiration (ET_o) and effective precipitation (P_{eff}) concept as outlined in FAO Irrigation Water Management Training Manual No. 3 (FAO, 2011). The reason for selecting this method is because there is no data available on the exact amount of water used for irrigation and farmers do not know how much water they use for irrigation purposes. Also, the method has been used by various scholars to calculate irrigation water requirement in many basins (Ndiiri 2011; Nolan and Shields 2000), and has worked well. Irrigation water requirement can be estimated on the basis of the crop with the highest water requirement or the crop that will ultimately occupy the largest area. Therefore, maize is the dominant crop in the sub catchment, hence was chosen for the purpose of computing crop water requirements. The formula for calculating Irrigation Water Requirement (IWR) is given by;

$$IWR = Kc.ETo - Pe \quad (1)$$

In the above equation, Kc is the Crop factor (unit less), ET_o is the reference crop evapotranspiration (mm) and P_e is the effective rainfall (mm). The maize growing days are ranging from 125- 180 and the average crop factors (Kc) is 1.15. The effective rainfall was determined by using the following formula based on the FAO manual number 3;

$$Pe = \left(0.8xP\right) - 75 \text{ If } P > 75\text{mm/month} \quad (2)$$

$$Pe = \left(0.6xP\right) - 75 \text{ If } P < 75\text{mm/month} \quad (3)$$

where; P_e is the effective rainfall (mm/month) and P is the mean monthly rainfall (mm/month). Therefore, total irrigation water requirement was calculated by subtracting the effective rainfall during the period from the crop water requirements as shown in Table 1.

RESULTS AND DISCUSSION

Adopted Practices of Irrigation Water Management

The results of the study showed that water from the Kiladeda River is diverted to the irrigated farms by using earth furrows which are poorly constructed and maintained. The study found seven open furrows in the sub catchment. Two furrows were located in the upstream part of Kiladeda River, two in the midstream and three in the downstream zone (Figure 3). The furrows are divided into primary, secondary and tertiary furrows. The primary furrows divert water direct from Kiladeda river, secondary furrows divert water from the primary (main furrows) while tertiary furrow diversions diverted water from secondary and by gravity convey the water to the farms. The primary furrow were constructed by using cement, stones and corrugated floor surface, and well protected with abstraction control gates at the intake immediately from the river. The secondary and tertiary furrows were constructed of unlined earth and normally lose large quantities of water. Therefore, farmers who have no direct access to tertiary furrows used plastic buckets to irrigate their crops.

Furrow and bucket irrigation were the two irrigation management practices adopted by the farmers in the sub catchment. However, no farmer uses both methods. Nonetheless, furrow irrigation was the dominant (94%) while 6% of the farmers use buckets irrigation methods in the upstream of the sub-catchment (Figure 2). The

Table 1. Monthly irrigation water requirements

Month	Jan	Feb	Mar	Apr	May	Jun	Jul	Aug	Sept	Oct	Nov	Dec
Monthly rainfall, P (mm)	32.8	33.3	110.7	230.7	127.6	33.5	6.8	20.3	15.6	34.7	68.2	68.2
Effective rainfall, Pe (mm)	9.7	10.0	63.6	159.6	77.1	10.1	0.0	2.2	0.0	10.8	30.9	30.9
Monthly Evaporation (mm	145	131	135	83	78	69	72	81	112	127	119	134
Crop water need, E.Tc (mm)	167	151	156	95	89	80	83	93	129	146	137	154
IWR (mm/ month)	157	141	92	0	12	70	83	91	129	135	106	123
IWR (mm/ day) (Kc =1.15)	**5.2**	**4.7**	**3.1**	**0**	**0.4**	**2.3**	**2.8**	**3.0**	**4.3**	**4.5**	**3.5**	**4.1**

Figure 2. Irrigation practices employed by farmers in Kiladeda sub catchment

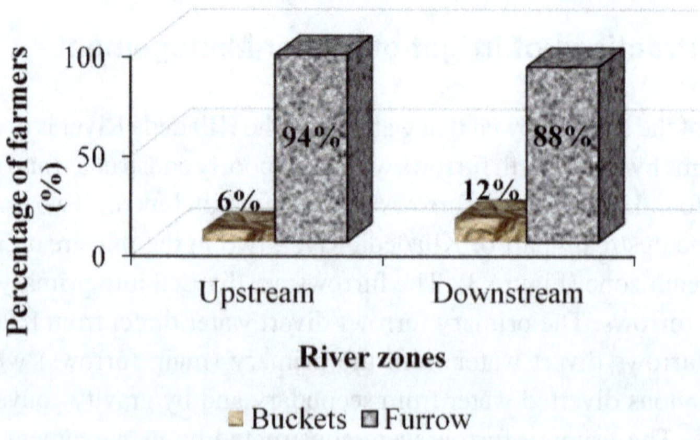

study also revealed that only 88% of farmers employed furrow irrigation methods in downstream of the river (cf. Mattias 2010; Komakech and van der Zaag, 2011). However, the system is less organised, technically inferior and water inefficient. This leads to water shortage due to several leakage points of lined and unlined furrow through the earthen materials and from breakages of cemented canals system.

Water Allocation for Irrigation

The average river discharges during wet and dry seasons were 0.44m³/s and 0.11m³/s, respectively. This implies that the river experiences very low flows during dry season from December to February. It was also noted that water abstraction for irrigation was critical during dry season with an average of 0.04m³/s (Figure 4). According to Pangani Basin Water Office (PBWO), the maximum allowable water for abstraction in irrigation furrow is 0.03m³/s per furrow. This means that farmers abstract water above permitted amount. Also, the study found that there was very low water abstraction rate during rainy season. The average abstraction rate was 0.02m³/s during wet season. This can be attributed to the fact that farmers are not relying on river water during rainy season. Moreover, it was noted that upstream farmers (Kiladeda A and B) abstract more water for irrigation then downstream farmers (Kiladeda D, E and Chekereni). These observations are in agreement with those of Mromba, (2012) who studied the impacts of water abstraction on the riparian vegetation. However, this situation brings water conflict between upstream and downstream farmers especially in dry season where by some downstream furrows are completely dry.

Figure 3. Location of spring and furrows in Kiladeda River
(Komakech et al., 2012)

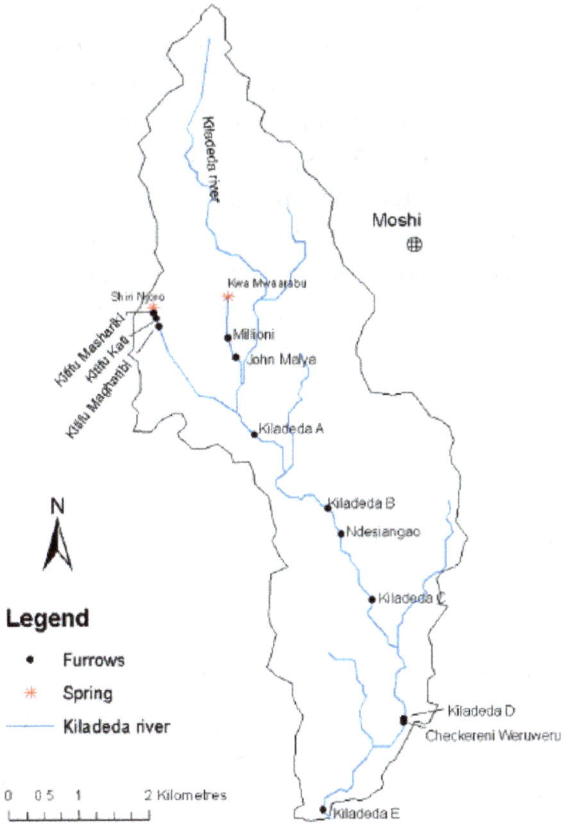

Figure 4. River discharge and abstraction rate of dry and rain season

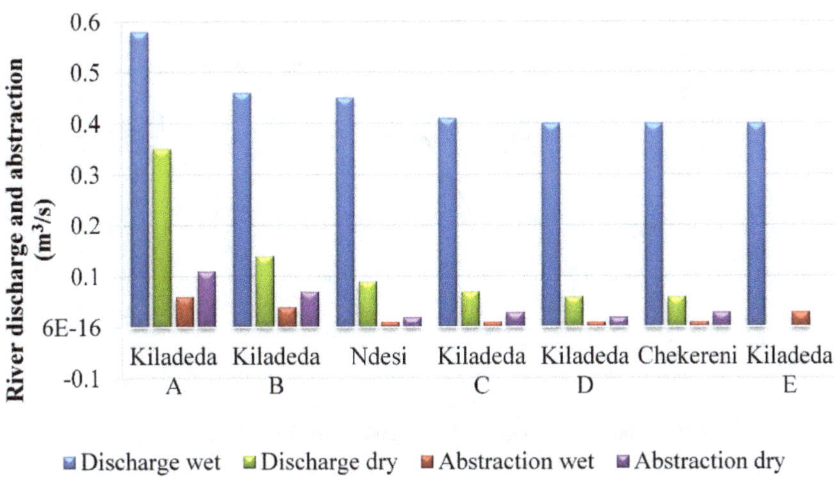

Results of Model Calibration and Validation

WEAP model was calibrated with discharge data by comparing observed and simulated discharge. The Figure 5 shows the results of the calibration process. The results display good agreement between simulated and observed discharge of Kiladeda sub catchment. The correction coefficient (R), Nash-Sutcliffe efficiency (NSE) and the percentages bias (PBIAS) were used to assess the model performance in simulating observed discharge at the spot gauges (Table 2). The NSE (0.73-0.78) and R^2 (0.85-0.88) values indicate that the ability of the model was satisfactory. This is because the model is deemed to perform satisfactory if the NSE values range from 0.5 to 1.0 and percentage bias of ± 25% (Kagoda and Ndiritu, 2009).

Scenario Analysis

Baseline Scenario

The water demand in 2014 in the entire Kiladeda sub catchment was calculated. The model results showed that the 2014 water demand in the sub catchment is

Table 2. Results of the model performance of the Spot gauging stations

Gauging station	R^2	NASH (NSE)	PBIAS
Upstream	0.88	0.78	25.4
Downstream	0.85	0.73	22.5

Figure 5. Observed and simulated discharge of Kiladeda sub catchment

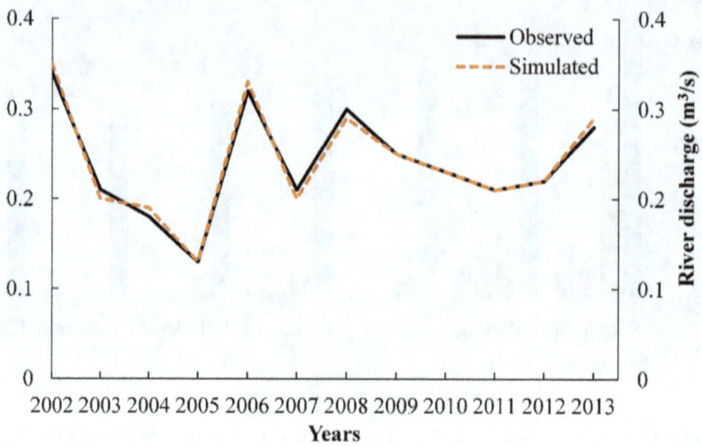

13.93Mm^3 annually (Table 3). The annual irrigation water demand for upstream and downstream were; 7.78Mm^{3}and 6.15Mm^3 respectively. The results imply that the upstream farmers use a higher volume of water than downstream farmers. The monthly irrigation water demand for all sub catchment areas are presented in Figure 6. On the other hand, unmet irrigation water demand for reference scenario was also computed. Unmet water demand was considered as the amount of water that is not provided from the river during a part of the year. Unmet water demand is important for sub catchment water management and planning purposes. The results showed that the current annual unmet irrigation water demand was 7.47Mm^3. This implies that only 46.4% of total irrigation water was met while 53.6% of irrigation water was unmet. Therefore, the results imply that sub catchment has water shortage of about 53.6% of the total irrigation water required. The unmet demand was distributed into 3.44Mm^{3}and 4.03Mm^3 for downstream and upstream respectively. The monthly distribution of unmet water demand for reference scenario is shown in Figure 7.

From the Figure 7, it is clear that Kiladeda River has low flows during dry season's months of January, August, September and December. During these months, the unmet irrigation water demand is high. Also, during these months, farmers depend

Table 3. Annual irrigation water demand distribution reference scenario

River zones	Water demand (Mm^3)	Percentage (%)
Upstream	7.78	55.85
Downstream	6.15	44.15
Total	**13.93**	**100.00**

Figure 6. Monthly irrigation water demand

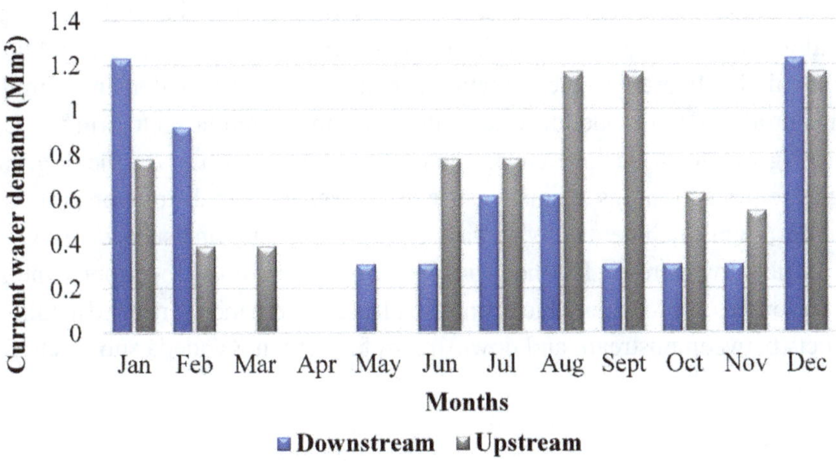

Figure 7. Monthly unmet irrigation water demand for reference scenario

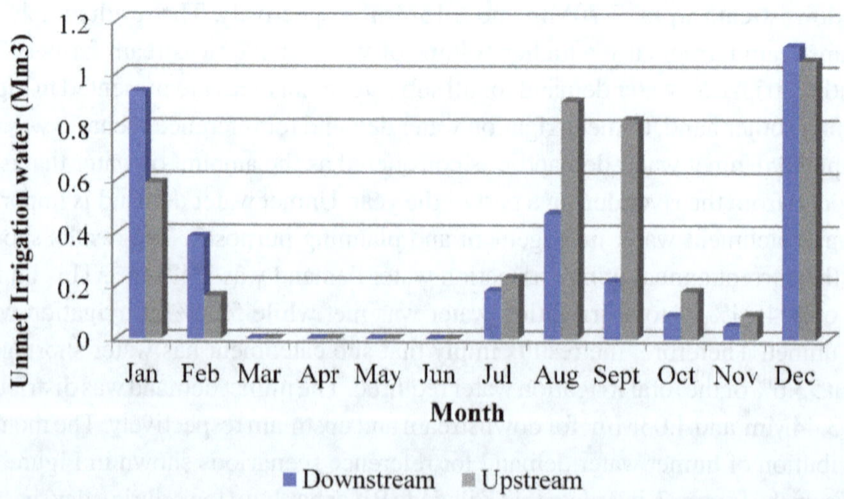

on rainfall for irrigation purposes. However, the results show that the irrigation water demand was almost fulfilled during the wet season months of March, April, May and June because heavy rainfall occurs. This also implies that farmers are not utilizing much river water for irrigation activities. Therefore, it can be concluded that despite the fact that farmers' abstract water above the permitted level during dry season, still the current irrigation water demand remains unmet in Kiladeda sub catchment. Hence, this situation accelerates water use conflicts between upstream and downstream farmers and will increase river degradation if other alternative sources of water will not be provided to the farmers to supplement the demand. The results are in agreement with those of Ndiiri, (2011) who modelled water demand for different demand sites in Mara River. She observed that agriculture sector was the least water consumer in the catchment. However, the demand for irrigated agriculture was unmet by 18%. Ndiiri, (2011) found that approximately 28 Mm^3 per year should be diverted to the irrigation schemes to satisfy the demand. Similarly, Mutiga et al. (2010) modelled water allocation for different sectors in the Upper Ewaso Ng'iro North Basin, Kenya. The past studies have observed that high water demand for irrigation was the main cause of excessive water abstraction particularly in the upstream catchments, giving rise to water shortages and subsequently, water use conflicts downstream. It is therefore, like in other catchment mentioned, integrate water resources approaches which consider all user's demand are needed to minimise conflicts between upstream and downstream farmers in Kiladeda sub catchment.

Irrigation Expansion Scenario

The study anticipated to answer the question "*What if irrigation area expands from 135 hectares to all 204 hectares that belong to agricultural land use*?" When irrigation expansion scenario was modelled, it was observed that average annual irrigation water demand in the sub catchment increased from 13.93Mm^{3}to 28.68Mm^3 per year. The total annual irrigation water distribution for sub catchment zones were 12.29Mm^{3}and 16.39Mm^{3}per year for downstream and upstream respectively (Figure 8). This implies that irrigation water demand will be almost doubled in 2020 as irrigation activities increase. Hence the situation will increase water shortage in the sub catchment.

Also, in the irrigation expansion scenario the model indicated high monthly irrigation water demand obtained in the dry season months of January, August and December while in the wet months of March to May, the irrigation water demand was low (Figure 9). However, the results show that the average annual unmet irrigation water demand was 18.78Mm^3 in irrigation expansion scenario (Figure 10). Indeed, it can be deduced from the Figure 10 that irrigation water demand for the wet season months of March, April and May will be covered for 100% in spite of increasing irrigation needs in irrigation expansion scenario.

The results of irrigation expansion scenario of this study conform to Ndiiri's (2011) findings in Mara basin that the demand will increase in the future when population increases and irrigated agriculture expand. As the result shortfalls in water supply will increase especially during dry periods. Similarly, Mutiga et al., (2010) observed that the changes in land use such as increasing the area under irrigation resulted

Figure 8. Annual irrigation water demand of irrigation expansion and reference scenario

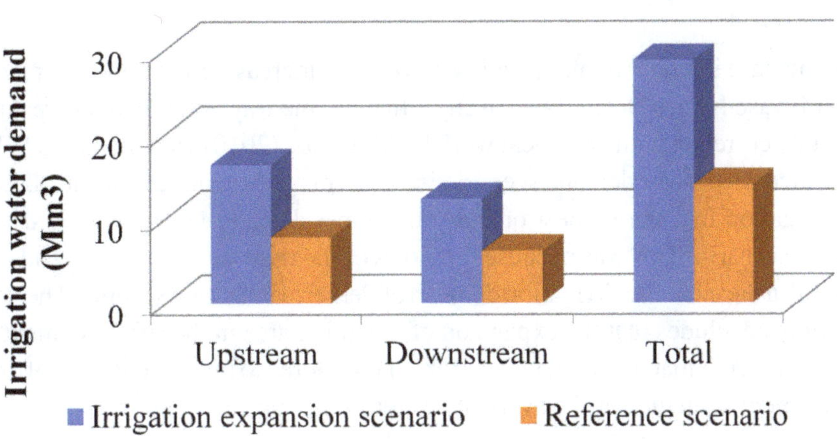

Figure 9. Monthly irrigation water demand of irrigation expansion scenario

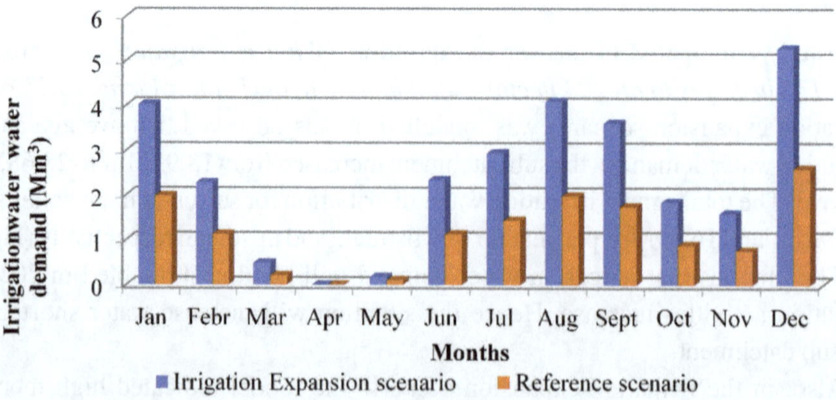

Figure 10. Monthly unmet water demand of irrigation expansion scenario

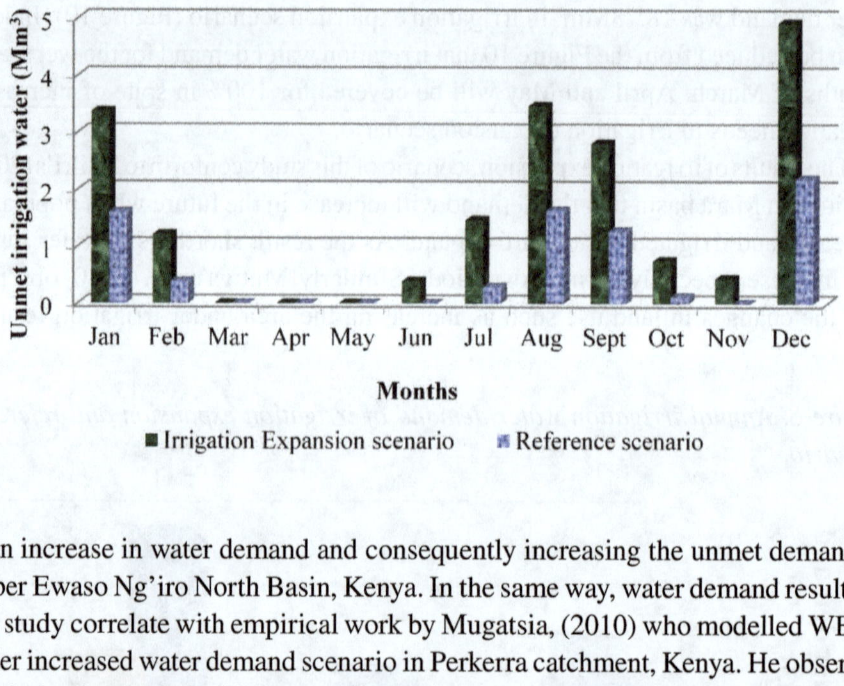

in an increase in water demand and consequently increasing the unmet demand in Upper Ewaso Ng'iro North Basin, Kenya. In the same way, water demand results of this study correlate with empirical work by Mugatsia, (2010) who modelled WEAP under increased water demand scenario in Perkerra catchment, Kenya. He observed that irrigation had the highest quantity of unmet demand under increased water demand scenario. This was because the flows of the river was decreasing and could not sustain the increased irrigation demand of Perkerra irrigation scheme. Therefore, this study concludes that the expansion of irrigation area in the sub catchment will increase water demand twice and therefore, it is required to implement more efficient management strategies of the available water in the sub catchment.

CONCLUSION

The study revealed that furrow irrigation was the main irrigation practice employed by farmers while few of them engaged on buckets irrigation. The findings also indicates high water abstraction rate during dry season in upstream then in downstream. WEAP results showed water shortages of about 46.4% in the sub catchment. For both scenarios tested, water demand increased and hence creates greater water shortage in the sub catchment. Therefore, improving irrigation efficiency would significantly improve water requirements and subsequently reduce the unmet water demands in the sub catchment. The study therefore recommends the use of improved irrigation water management practices in the sub catchment, particularly drip irrigation would be more efficient and economical in the use of irrigation water compared to bucket and furrow irrigation practices. It further emphasizes there viewing of the existing laws and regulations on water resources management in order to conform to the current irrigated area and water requirements.

ACKNOWLEDGMENT

The authors would like to acknowledge all lecturers and non-teaching staff of the Department of Geography, Kenyatta University. More so, the authors wish to acknowledge the assistance from the Pangani Basin Water Office and the Kiladeda Sub catchment committee.

REFERENCES

Booker, J. F., Howitt, R. E., Michelsen, A. M., & Young, R. A. (2012). Economics and the modeling of water resources and policies. *Natural Resource Modeling*, *25*(1), 168–218. doi:10.1111/j.1939-7445.2011.00105.x

Chiwa, R. (2012). *Effects of land Use and Land Cover Changes on the Hydrology of Weruweru-Kiladeda Sub-catchment in Pangani River Basin Tanzania* (Unpublished Master's Thesis). Kenyatta University, Nairobi, Kenya.

IUCN Eastern and Southern Africa Programme. (2009). *The Pangani River Basin: A Situation Analysis* (2nd ed.). Nairobi, Kenya: International Union for Conservation of Nature.

FAO. (2011). *AQUASTAT online database*. Rome: Food and Agriculture Organization. Retrieved from http:// www.fao.org/nr/water/aquastat/data

Hussain, I., Hussain, Z., & Sial, M. H. (2011). Water balance, supply and demand and irrigation efficiency of Indus basin. *Pakistan Economic and Social Review*, *49*(1), 13–38.

IFAD. (2012). *Challenges and opportunities for agricultural water management in West and Central Africa: lessons from IFAD experience*. International Fund for Agricultural Development.

Kagoda, P. A., & Ndiritu, J. G. (2009). *Forecasting of Daily Streamflow in the Luvuvhu River Catchment Using Artificial Neural Networks*. 10th WARFSA/WaterNet Symposium, Entebbe, Uganda.

Komakech, H. C., & van der Zaag, P. (2011). Understanding the emergence and functioning of river committees in a catchment of the Pangani basin, Tanzania. *Water Alternatives*, *4*(2), 197–222.

Komakech, H. C., van der Zaag, P., & van Koppen, B. (2012). The Last Will Be First: Water Transfers from Agriculture to Cities in the Pangani River Basin, Tanzania. *Water Alternatives*, *5*(3), 700–720.

Kothari, C.R. (2004). *Research Methodology Methods and Techniques*. New Age International (P) Limited, Publishers.

Mattias, T. (2010). *Studies of the waterscape of Kilimanjaro, Tanzania Water management in hill furrow irrigation* (Doctoral Thesis). Norwegian University of Science and Technology.

Mounir, Z. M., Ma, C. M., & Amadou, I. (2011). Application of Water Evaluation and Planning (WEAP): A Model to Assess Future Water Demands in the Niger River (In Niger Republic). *Modern Applied Science*, *5*(1), 38–49. doi:10.5539/mas.v5n1p38

MOWI. (2010). *Water sector status report*. Dar es Salaam, Tanzania: Ministry of Water and Irrigation.

Mromba, C. (2012). *Effects of water abstraction on the riparian vegetation of Kiladeda River in Pangani river basin, Tanzania* (Unpublished Master's Thesis). Kenyatta University, Nairobi, Kenya.

Mugatsia, E. A. (2010). *Simulation and Scenario Analysis of Water Resources Management in Perkerra Catchment Using WEAP Model (Master's Thesis)*. Department of Civil and Structural Engineering, School of Engineering, Moi University. Retrieved from http://www.weap21.org

Mutiga, J. K., Mavengano, S. T., Zhongbo, S., Woldai, T., & Becht, R. (2010). Water Allocation as a Planning Tool to Minimise Water Use Conflicts in the Upper EwasoNg'iro North Basin, Kenya. *Water Resources Management, 24*(14), 3939–3959. doi:10.100711269-010-9641-9

Ndiiri, J. A. (2011). Application of the Water Evaluation and Planning (WEAP) Model to assess and plan future water demands in the Mara River Basin, Kenya. University of Dar es Salaam.

Nolan, K. M., & Shields, R. R. (2000). *Measurement of stream discharge by wading*. US Department of the Interior, US Geological Survey. Retrieved from http://www.esd101.net/cms/lib

Ntale, J. F., & Litondo, K. O. (2013). An investigation into the entrepreneurial behaviour and human capital formation among small scale farmers in Kenya. *African Journal of Social Sciences, 3*, 122–134.

Qadir, M., Wichelns, D., Raschid-Sally, L., McCornick, P. G., Drechsel, P., Bahri, A., & Minhas, P. S. (2010). The challenges of wastewater irrigation in developing countries. *Agricultural Water Management, 97*(4), 561–568. doi:10.1016/j.agwat.2008.11.004

SEI. (2011). *Water Evaluation And Planning System: User Guide*. Stockholm Environment Institute, U.S. Center. Retrieved from http://www.weap21.org

Shah, T., Verma, S., & Pavelic, P. (2013). Understanding smallholder irrigation in Sub-Saharan Africa: Results of a sample survey from nine countries. *Water International, 38*(6), 809–826. doi:10.1080/02508060.2013.843843

UN-Water. (2013). *United Republic of Tanzania Water Country Brief (Water country report)*. United States Department of State.

United Republic of Tanzania (URT). (2012). 2012 Population and Housing Census. Dar es Salaam, Tanzania: National Bureau of Statistics.

URT. (2009). The national irrigation policy. Ministry of Water and Irrigation, Tanzania.

WWAP. (2012). *The United Nations World Water Development Report 4: Managing Water under Uncertainty and Risk (Executive Summary)*. Paris, France: United Nations Educational, Scientific and Cultural Organization.

Yates, D., Sieber, J., Purkey, D., & Huber-Lee, A. (2005). WEAP21 – A Demand, Priority, and Preference-Driven Water Planning Model: Part 1, Model Characteristics. *Water International, 30*(4), 487–500. doi:10.1080/02508060508691893

Chapter 6
Probabilistic Studies of Hydrologic Drought Events in Juba River in Somalia

Anne Njeri Njoroge
Kenyatta University, Kenya

ABSTRACT

Juba River runs from its headwaters in Ethiopia through Somalia and discharges into the Indian Ocean near Kismayo. Severe droughts recur in Juba River Catchment. This chapter examines the time series of hydrologic droughts and the factors influencing their occurrence in the catchment. The author further demonstrates the application of probability distributions to compute the exceedance probabilities of the hydrologic drought experienced in the catchment in the past, from 2001 to 2014. The fits of the probability distributions are compared to show the best fitting of all distributions. The probability distribution that will reasonably fit well to the data set is adapted to generate graphical plots for forecasting the return periods of drought events in the catchment. Reconstructing past drought episodes and future drought predictions is useful in drought risk management in Baardheere region of Somalia.

INTRODUCTION

Probability distributions are used in frequency analysis of hydrological data. A probability distribution is a continuous mathematical expression that determines the probability of a particular event, for example, drought or flood. A probability distribution gives important information about the data, for instance, the flows, how the flow values are changing, whether they are bunched together or spread out, and

DOI: 10.4018/978-1-7998-0163-4.ch006

whether they are symmetrically disposed on the X-axis or not. The distribution also tells the relative frequency of plotted flow values in the population on the X-axis. Knowledge of probability distributions is fundamental to water and environmental professionals (hydrologists, water users, scientists and decision makers) in such tasks as water uses, the operation of water resources and drought risk management. This chapter covers fundamental probability distributions in hydrology: 1. Normal distribution (symmetric), 2. Log-Normal distribution (skew to right), 3. Gumbel (Extreme value type I) distribution (skew to right), 4. Generalized Extreme value (GEV) (any skewness), 5. Weibull distribution (any skewness). Frequency analysis is performed by fitting these probability distribution functions to the discharge variability of Juba River at Baardheere River Gauge Station. The results will be useful in drought risk forecasting and management in Baardheere town, Somalia.

Droughts are a natural phenomenon and can be categorized as meteorological, hydrological and socio-economic droughts. Hydrological drought occurs when low water supply becomes evident, especially in streams after many months of low or no rainfall. The monthly flow sequence of Juba River at Baardheere gauging station was developed using a time series approach to identify past hydrologic droughts events in the period 2001 to 2014. The flow sequence was truncated at a desired truncation level, the mean level. All positive and negative departures from the mean for some selected months were examined using Figure 1. The negative departures from the mean were associated with hydrological droughts; two important parameters were identified from the graph, duration and the anomalies. Generally, the months of February and March were noted to have the greatest drought severity in terms of magnitude of deficit. Major droughts events in the juba river at Baardheere gauge station for the study period occurred in 2002, 2003, 2008 and 2011. In 2011, the severe drought felt in the whole country triggered a famine that resulted in the deaths of about 260,000 people (FAO, 2011); this has made increased emphasis on drought forecasting in the region.

Factors Influencing Drought Occurrence

Juba River typically results from low rainfall characteristics, often in combination with flow across a dry land and increased watershed developments. The factors are outlined as follows:

- **Topography and Elevation:** Somali is located on the North Eastern part of Eastern Africa; it's the horn of Africa. About 85% of Somali land is Arid and Semi-Arid Lands. The elevation ranges from 0 meters at the mouth of Juba River to 300 m above the mean seal level, as the river enters Somali from Ethiopia as shown in Figure 2. The main perennial rivers are Juba River

Figure 1. Time series model of hydrologic drought events in Juba River
(Author, 2016)

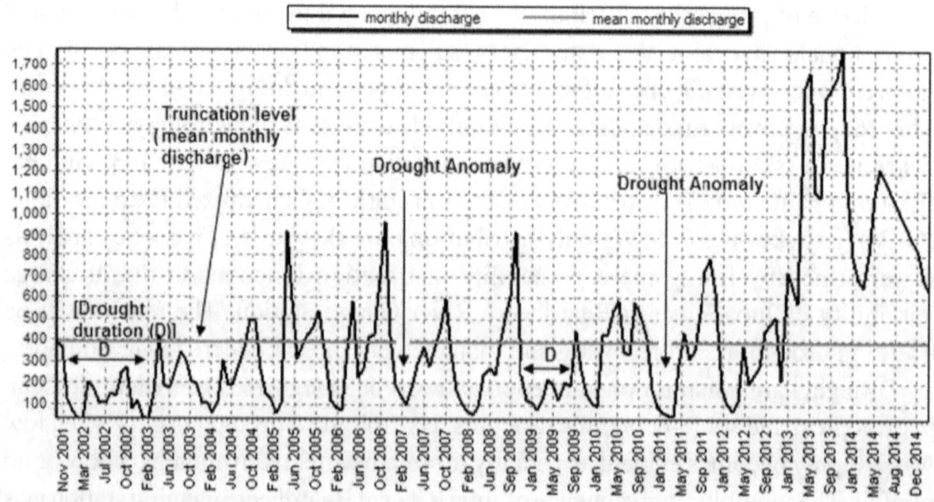

and Shabelle Rivers. The low relief of the region influence rainfall formation by enhancing high temperatures which reduce humidity levels. With low moisture levels, the region consequently has high spell most times of the year.

- **Rainfall Characteristics**: Juba River Catchment experiences arid and semi-arid type of climate. The major wet season is in the months of April to June while the minor wet season is in the months of October to November. The major dry season is in the months of December to march while the minor dry season is in the months of July to September (Muchiri, 2007).
- **Watershed Development**: Juba river flow is affected by human developments. The river being a perennial stream attracts development of major towns, for instance, Baardheere town, and villages along its riparian banks in order to utilize its waters for domestic, livestock and irrigation purposes. Baardheere town is a major agricultural centre; its climate is ideally suited for year-round crop production and has large farms of sorghum, corn, onions, beans, sesame plants, tobacco, and fruits such as bananas, watermelon, oranges, papayas, and mangoes along the river banks.

Figure 2. Surface water resources and topography of Somalia mapped from Natural Earth and Shuttle Radar Topography Mission 1 Arc-second Global respectively (Author, 2016) Source: USGS, 2016

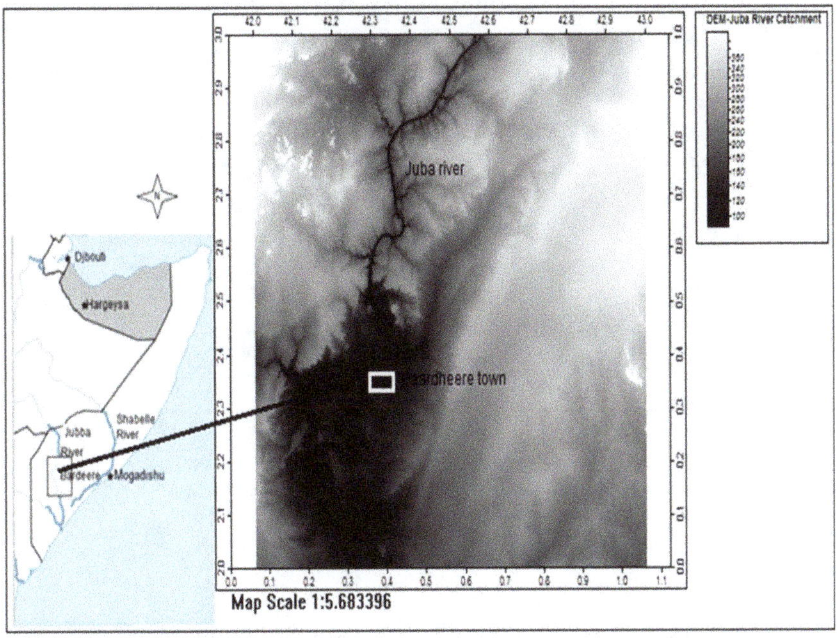

METHODOLOGY

Data Sources, Quality Checks, and Hydrological Models

The daily discharge data of Juba River at Baardheere station, altitude 90 m above mean sea level for 14 years (Oct, 2001- March, 2015) were obtained from FAO-SWALIM Data and Documents Repository at URL: http:sddr.faoswalim.org/. The daily discharge data was aggregated into monthly time steps.The mean monthly discharge at Baardheere is estimated at 390 and a monthly standard deviation of 358 m^3/s. The maximum width of Juba River is 100m, maximum depth is 8m and maximum flow is 1800 m^3/s (Artan, Gadain, Muthusi and Muchiri, 2007). To check the homogeneity of the discharge data, double mass curve analysis was carried out as shown in Figure 3.

A double mass curve is used to check the consistency of many kinds of hydrologic data by comparing date for a single station with that of a pattern composed of the data from neighbouring stations in the area. However, when the reference and tested time series comes from the same parent population to test the homogeneity

Figure 3. Double mass curve for Juba River at Baardheere station
(Author, 2016)

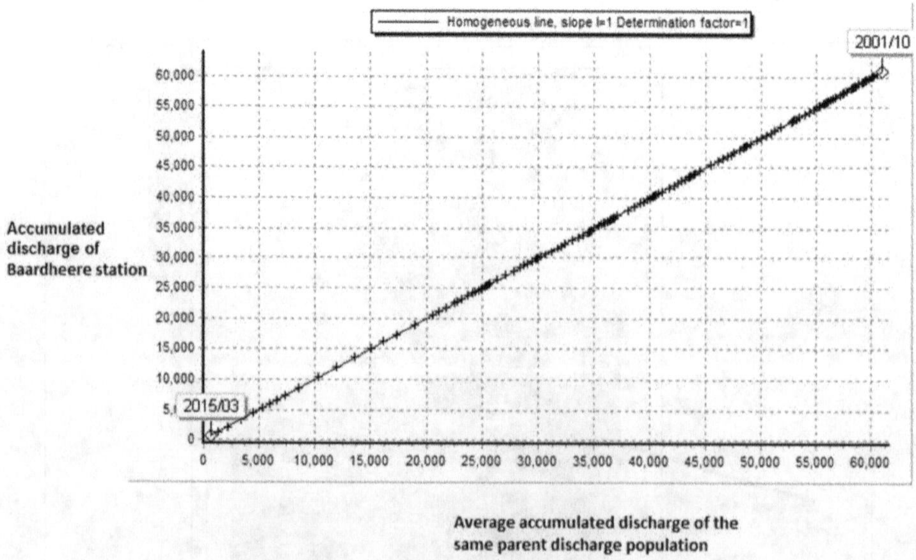

of the data over the period of record, the time series are consistent, and the graph is a straight line.

Hydrognomon, an open source hydrological data management and processing software tool is used for all data analysis herein. The programs statistical module provides tools for sampling analysis, distribution functions, statistical forecast, and Monte Carlo simulation, analysis of extreme events and construction of intensity-duration curves. Hydrognomon software can be downloaded at URL: http://hydrognomon.org/download.html

Procedure of Fitting Probability Distribution Functions

There are two approaches of fitting the probability distribution functions to hydrological data. The first method is use of graphical plots. Weibull, Gringorten or Blom points are initially used as plotting positions of ranked data series on suitable probability papers, thereafter; the best-fitting line is constructed. The second method is the parametric method. This involves determining the parameters of the probability distribution (for example, the mean and standard deviations) from the data.

In this chapter, the graphical method will be emphasized because of the visual assessment of the line of fit and the performance of each distribution can be evaluated through goodness of fit tests. The fitted distributions can be used to interpolate and extrapolate the return periods of the discharge values observed at Baardheere gauge station.

Computation of Probability of Exceedance

To determine the probability of exceedance, the data points are ranked in descending order. If N is the total number of data points and z is the rank of an individual point, the exceedance probability of the zth largest value, x_z is expressed in Equation (1) (Weibull plotting positions formula):

$$p_z = p\left(X \geq x_z\right) = \frac{z}{N+1} \tag{1}$$

To apply the formula, for example, with a 100 year events record, the largest event is ranked 1, hence its exceedance probability, $P_1 = 1/100+1$. The smallest event is ranked 100 and its exceedance probability, $P_{100} = 100/100+1$.

Return Periods

The return period of the hydrological drought is the time interval (T years) in which that particular drought event can be expected to occur once on the average. The reciprocal of the return period is the probability that the drought event is exceeded once, on the average, every T years. The return period T of the event $X \geq x_T$ and the probability of exceedance are linked by (Willems, 1998):

$$P\left(X \geq x_T\right) = \frac{1}{T} \tag{2}$$

In this chapter, the return periods of past hydrologic drought events were estimated by fitting the best performing probability distribution to the discharge data set using the hydrologic model, Hydrognomon.

To mention, Normal and Log-Normal probability distributions of the Normal family and Gumbel (Extreme Value Type I), Weibull (Extreme Value Type III) and Generalized Extreme Value (GEV-min) distributions of the Extreme Value family (moment) were selected for analyses. The most frequently used probability density functions used in the hydrologic analyses are given below. The equations and graphical plots of these most frequently used probability distributions in hydrology are discussed hereafter.

Normal Distribution

The probability distribution is also called Gaussian distribution. Many hydrologic variables are not normally distributed, but transformation can, in many cases, make them approximately normally distributed. When a hydrologic variable as well is integrated over a large time period is used in analysis, the variable is expected to follow a normal distribution because the number of causative effects increases. The normal distribution has a symmetrical bell-shaped probability density function. The distribution has two parameters; mean (μ) and standard deviation (σ) (Stedinger et al., 1992).

Its Probability Density Function (PDF) is expressed in Equation (3):

$$f(x)=\frac{1}{\sigma\sqrt{2\pi}}\exp\left[-\frac{(x-\mu)^2}{2\sigma^2}\right], -\infty < x < \infty \tag{3}$$

Its Cumulative Distribution Function (CDF) is expressed in Equation (4):

$$F(x)=\frac{1}{\sigma\sqrt{2\pi}}\int_{-\infty}^{\infty}\exp\left[-\frac{(x-\mu)^2}{2\sigma^2}\right]du \tag{4}$$

The reduced variate of the normal distribution is defined in Equation (5):

$$Z=(x-\mu)/\sigma \tag{5}$$

The properties of the reduced variate are: $\mu = 0$, $\sigma = 1$, and the coefficient of skewness is equal to 0.

From Figure 4 discharge values below (390 m^3/s) the mean value signifying hydrological droughts, shown on the vertical axis can be read. For instance, following the fitted normal probability line, 200 m^3/s and 400 m^3/s discharge values has 70% and 48% probability of exceedance respectively.

Log-Normal Distribution

This is the probability distribution of a random variable whose logarithm is normally distributed. A hydrological variable X with a normal distribution has its log-normal

Figure 4. Weibull Plotting positions and Exceedance probability (%) of discharge values at Baardheere gauge station plotted on normal probability paper with a fitted normal probability line
(Author, 2016)

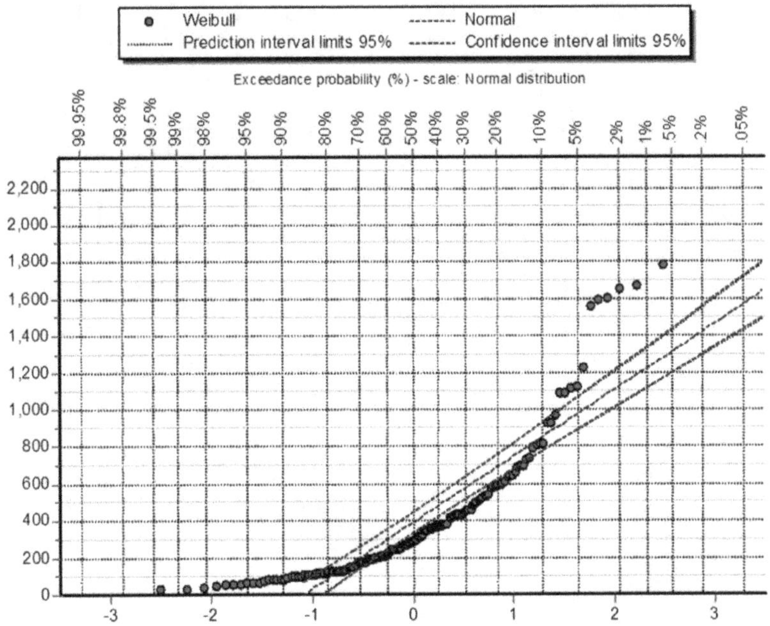

distribution as $Y = \exp\left(X\right)$ (Stedinger et al., 1992). The PDF of log-normal distribution is expressed in Equation (6)

$$f\left(x\right) = \frac{1}{x\sigma_y\sqrt{2\pi}}\exp\left[-\frac{\left(\ln x - \mu_y\right)^2}{2\sigma y^2}\right], x > 0 \tag{6}$$

The log-normal distribution has two parameters: Mean $\left(\mu_y\right)$ and standard deviation $\left(\sigma_y\right)$ which can be estimated by transforming all x_i's to y_i's as in Equation (7):

$$y_i = \ln x_i \tag{7}$$

From Figure 5, following the fitted Log-Normal probability line, 200 m^3/s and 400 m^3/s discharge values has 70% and 32% probability of exceedence respectively.

Figure 5. Weibull Plotting positions and Exceedance probability (%) of discharge values at Baardheere gauge station plotted on normal probability paper with a fitted lognormal probability line
(Author, 2016)

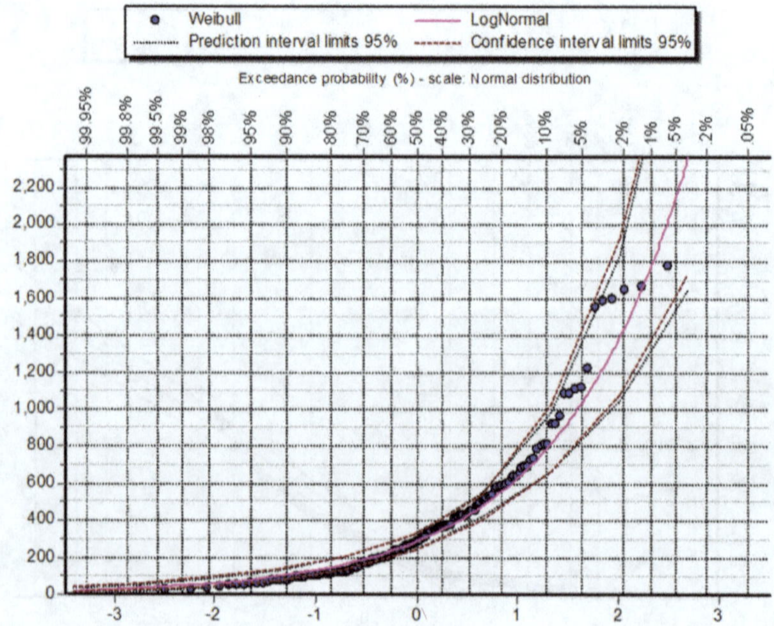

Gumbel (Extreme Value Type I) Distribution

Extreme Value (EV) distributions are distributions of *m* and *n* extreme values *(maximum and minimum respectively),* each value *(x)* of *m* or *n* category is selected from a large number of *N* observation values. The *x* values of both categories tend to approach an asymptotic limit as *N* is increased indefinitely. Mostly, the *maximum* and *minimum* categories correspond to, for example, floods and droughts respectively. Depending on the initial distribution of *m* and *n* values, three asymptotic (types) extreme distributions are derived, namely, Gumbel (Extreme Value type I), Generalized Extreme Value (GEV) and Weibull (Extreme Value type III) distributions.

Gumbel (Extreme Value type I) distribution is also called Gumbel (max) distribution. The Gumbel distribution has often been used to describe the *maximum* flows (Yevjevich, (1972); Stedinger et al., 1992; Hosking & Wallis, 1997). Its CDF and PDF are expressed in Equations (8) and (9):

$$f(x) = \alpha \exp\left\{-\alpha(x-\beta) - \exp\left[-\alpha(x-\beta)\right]\right\}, -\infty < x < \infty; -\infty < \beta < \infty; \alpha > 0 \quad (8)$$

$$F(x) = \exp\left\{-\exp\left[-\alpha(x-\beta)\right]\right\} \quad (9)$$

where, α and β are scale and location parameters respectively.

From Figure 6, following the fitted Gumbel (max)probability line, 200 m^3/s and 400 m^3/s discharge values has 68% and 42% probability of exceedance respectively.

Unlike the other EV distributions, Gumbel (Extreme Value type I) distribution third parameter (the estimator of the shape parameter, kappa (k) is equal to 0. When $k < 0$, the distribution is Extreme Value type II and when $k > 0$, the distribution is Extreme Value type III (Weibull) distribution, and the distribution is Generalized Extreme Value (GEV) distribution when $k \neq 0$.

Figure 6. Exceedance probability (%) of discharge values at Baardheere gauge station plotted on Gumbel probability paper with a fitted Gumbel (max) probability line (Author, 2016)

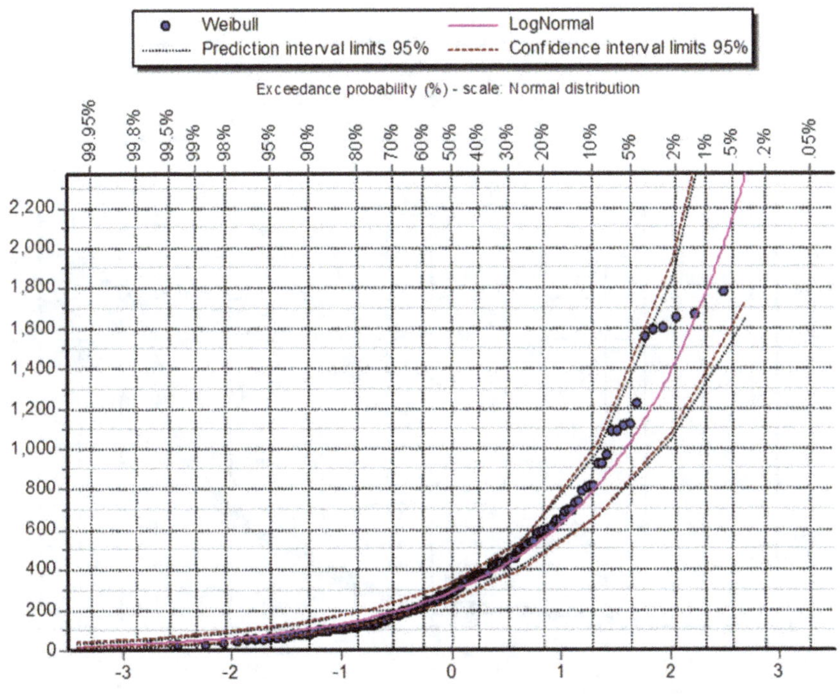

Generalized Extreme Value (GEV) distribution. The probability distribution is a general mathematical expression that incorporates the type I, II, and III extreme value (EV) distributions (Willems, (1998); Yevjevich, (1972); Stedinger et al., 1992; Hosking & Wallis, 1997). The GEV (min) distribution has the Probability Distribution Function (PDF) expressed in Equation (10):

$$F\left(x\right) = \exp\left[-\left\langle 1 - k\frac{x-\xi}{\alpha}\right\rangle^{1/k}\right], \tag{10}$$

where, k is the shape parameter, α is the scale parameter and ξ is the location parameter and $k \neq 0$. When $k > 0$, $x < \left(\xi + \frac{\alpha}{k}\right)$ and when $k < 0$, $x > \left(\xi + \frac{\alpha}{k}\right)$.

From Figure 7, following the fitted GEV (min) probability line, 200 m³/s and 400 m³/s discharge values has 60% and 35% probability of exceedence respectively.

Figure 7. Exceedance probability (%) of discharge values at Baardheere gauge station with a fitted GEV (min) probability line
(Author, 2016)

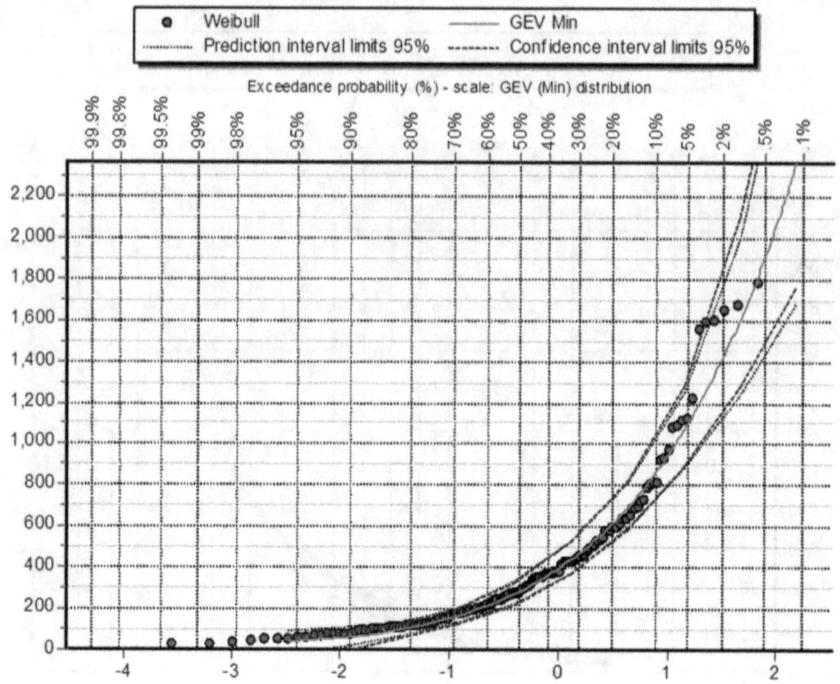

Weibull (Extreme Value Type III) Distribution

The probability distribution is also called Extreme Value Type III distribution. Weibull distribution has often been used to describe the *minimum* values from the *N* observation values (Yevjevich, (1972); Stedinger et al., 1992; Hosking and Wallis, 1997). The CDF and PDF of Weibull distribution are expressed in Equations (11) and (12):

$$f\left(x\right)=\left(\frac{k}{\alpha}\right)\left(\frac{x}{\alpha}\right)^{k-1}\exp\left[-\left(\frac{x}{\alpha}\right)^{k}\right], x>0;\alpha,k>0 \tag{11}$$

$$F\left(x\right)=1-\exp\left[-\left(x/\alpha\right)^{k}\right] \tag{12}$$

where k is the shape parameter and α the scale parameter.

From Figure 8, following the fitted weibull probability line, 200 m^3/s and 400 m^3/s discharge values has 63% and 37% probability of exceedence respectively.

Performance of the Probability Distributions

The fit of different distributions using the same data set of Baardheere River Gauge Station are compared to show the best fitting of all distributions. Graphical plots, Parameter values and Goodness of fit method are used to assess the distribution fit. The probability distributions that will reasonably fit well to the data set, is used to interpolate and extrapolate the return periods of extreme drought events.

Graphical Plots

The probability distribution functions plots figures (4-8) allowed visual assessment of the line of best fit. Comparing interpolations of different exceedance probability percentages drawn from the graphical plots figures (4-8), GEV (min) probability distribution fits the data well and is the best fit at 50% exceedance probability, followed by Weibull, Log-Normal, Gumbel (max) and then the Normal distribution. At 75% exccedance probability, Weibull distribution was the best fir and at 25% exceedance probability, Log-Normal distribution was the best fit of all the distribution functions as shown in Table 1.

Figure 8. Exceedance probability (%) of discharge values at Baardheere gauge station with a fitted Weibull probability line
(Author, 2016)

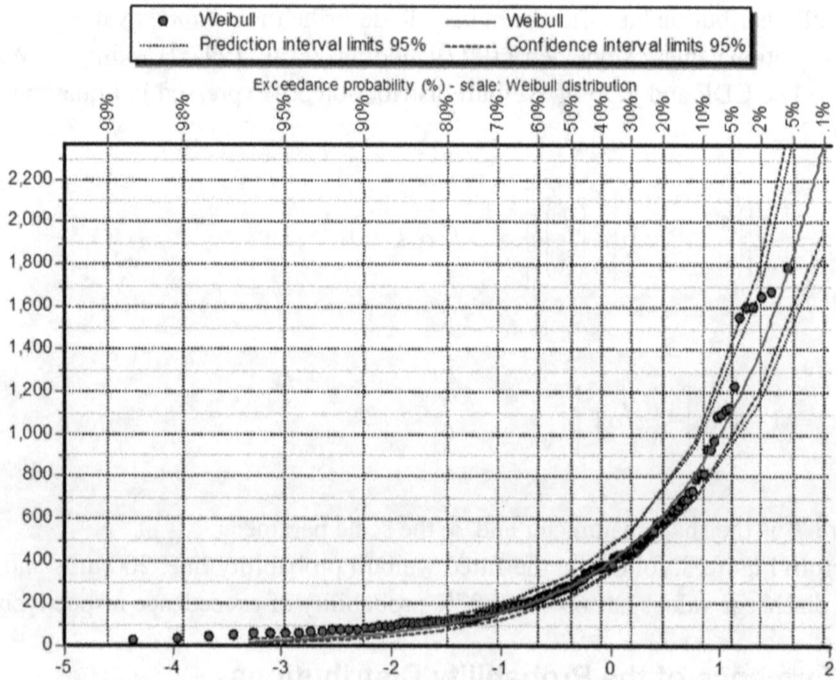

Table 1. Interpolating probability of exceedance

To an exceedance probability value (%)	Normal family		Extreme Value family (moment)		
	Normal	Log-Normal	Gumbel (max)	Weibull	GEV(min)
	(Discharge in Cumecs)				
75	148.5	169.6	137.6	128.4	132.5
50	390.5	287.6	331.5	288.0	283.0
25	632.5	487.4	577.6	544.4	534.6

(Author, 2016)

Parameter Method

The probability distribution parameters were estimated using the hydrological model as shown in table 1. The estimator of the shape parameter (kappa-k) belonging to the EV family (moment Method) distributions was used to define the best fitting of all distribution. From Table 2, GEV (min) emerged a very adequate distribution

Table 2. Parameter values of the probability distributions

Distribution model	GEV (min)	Weibull	Gumbel (max)	Log-Normal	Normal
Statistical distribution parameters	Kappa $(k) = 0.83$, Lambda $(\alpha) = 6672$ psi $(\xi) = 1.34$	$(k) = 0.74$, $(\alpha) = 6810.6$	$(k) = 0$ $(\alpha) = 4917$ $(\xi) = 1.14$	$(\mu_y) = 8.8$ $(\sigma_y) = 0.67$	$(\mu) = 8447.4$ $(\sigma) = 6303.3$

(Author, 2016)

model, its k parameter was, $k = 0.83$, followed by Weibull distribution, $k = 0.74$ and Gumbel (max) distribution, $k = 0$.

Goodness of Fit Test

A goodness of fit test measures the compatibility of a random sample with a theoretical (fitted) probability distribution function. In other words, the test shows how well the distribution one select fits to the data. Kolmogorov-Smirnov (K-S) test was selected as goodness of fit test. Kolmogorov-Smirnov test measures the largest vertical difference (Dmax) between the fitted and the observed cumulative distribution function (Chowdhury and Stedinger, 1991). The K-S test procedure can be summarized as follows:

1. Sorting the samples $X\left(x_1, x_2, \ldots x_n\right)$ by ascending order, and storing it to a new vector $X'\left(x'_1, x'_2, \ldots x'_n\right)$, and calculating the observed distribution function, see Equation (13):

$$F_n\left(x'\right) = \begin{cases} 0, x' < x'_1 \\ \frac{k}{n}, x'_k \leq x' < x'_{k+1} \\ 1, x' \geq x'_n \end{cases} \tag{13}$$

2. Calculating the K-S statistics $(D\max)$, Equation (14) and (15): $(D\max)$

$$D\max = \max_{x'_1 \le i \le x'_n} \left| F_n\left(x'\right) - F_0\left(x'\right) \right| \tag{14}$$

$$D\max = \max_{1 \le i \le n} \left\| \frac{i}{n} - F_0\left(x'_i\right) \right|, \left| \frac{i-1}{n} - F_0\left(x'_i\right) \right\| \tag{15}$$

3. Hypothesis Testing

The null and the alternative hypotheses are:

H0: The monthly discharge variability at Baardheere River gauge station do not follow the specified distributions;
H1: The monthly discharge variability at Baardheere River gauge station do follow the specified distributions.

Specifying a significance level α, if $p\left(D\max \ge D\max\left(1-\alpha\right)\right) \ge \alpha$, then accept the alternate hypothesis, otherwise reject it.

The K-S goodness-of-fit test was used to check whether monthly discharge variability at Baardheere River gauge station is consistent with GEV-min, Weibull, Gumbel (max), Log-Normal and Normal distributions. From the measure of discrepancy $\left(D\max\right)$ values for each fitted distribution, GEV (min) had the least vertical difference between the fitted and the observed cumulative distribution function, hence, GEV (min) performed better than the other four distributions. It was also noted that the difference between observed cumulative frequency values of the monthly discharge variability and the fitted distribution cumulative values, only Weibull and GEV- min distributions were acceptable at all level of significance of $\alpha = 1\%$, $\alpha = 5\%$ and $\alpha = 10\%$. The GEV-min distribution was statistically the best fit of all distributions at all significant levels. Table 3 shows Kolmogorov-Smirnov Test statistic (Dmax) values and attained significance level α for the fitted probability distributions functions.

Forecasts

The best fitting of all probability distribution functions GEV (min) was used to interpolate the return periods of hydrological drought based on the past discharge observations. Figure 8 show return period curve of past drought discharge observations taken at Baardheere station from 2001 to 2014. From the curve, discharge values

Table 3. Kolmogorov-Smirnov Test statistic (Dmax) values and attained significance level α for the fitted probability distributions functions

K-S test	$\alpha = 1\%$	$\alpha = 5\%$	$\alpha = 10\%$	Attained α	$D\max$
Normal	REJECT	REJECT	REJECT	0.19012%	0.14934
Log-Normal	ACCEPT	ACCEPT	REJECT	6.92846%	0.10382
Gumbel max	ACCEPT	REJECT	REJECT	1.97598%	0.12165
Weibull	ACCEPT	ACCEPT	ACCEPT	46.9475%	0.06784
GEV-min	ACCEPT	ACCEPT	ACCEPT	86.1475%	0.04821

(Author, 2016)

below the mean discharge level of 390m³/s signifying hydrological drought episodes have estimated return period of between 1.5 to 200 years.

A Summary of selected hydrologic droughts and their return periods drawn from the return period curve, Figure 9, is shown in Table 4.

Figure 9. Return period curve of past drought discharge observations taken at Baardheere station from 2001 to 2014
(Author, 2016)

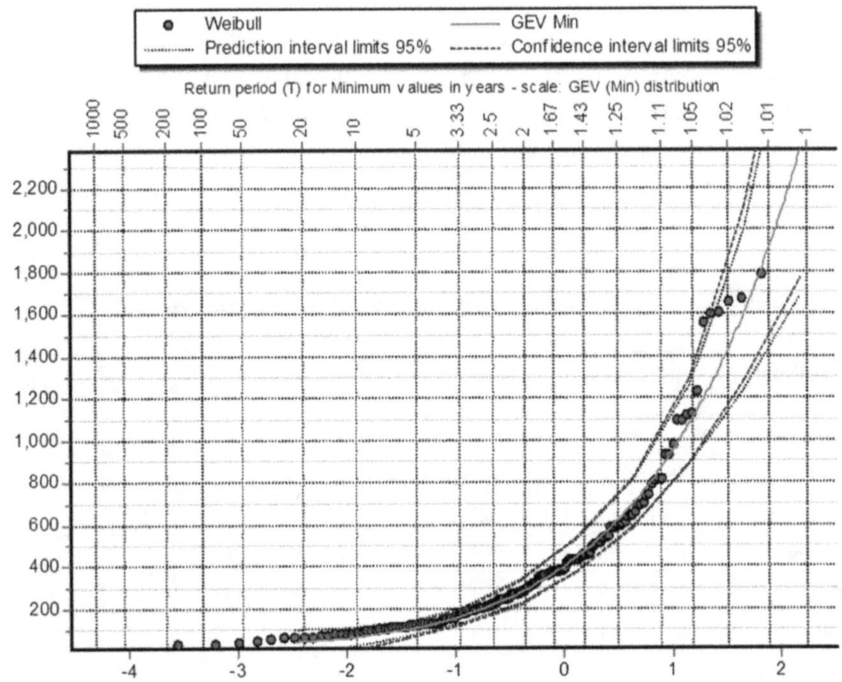

Table 4. Summary of selected discharge (cumecs) of past hydrologic droughts and their return periods

Discharge signifying extreme hydrological drought (M^3/s)	Return periods (years)
50	10
100	5
200	3.33
300	2
400	1.60

(Author, 2016)

The forecasts of hydrological droughts using different probability distributions are shown in Table 5.

CONCLUSION

This chapter examines the application of probability distributions most common in hydrological studies to describe exceedance probabilities and forecasting hydrological drought. Normal, Log-Normal and Gumbel (Extreme Value Type I), Weibull (Extreme Value Type III) and Generalized Extreme Value (GEV-min) distributions were successfully fitted for observed Baardheere gauge station data. The best fitting probability distribution was tested using graphical plots, parameter method and Kolmogorov-Smirnov goodness of fit tests. Generalized Extreme Value (GEV) distribution belonging to the Extreme Value distributions emerged the best fitting of all distributions considering the outcome performance in all the tests. Hence, the

Table 5. Forecast of hydrological droughts at different return periods

Return Periods (T) in years	T2	T5		T10		T25		T50		T100	
Discharge (m3/s)	Max	Min	Max	Min	Max	Min	Max	Min	Max	Min	Max
GEV-Min	300.9	107.6	416.3	50.1	536.6	39.1	681.3	27.8	783.2	21.9	880.3
Weibull	233.2	106.1	418.9	62.9	537.4	32.6	678.3	19.9	776.8	12.3	870.1
Log-Normal	223.2	130.8	381.1	98.9	503.9	73.4	678.9	60.5	772.6	50.9	978.6
Gumbel	304.8	114.9	411.7	61.1	524.5	10.8	666.9	-18.5	822.9	-42.9	877.5
Normal	273.2	111.1	435.2	26.3	519.9	-64.1	610.3	-122.4	668.7	-174.9	721.2

(Author, 2016)

probability of exceedance (occurrence) of GEV (min) can be used in drought risk management in the study area. In addition, these methods can be used for any set of hydrological data and probabilistic studies of flood events. Furthermore, the methods can be applied in different hydrological regions as potential areas for research.

ACKNOWLEDGMENT

The author acknowledges FAO-SWALIM Organization for the hydrologic data. The author would also like to thank the reviewers for their constructive comments that have helped to improve the Chapter contribution.

REFERENCES

Artan, G., Gadain, H. M., Muthusi, F. M., & Muchiri, P. W. (2007). *Improving Flood Forecasting and Early Warning in Somalia, Feasibility Study, Technical Report No W-10*. Nairobi, Kenya: FAO-SWALIM.

Chowdhury, J. U., Stedinger, J. R., & Lu, L.-H. (1991). Goodness of fit tests for regional generalized extreme value flood distributions. *Water Resources Research*, *27*(7), 1765–1776. doi:10.1029/91WR00077

Food and Agriculture Organization of the United Nations (FAO). (2011). *FAO's Preparedness and Response to the Horn of Africa Drought*. Retrieved from URL: http://www.fao.org/emergencies/emergency-types/drought/en/

Hosking, J. R. M., & Wallis, J. R. (1997). *Regional frequency analysis: an approach based on L-moments*. Cambridge, UK: Cambridge University Press. doi:10.1017/CBO9780511529443

Krause, P., Boyle, D. P., & Base, F. (2005). *Comparisons of Different Efficiency Criteria for Hydrological Model Assessment*. Advances in Geosciences. doi:10.5194/adgeo-5-89-2005

Muchiri, P. W. (2007). *Climate of Somalia. Technical Report No W-01*. Nairobi, Kenya: FAO-SWALIM.

Stedinger, J. R., Vogel, R. M., & Foufoula-Georgiou, E. (1992). Frequency analysis of extreme events. In D. R. Maidment (Ed.), *Handbook of Hydrology*. New York: McGraw-Hill.

Willems, P. (1998). Hydrological Applications of Extreme Value Analysis. In H. Wheater & C. Kirby (Eds.), *Hydrology in a Changing Environment* (Vol. 3, pp. 15–25). Chichester, UK: John Wiley & Sons.

Yevjevich, V. (1972). *Probability and Statistics in Hydrology*. Fort Collins, CO: Water Resources publications.

Section 2

Hydrological Data Systems and Land Use Change Dynamics and Impacts

Chapter 7
Infrastructures for Data in the Context of Flow Forecasting Using Artificial Neural Network Model for Okavango River in Namibia

Jacobine Taukondjele Amutenya
Ministry of Agriculture, Water, and Forestry, Namibia

Gerald (Augusto) Corzo Perez
UNESCO-IHE, The Netherlands

ABSTRACT

A number of evolutions on data collection and sharing have been published. Countries have collected data, but lack of access and complexity to implement these technologies has limitations. HydroServer Lite, a web-based server for sharing water data, helps to address the need of data sharing and storing in a standard format. Namibia Hydrological Services has no common online system for storing and sharing of water data. This study extends the research on HSL features as data system linked to online ANN forecasting model. This is done by implementing a Namibian HSL using real-time connection to the database to operate in real-time tools developed to visualize and fill in missing data. Lastly, a model was build using Waikato Environment for Knowledge Analysis. Results of the best model obtained are coded in hypertext preprocessor with near real-time data to provide continuous forecast. Linking data system for water resource management in a standard format is practical and promising.

DOI: 10.4018/978-1-7998-0163-4.ch007

INTRODUCTION

Water is a critical component of the world's natural resources. It is therefore imperative to understand how much water resources are available and where it is available. Recent extreme events have added an important motivation to the need of better flood and drought management. This requires knowledge on the behavior of hydrological phenomena and nowadays this can be realized by the use of hydrological models which run on water data such as water level and discharge (Albert *et al.* 2009). Currently, information technology has not been widely applied on the large amount of data that is being produced. Attempts of organizing water data that will help in effective water resource management are not formally archived in a central place in one standardized format, but it is spread over databases at different locations, in different formats. Although water resource managers do have access to this data, the variation in this data makes it difficult to locate the appropriate sources to support water resource management. Researchers spend their time on collecting data and transforming this data instead of scientifically analyzing hydrological models and making the required decisions (Goodall *et al.* 2008).

Thus, there is a need to build infrastructures to manage, archive and share this data. One such technology is in the context of Cyber infrastructures for hydrological data management. Cyber infrastructures is defined as "a combination of data resources, network protocols, computing platforms, and computational services that brings people, information, and computational tools together to perform science or other data-rich applications in this information-driven world"(Yang *et al.* 2010). The Consortium of Universities for the Advancement of Hydrological Science (CUAHSI) has developed a Hydrological information system known as the CUAHSI-HI. This university initiative has developed the first standards for the exchange and storage of water data on the internet (Botts*et al.* 2006, Tarboton*et al.* 2010). HydroServer Lite (HSL) which is also part of CUAHSI used in this study is a web based system for storing and sharing time series hydrological data using standards. However, although these data management infrastructure is already available for use, it still needs to be further explored, tested and improved to overcome the roadblock for effective water resource management (Kadlec *et al.* 2011, Kadlec 2014).

Many studies such as those by (Wu *et al.* 2005, Corzo 2007, Corzo 2009, Hung *et al.* 2009) relating to ANN in stage forecasting have emerged. These studies claim that ANN has been accepted as a good alternative to forecasting with hydrological and hydrodynamic models. This has been proven in several cases to be superior in terms of accuracy, performance and flexibility. However, in experience of the HydroServer data infrastructures, the development of such models has not been tested. By integrating the model to the data infrastructure to feed the model will be valuable in terms of generating reliable prediction. This study, therefore, seeks to address this gap by testing the applicability of ANN models in integrating the model with new emerging data infrastructures. Thus, by exploring this line of work comprehensively, the study contributes to the knowledge of data driven modeling in forecasting stages in an online environment.

The Namibia Hydrological Services in the Ministry of Agriculture Water and Forestry is the responsible body that has the mandate to provide state services in the field of hydrological monitoring including hydrological regime of surface water bodies. However, the data in the organization comes from diverse sources with different formats. There are currently a lot of data sources but for the purpose of this study consideration is given to only 3 data sources. This includes Hydstra database with historical data, South African Development Country Hydrological Cycle Observing System (SADC-HYCOS)as well as the Namwater Telemetry System. Rainfall data is provided by a different government body known as the Namibia Meteorological services. For all the data sources defined above the observations are rarely available. Moreover, all this data sources are not linked. Data from the NHS is disseminated per request mainly done through email. The process of obtaining the data Figure 1 includes finding the responsible data manager who then sends the request to the hydrologists responsible to export the data from the database. The hydrologists forward the data to the data manager who then sends out the data to the user.

Many organizations have tried to help the Hydrological Services of Namibia with water resources management through the development of data infrastructures as well as forecasting models. This includes organizations such as National Aeronautics and Space Administration (NASA) which developed a Coupled Routing and Excess STorage (CREST) model to assess water resource data at the Rundu gauging station and TigerNet which developed a model to forecast flow at Rundugauging station using

Figure 1. Obtaining stage/water level data from Namibia hydrological services

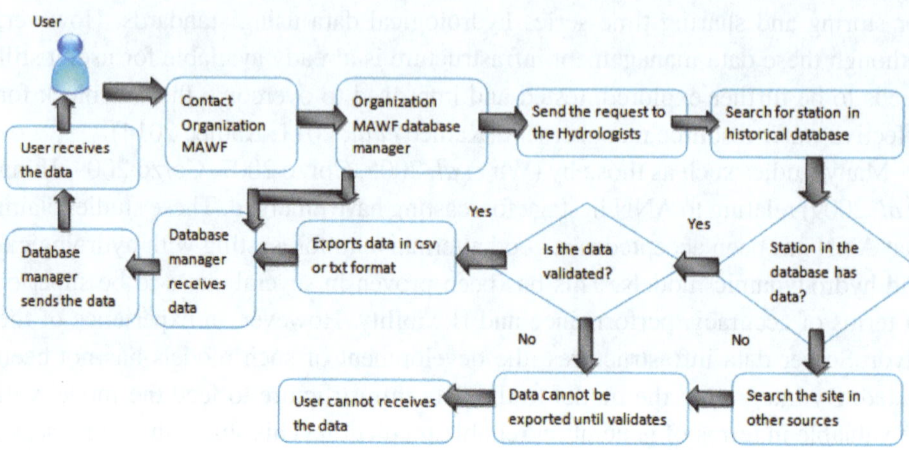

data from their Water Observation System (Walli *et al.* 2013). Another system is the Namibia Flood dashboard developed for sharing hydrological regimes in Namibia. These models and systems that are running mainly from their organizations that developed them obtain their data through File Transfer Protocol (FTP) servers or emails such as Global Network on Water and Development Information (G-WADI). Due to this sharing of data files using FTP, these systems and models are not operational because there is no standard online data management platform that stores and shares this data in a standardized format. Moreover, there are no tools or methods to automate the collection of hydrological data from the different sources allowing more accurate and more frequent measurements.

Okavango River Basin

The Model system is applied for the Okavango River basin in Namibia. The Okavango river basin (Figure 2) is located between latitude 18.9619° S and longitude 22.4844° E. This is a shared river between Namibia, Botswana and Angola. The total area of the basin in Namibia is 106,798 km^2. Rainfall is not equally distributed across the basin as Angola receives the highest annual rainfall of 1320 mm and Namibia the lowest which is 595 mm. Namibia is also the driest country in the sub-Saharan Africa, benefiting from an average annual rainfall of only 250 mm, of which 42.5 mm is lost to evapo-transpiration (Mendelsohn *et al.* 2004). This region is prone to floods that sometimes occur during the rainy season observed between October to early May each year.

METHODOLOGY

The methodology applied has three main components. Initially, the HydroServer Lite was enhanced and scripts were developed to parse data from different sources into the server. Then, this information was used as input to a *Waikato Environment for Knowledge Analysis* (WEKA) based Artificial Neural Network (ANN) model that is developed in (PHP).

Improving and Setting up the HydroServer

HydroServerLite is a freely available hydrological data hosting server. This server can be run on any webhosting services that support PHP and MySQL such as Wamp or Xamp. The Hydroserver has its own standard database known as the Object data model (ODM) shown in Figure 3 (Conner 2017). For the HSL, the essential metadata attributes for each data value include the source of the data, the monitoring

Figure 2. Overview of Okavango River
(OKACOM, 2015)

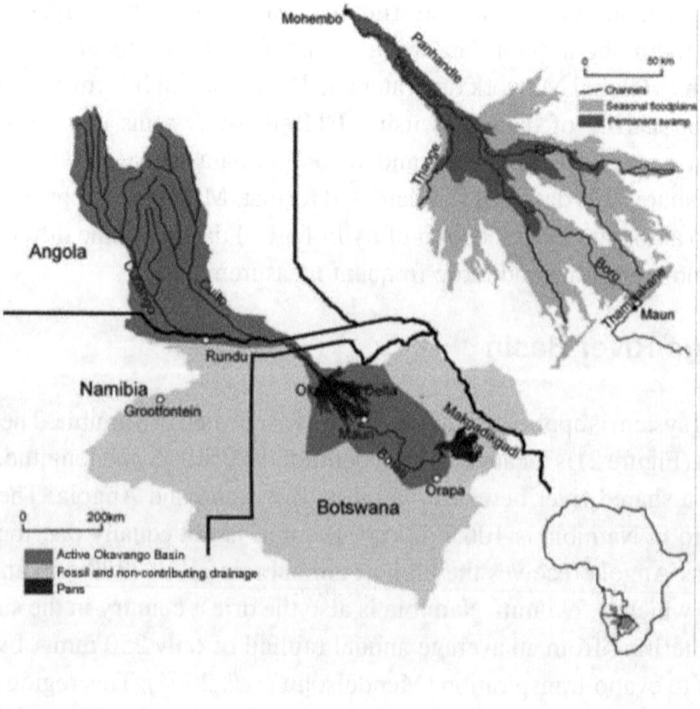

Figure 3. Simplified data model for HydroServerLite relational database structure
(Conner et al. 2013)

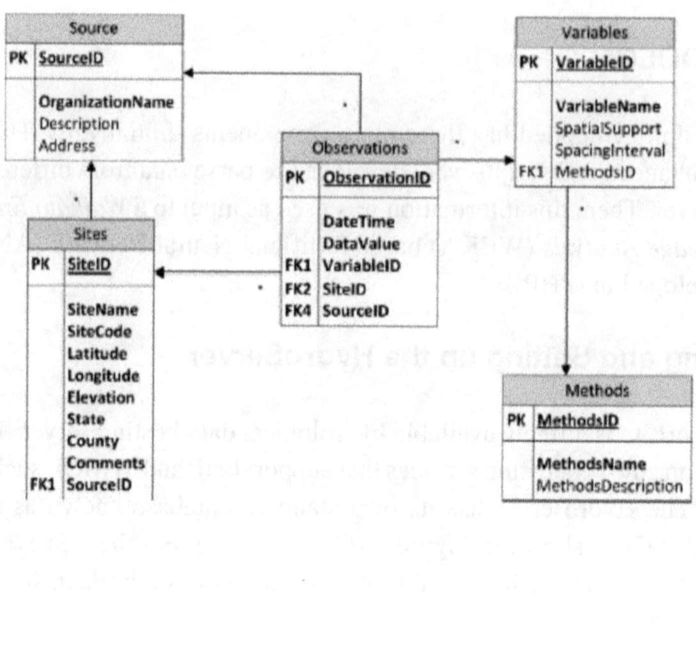

site, type of variable representing by the data value as well as the method to obtain data measurements (Horsburgh *et al.* 2008). In order to be able to use the HSL, a copy of the HydroServer was setup on a Personal computer. This was done by first installing a freely available open source Relational Database Management System that uses the Structural Query Language known as the MySQL Workbench 5.2 and then creating an empty database. This is the database that was used to populate the data tables required by the HSL during installation in order for the ODM to function. Changes were made to the HSL interface. Once the modification was done the Meta data supporting the data values was defined.

Hypertext Preprocessor (PHP) is as a server-side scripting language designed primarily for web development but also used as a general-purpose programming language (Lengstorf *et al.* 2009).For real time data, software tools were developed using (PHP) to automate data extraction from the different data sources and insertion into the improved HydroServer. However, for historical data the data insertion was done manually through the HSL interface. With the availability of this data in the server, the next step was to create another web page known as Ministry Of Agriculture Water and Forestry Forecasting Portal (MAWF-FP) that was to be fed with data from the improved HSL as this is where the Model will run from.

Artificial Neural Network in WEKA

WEKA is abbreviation of Waikato Environment for knowledge Analysis. This is a popular suite of machine language software that contains a collection of visualization tools and different algorithms for data analysis and predictive modeling. The scope of this research only focuses on the ANN building functionality. For the Okavango river basin, 5 years of Stage dataset was prepared and used to build the model at Rundu gauging station. Both low and high stage events were covered. The idea was to use the stage at Nkurenkuru gauging station which is upstream of Rundu gauging station as input to the model. However, the Nkurenkuru gauging station has been re-calibrated. Due to these unseen events, there was no correlation between these two stations. The ANN model was built offline. The function used in training the network was the Multi-Layerperceptrons (MLP). The default parameters were used to train and test the network. For training the model the training dataset was used. The input summary selected includes $((H_t,H_{t-1}),(H_t,H_{t-1},H_{t-2}),(H_t,H_{t-1},H_{t-2,}\ H_{t-3}),(H_t,H_{t-1},H_{t-2,}\ H_{t-3,}\ H_{t-4}),(H_t,H_{t-1},H_{t-2,}\ H_{t-3,}\ H_{t-4,}\ H_{t-5}))$ where H_t is current stage, H_{t-n} is antecedent stage (at t–1, t–2 . . . t–n time steps).Testing of the model was done with unseen data points which were used with the input

Programming the Neural Network in PHP

Since this model is expected to be online, the next step was to program the Network Structure in PHP for the 5 selected input variables built using WEKA. PHP was used in this study to program an adaptive WEKA based ANN model. In order to program the WEKA based ANN model in PHP there is a need to be acquainted with the network structure. By understanding the network means knowing the number of nodes making up the hidden layer and the output layer as well as the weights associated with these nodes.

To develop the Multi-Layer Perceptron (MLP) model the values associated with these nodes were extracted from WEKA. The sigmoid node defines the hidden nodes for every input as well as the threshold applied at each node. The linear node defines the output node with their weight as well as the threshold value. These are the values that were used to build the network. The steps followed are presented below.

Input Values

Getting Data From HydroServer Lite: Web Services

WaterOneFlow method is a web service for publishing the data on the internet. WaterML which stands for Water Markup Language is a type of XML developed specifically for the sharing of water data using the web (Cox 2007, Taylor *et al.* 2012). This was observed when data was collected and stored into the Object Data Model with their associated metadata. Once the data was available in the database, it was made available online using the REST WaterOneFlow web services which returns the data in a consistent format which is WaterML 1.0.

With the HydroServer acting as the main data source that receives data in real time, a script was developed to read data from the HSL in the Comma Separated Values (CSV) format using the HyperText Markup Language (HTML) request defined. In this link there is a need to specify the start date as well as the end as the main important data points. The HTML reply will be the HTML data set. To be able to extract this data the current date function gets the data from the database where by the current date and time is at 08:00 AM for the current date. However, if the CSV file does not return any value which means, the value of -999 will be returned to the database. This is the default value that specifies that the current date is null. This data is inserted into the forecasting portal using the SQL query shown in Figure 4.

Since this script does need to run automatically, windows scheduler will be run at 08:00 o`clock every morning to download the data and insert it into the Object Data Model (ODM) where the model will be running from.

Figure 4. WaterOneFlow webservice method used to publish data in a standardized format

```
$CSV=file_get_contents("
http://localhost/hydro/server/client/data_export.php?siteid=61&varid=6&meth=3&startdate=
2010-03-29%20%2000:00:00&enddate=2012-03-19%2000:00:00");


$query = "insert into datavalues(DataValue,
LocalDateTime, DateTimeUTC, SiteID, UTCOffset, SourceID,
MethodID, VariableID, QualityControlLevelID)";
$query .= "values
('$DataValue', '$LocalDatTime', '$LocalUTCTime', '$SiteID', '$UTCOffset', '
$SourceID', '$method_id', '$VariableID', '$QualityControlLevelID')";
```

Getting Data From ODM Linked to Model

There is a need to input the data in real time into the models. In order to do this an SQL statement is used to get the data from the database. However, since the data required is only for 08:00 in the morning, the SQL statement was restricted to only pick the 08:00 reading daily based on the date and time. Here the Siteid defines the station to be forecasted while the Varid is the water level in this case being forecasted. The data is also ordered based on the date and time.

Create Input Array

For forecasting the time series data can be available for many years, the forecasting date and values can be restricted to certain days. In this study it was restricted to forecast only for 160 days. Thus, with PHP Arrays it is easy to get only the last elements of the data array returned from the database. The data used for producing the forecast is from 19th. March 2012 is the current day. For forecasting 1 day ahead we will need to forecast a value for the 20, two days if for the 21, 3 days for the 22 and 4 days will be the 23 and 24 for 5 days.

With this PHP array containing the limited data values, the data is then shifted based on the lag time in order to create the input arrays. In this case the date was also considered and shifted and new dates were popped to the end of the array. Since PHP does not have the build in ANN functions like other software tools such as MATLAB, this function was developed and called Normalize. The Reverse of this function was also use to de-normalize the final output.

Apply Weight Function to Sigmoid Nodes

Apply Weight Function: In this stage the developer defines the number of hidden layers based on the weight, neuron as well as the threshold value obtained at that specific hidden node as they can be observed. By applying this to the input array the results from each node is then transformed by calling sigmoid function. To obtain the final output there, Apply Final Weight function was called to calculate the output. The Output was then de-normalized. By supplying the model with the testing data the model will then generalize the output that can be visualized using the High Chart Plot.

Integrating the System

Using data management tools data is downloaded from their sources into one common data storage which is the HydroServer Lite instance that was setup. Through web services users such as the ANN stage forecasting model can access and parse the data. This stipulates the data management system that was setup. The scripts require window scheduler to automate them without user intervention.

RESULTS

Developing and Enhancing the Data Management System

By exploring the HSL instance setup known as the MAWF HSL, it was determined that the HSL approach to managing data provides an effective and efficient way to implement a standardized data storage as well as dissemination plan using Water Markup Language. These findings are important in the sense that the introduction of this approach to be specific to the MAWF will allow them to facilitate data sharing with intended organizations such as NASA, TigerNet and the Hydrological community in SADC. Software bugs observed in the HSL is that when inserting metadata for the same site with different instruments which are the case for most gauging sites in Namibia, it does not display this site on the map viewer. A better option would be to link the HSL to a statistical tool such as MATLAB or R to improve data analysis. In order to have data in one common source with the same data format, software tools were developed to download and parse the data into the ODM database. The finding to this was that, with these tools the ODM database has the capability to handle both manual and automatic station data. Despite the common data storage, these findings are also important in the sense that automating real time data using tools is effective for the timely data collection for the developing countries such as Namibia mostly relying on manual observations.

Customization of the HydroServer has proved that it is possible to visualize data stored on the HydroServer with missing data as well as without data outliers. There was a need for a tool to mediate between the database as well as the HighChart API to fill and detect data outliers. During this research this tool was developed, and the user has a choice to visualize the raw data as well as missing data and outliers using the HydroServer interface.

This is an important functionality when dealing with data provided in real time as there is no time to validate the data. One might expect to provide their data in real time and at the same time preserve their raw data in the data storage but as well be able to allow users to visualize the data as it comes in real time but without outliers and as well with the missing data. Comparing the two plots, we see a difference in the sense that data plotted with missing data is simpler to visualize and compare which is important for the staff, users, researchers and all interested public in Namibia. The Chavelot criterion method used in this research has also proven itself to be a good method for filtering outliers. However, the problem with it is that it does not filter out low values. This research only puts a value of null for all the data that are missing as well as for the outliers. Future research should look into filling this data gaps with a value based on statistical analysis.

Developed WEKA Based Stage Forecasting Model

Without the availability of rainfall data in the study area, the WEKA ANN model was able to use the water level at Rundu to replicate its own stage up to 5 days ahead with minimal Root Mean Squared Error (RMSE) of 0.0115. The results of the experiment have shown that the ANN model is able to identify the stage to stage

Figure 5. Improved hydroserver interface

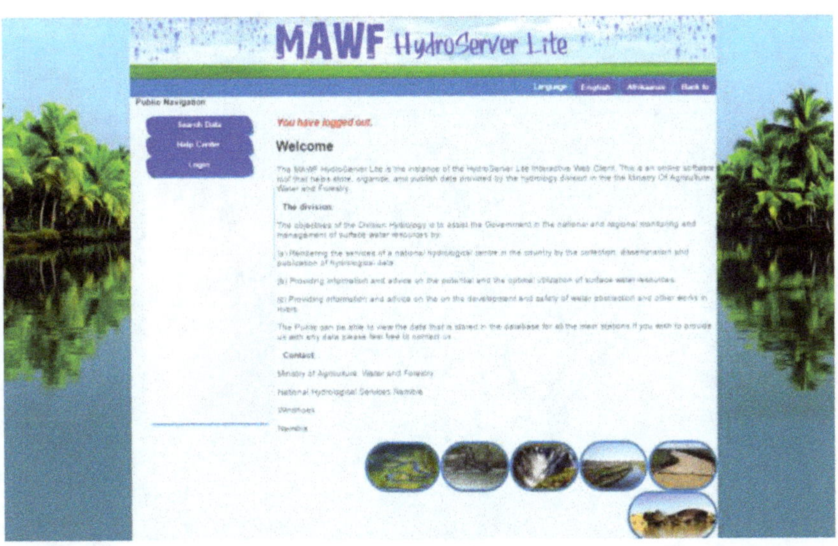

relationship for a specific site. The error statistic supports this with very low values for the RMSE for up to 3 days ahead and increases as the lead time is increased from 3 to 5 days as shown in Figure 6. This forecast produced after the model was trained was satisfying enough to allow use of the structure of the network to build an adaptive WEKA model in PHP.

The network structure in PHP was tested based on the same data set used to train the model built in WEKA. Overall the forecast is still significant enough to provide water level as the predicted and observed water level through the MAWF forecasting portal. Although WEKA already provides all this functionality by building WEKA-based ANN model algorithms, it gives the modeler more control of the system. The control to this model in PHP is linked to the MAWF HSL instance that was forecast water level online.

Standards for Storing and Sharing Time Series Data

The prospect of using the HSL for storing and sharing hydrological data is promising. The ODM database provides a standardized relational format of storing the data as well as WaterOneFlow method for publishing the data on the internet. This was observed when data was collected and stored in the ODM with their associated

Figure 6. Prediction of H_{t+5} using PHP based WEKA ANN

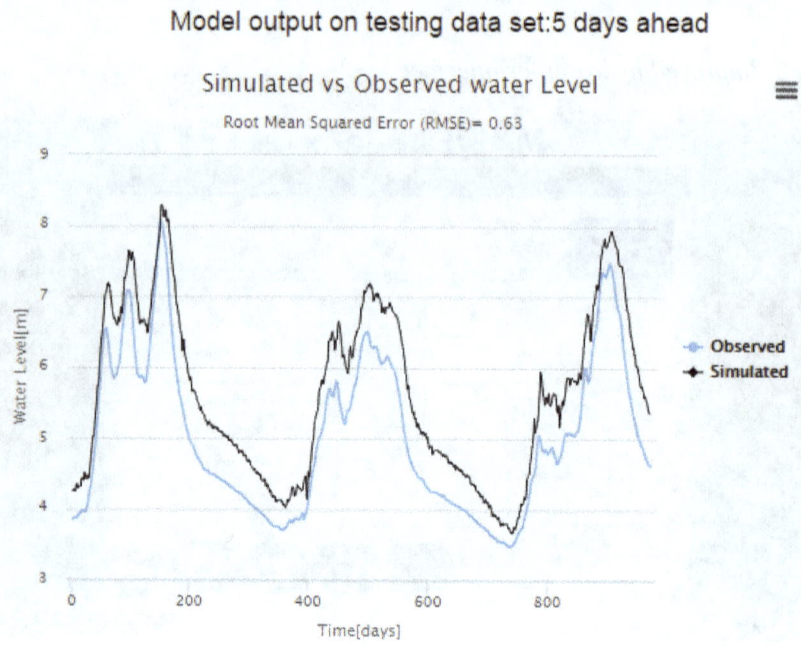

metadata. Once the data was available in the database, it was made available online using the REST WaterOneFlow web services which return the data in a consistent format which is in Water MarkUp Language (WaterML 1.0). Another format returned is by using the Rest Style Universal Resource Locator (URL) which returns the data in a consistent format which is Comma Seperated Values. Feedbacks from other agencies such as the United States Geological Survey (USGS) Goodall *et al.* (2008) and Czech Republic Jenicek (2011) whose operational systems are currently using these standards have proved that this method is a success.

Although this data is available, there was a need to parse this Extensible MarkUp Language (XML) and CSV data format in order to obtain the information. However, for this data the CSV data format was used as the XML data file especially the one returned by the GetValue method was difficult to parse. Since not every user of this system is a programmer there is a need to develop a tool for parsing the data returned by these methods in PHP. WaterML 1.0 returned by this is not in the international OGC standard. According to Almoradie *et al.* (2013) who used the WaterML 2.0, to use this encoding format, there is a need for a customized GeoServer package that will setup WaterML 2.0 accompanied by a database structure. Based on the result of this research, the WaterML 1.0 is more convenient to use since it is accompanied by the standardized ODM database structure with simple configurations.

Integrating the ANN Model With the Enhanced HydroServer

With the difficulty of locating of input data for forecasting models, this research looked into integrating the developed PHP based ANN model with the developed system. The evaluation of this integration has shown that the integration is possible. The integration was between the ANN model to the MAWF HSL instance using data sharing (CSV format) standards to facilitate the transfer of data between the model and the MAWF HSL. This was in order to feed the model with observational data and be able to visualize the model output with the MAWF forecasting data portal. From this integration, the results of the forecast for up to 5-day lead time are observed in data used for this testing as the current data value observed was the 19 March 2012. This time series data is visualized on the MAWF data portal. Since the development of the Consortium of Universities for the Advancement of Hydrological Science-Hydrological Information System (CUAHSI-HIS), many researchers supported the use of data sharing standards as a mediator between the models and the HI. This includes scientists such as Tarboton *et al.* (2011) who used web services to search and download hydrological data that they fed in their model. Their results showed that Hydrological Information System (HIS) web services can indeed be used to query, download and feed data directly into the model from

HIS. However, Tarboton *et al.* (2011) research did not use HSL that has been used in this research.

MAWF HSL can support any model that requires data served through the HSL. The level of integration between the model setup and the HSL is dependent on the model developer. Thus, there are still a lot of rules to define for inputting the data in the model. Since the software tool developed to filter data only filters extreme high data values and replace the data value with null, a great option will be to interpolate this values as well as being able to visualize the interpolated value with a different line in HighCharts. Additional work is needed to use the WebOneFlow web services for querying hydrological data directly into the model instead of the CSV data format used. The Ministry Of Agriculture Water and Forestry (MAWF) do not have an operational forecasting model and for water resource management. Daily water level is used in the daily flood bulletin and distributed through email as a way of informing the Lodge owners, community members and the public at large as to what the situation is in the rivers especially at the town of Rundu. This informs them of what the situation in the river is and when it will cause risks such as flooding. By having this PHP forecasting model integrated to the HSL the manual work of collecting data and changing the format to input the data in the model can be improved as well as minimize human errors. Water level forecasting is important mainly for flood control in the study area. When flood events occur, the reliable water level forecast to mitigate the floods.

CONCLUSION

This research has proved that it is indeed possible to apply data sharing and storing standards for integrated water information management through an operational ANN stage forecasting model. With these standards, users can establish the communication to the common time series server known as the HydroserverLite, search and obtain the appropriate input time series and retrieve the data through the internet directly into another source or modelling application. Data quality was a major drawback in this research as for inputting the data into the model data was screened for quality of the data in order to obtain information that can support water resource management with high accuracy. Further research should research into developing an interface between the model and data parser that will screen the data for quality based on the type of model is being considered. Overall, the integration and setting up of this system required backend configuration. However, for this method to be more applicable to other study areas it is advisable to link the HydroServer database directly to the sensors in the field as well as filling missing data without giving it a value of null. Regarding the ANN model more work need to be to allow WEKA to directly obtain its input from the database. The method of linking data sources to a standardized data source for sharing data with a model is practical and promising.

The HydroServer Lite infrastructure was setup, configured with Web Services and modified to plot missing data as well outliers. An ANN model was also developed in PHP based on the WEKA structure. This model uses the CSV format to input data in the model from the HydroServer and displays it on a web interface that was developed. This is a good way for the National Hydrological Service the case study in Namibia to allow user's access to their data this is normally through a daily bulletin that someone prepares manually. By having this data online the farmers and lodge managers can view the stage on the website without waiting for an email.

REFERENCES

Albert, M. G. K., Francis, W. Z., & Xuebin, Z. (2009). *Guidelines on analysis of extremes in a changing climate in support of informed decisions for adaptation*. World Meteorological Organization.

Ames, D. P., Horsburgh, J., Goodall, J., Whiteaker, T., Tarboton, D., & Maidment, D. (2009). Introducing the open source CUAHSI Hydrologic Information System desktop application (HIS Desktop). In *18th World IMACS Congress and MODSIM09 International Congress on Modelling and Simulation, Modelling and Simulation Society of Australia and New Zealand and International Association for Mathematics and Computers in Simulation* (pp. 4353-4359). Academic Press.

Botts, M., Percivall, G., Reed, C., & Davidson, J. (2006, October). OGC® sensor web enablement: Overview and high level architecture. In *International conference on GeoSensor Networks* (pp. 175-190). Springer Berlin Heidelberg.

Conner, L. G., Ames, D. P., & Gill, R. A. (2013). HydroServer Lite as an open source solution for archiving and sharing environmental data for independent university labs. *Ecological Informatics*, *18*, 171–177. doi:10.1016/j.ecoinf.2013.08.006

Cox, S. (2007). *OGC Implementation Specification 07-022r1: Observations and Measurements-Part 1-Observation schema*. Open Geospatial Consortium. Tech. Rep.

Horsburgh, J. S., Tarboton, D. G., Maidment, D. R., & Zaslavsky, I. (2008). A relational model for environmental and water resources data. *Water Resources Research*, *44*(5). doi:10.1029/2007WR006392

Hung, N. Q., Babel, M. S., Weesakul, S., & Tripathi, N. K. (2009). An artificial neural network model for rainfall forecasting in Bangkok, Thailand. *Hydrology and Earth System Sciences, 13*(8), 1413–1425. doi:10.5194/hess-13-1413-2009

Kadlec, J. (2014). *Hydrodata. Info: A Web Service For Hydrological Time Series Visualization*. Academic Press.

Kadlec, J., & Ames, D. P. (2011, June). Design and development of web services for accessing free hydrological data from the Czech Republic. In *International Symposium on Environmental Software Systems* (pp. 581-588). Springer Berlin Heidelberg. 10.1007/978-3-642-22285-6_63

Lengstorf, J., & Hansen, T. B. (2009). *PHP for absolute beginners* (Vol. 1). Apress.

Mendelsohn, J., & El Obeid, S. (2004). *Okavango River: The flow of a lifeline*. Struik.

OKACOM. (2015). *Overview of the Okavango River Basin*. The Okavango River Basin.

Tarboton, D. G., Maidment, D., Zaslavsky, I., Ames, D. P., Goodall, J., & Horsburgh, J. S. (2010). *Cuahsi hydrologic information system 2010 status report*. Consortium of Universities for the Advancement of Hydrologic Science, Inc.

Taylor, P., Cox, S., Walker, G., & Valentine, D. (2012, July). Adapting standard information models for water data. In *10th International Conference in Hydroinformatics, HIC2012*. Hamburg, Germany: IWA Publishing.

Walli, A., Tøttrup, C., Naeimi, V., Bauer-Gottwein, P., Bila, M., Mufeti, P., & Koetz, B. (2013, September). TIGER-NET–enabling an Earth Observation capacity for Integrated Water Resource Management in Africa. In *ESA Living Planet Symp.*, Edinburgh, UK.

Wu, J. S., Han, J., Annambhotla, S., & Bryant, S. (2005). Artificial neural networks for forecasting watershed runoff and stream flows. *Journal of Hydrologic Engineering, 10*(3), 216–222. doi:10.1061/(ASCE)1084-0699(2005)10:3(216)

Yang, C., Raskin, R., Goodchild, M., & Gahegan, M. (2010). Geospatial cyberinfrastructure: Past, present and future. *Computers, Environment and Urban Systems, 34*(4), 264–277. doi:10.1016/j.compenvurbsys.2010.04.001

Chapter 8
A Spatial Database of Hydrological and Water Resources Information for the Nyangores Watershed of Kenya

Luke O. Olang
Technical University of Kenya, Kenya

Mathew Herrnegger
University of Natural Resources and Life Sciences, Austria

Doris Wimmer
University of Natural Resources and Life Sciences, Austria

Josef Fürst
University of Natural Resources and Life Sciences, Austria

ABSTRACT

Advances in environmental remote sensing have provided an opportunity to monitor water resource systems in vulnerable regions with data scarcity. The spatial datasets can be used to build spatial models of reality to enable derivation of catchment-based characteristics, also often required by models in hydrology. The derived estimates can then be mapped and cartographically presented to support water resources planning within the concerned developing regions. This contribution presents a database of water resources information for an upstream catchment of the Mara River Basin of Kenya developed from freely available spatial datasets. Additionally, water quality parameters (pH, electrical conductivities, and total dissolved solids) selected as essential indicators of the suitability of the water resources for domestic applications were measured and mapped. The database, packaged as spatial maps, has been presented to the local stakeholders for developing appropriate catchment management strategies within the important watershed.

DOI: 10.4018/978-1-7998-0163-4.ch008

INTRODUCTION

Many water resource systems in the developing countries are continually faced with formidable water quantity and quality challenges consequent of environmental change effects (Dessu *et al* 2014). An understanding of the environmental changes and their interactions within the complex water systems is necessary for developing considerate planning options required to support an integrated and sustainable management. However, there exists inadequacy of observed hydrological data required to understand the impact of the changes within the vulnerable watersheds. This constraint not only hampers extended hydrological predictions, but also the design of appropriate water management structures required to safely control surface water flows within the catchments during storm events (Onyando *et al.*, 2005; Juston *et al.* 2014). In such regions with data limitation, physically based modeling approaches can be employed to define the interactions of concerned hydrological variables using parameters that can be obtained from the physical conditions of the basins (McColl and Aggett, 2007; Baldyga *et al.* 2007). The application of such an approach has been boosted by the recent advances in environmental satellite remote sensing techniques with the capacity to provide spatial datasets that exemplify the physical conditions. However, considering the difficulty of validating the model estimates due to lack of *in-situ*hydrological observations, the remotely sensed spatial datasets require adept processing based on hydrological experience in order to adequately represent the land surface terrestrial processes (Refsgaard and Hans 2004; Ayuyo and Sweta 2014). Furthermore, the geographic entities should be topologically defined to create a conforming virtual model of reality from which correct hydrological parameters can be derived.

Today, a large amount remotely sensed datasets are available for many regions around the globe due to the need for integrated environmental monitoring and assessment (Githui 2009; Olang *et al.* 2012). To select the right sets of data that address a unique water resource challenge of a region, with respect to the socio-economic demands, it is essential to comprehensively review the metadata to understand the generation process and recommended overall applicability of the data (Olang *et al.* 2011; Kilonzo *et al.* 2014). It is also necessary, where possible, that the derived estimates are authenticated through short-term measurements and validation achieved through on the ground-truthing of selected geographical entities. In cases where detailed analysis is supported by a spatial information system, it is further vital that the spatial maps are cartographically and visually presented in an elaborate manner that clearly and easily communicates the results of study to the concerned decision makers (Hake et al. 2002; Zlinszky & Timár 2013). In most river basins of East Africa, the use of spatial maps to present hydrological and water resources information is rapidly gaining prominence due to ease of interpretation by the local stakeholders charged with day to day planning and management of water resources (Olang and Fürst 2011; Oloo et al. 2014). The production of such maps have been promoted by developments in geographical information science that couples digital geo-referencing and data integration techniques to provide an opportunity to package analytical results into spatial maps that are easy to comprehend at the local levels. Though Geographical Information Systems (GIS) has increasingly been used, a majority of the maps developed from most studies in Kenya notably lack detailed spatial cartographic information on the hydrological and water resource challenges in a straightforward visual manner (Fürst 2004; Olang, 2009). It is hence imperative that the available remotely sensed data for water resources management are assembled, consistently processed and cartographically presented to provide databases that can support resource management (Fürst et al. 2015; Wimmer et al. 2015).

The objective of this study was to explore water resources planning and management options in the upstream catchments of Mara River Basin of Kenya, with a view of enhancing environmental sustainable planning and management decision makingas per the country development strategy (County Government of Bomet 2014; MRWUA 2011). The study region has in the recent past witnessed rising water insecurity, consequent of competing users, and deteriorating water quantity and quality. Despite this, however, the upstream catchments are data scarce with respect to observed hydro-meteorological data (Kilonzo *et al.* 2014). Consequently, the study collated and analyzed the major water sources using tools and spatial datasets that are easily accessible and readily available from remote sensing sources. This contribution presents the geo-database of cartographic maps developed for the region as an illustration of how best practices in spatial mapping can generally support water resources development and management. The maps have been presented to the local community stakeholders and proved important in visually communicating the concerned water related challenges, thereby promoting improved decision making within the River Basin that feeds the Mara ecosystem classified as a world heritage site (Ogutu et al. 2012; Olang & Kundu 2011).

MATERIALS AND METHODS

The Study Area

The Nyangores Watershed of study is located between Latitudes 0° 22' and 1° 02' South of the equator and Longitudes 35° 12' and 35° 47' East. The basin is drained by River Nyangores, which originates in Keringet area of the Mau forest located in the Southwest of Kenya. The Nyangores River runs over a distance of about 94 km in southwest direction through the Counties of Nakuru, Narok and Bomet before joining River Amala at Kaboson area to form the main Mara River (MRWUA, 2011; Mango et al. 2011). The highest altitude of the basin is located at about 2970m above mean sea level (*a.m.s.l.*). This gradually falls to about 1900m *a.m.s.l.* at the Bomet Gauging Station, and finally to 1695m *a.m.s.l.* at the confluence of the Nyangores and Amala Rivers. The basin has a total catchment area of about 930 km^2. The major hydro-meteorological processes of the Nyangores Watershed are precipitation, evapotranspiration, water infiltration and surface runoff. Infiltration is mainly influenced by soil characteristics, surface conditions, vegetation and the rainfall amount and intensity (Dessu and Melesse 2012; Dessu et al. 2014). The basin characteristics, including some of the sample points of study are illustrated in Figure 1.

The study watershed generally typifies a tropical climate that changes greatly with the change in altitude. Seasonal variations in rainfall are influenced a lot by the Inter-Tropical Convergence Zone (ITCZ). The rainfall seasons are bimodal, with the long rains experienced between March and May and the short rains being experienced in September and December (Juston et al. 2014; Kiragu 2009). Though there are many meteorological stations in and around the Nyangores Watershed, very few (i.e. less than eight) stations provide more or less reliable data that can be used for analysis. A typical hyetograph indicating the mean monthly rainfall distribution for the Bomet Water Supply rainfall station is shown in Figure 2 based

Figure 1. Physical Characteristics including the selected field sampling points for the Nyangores Watershed

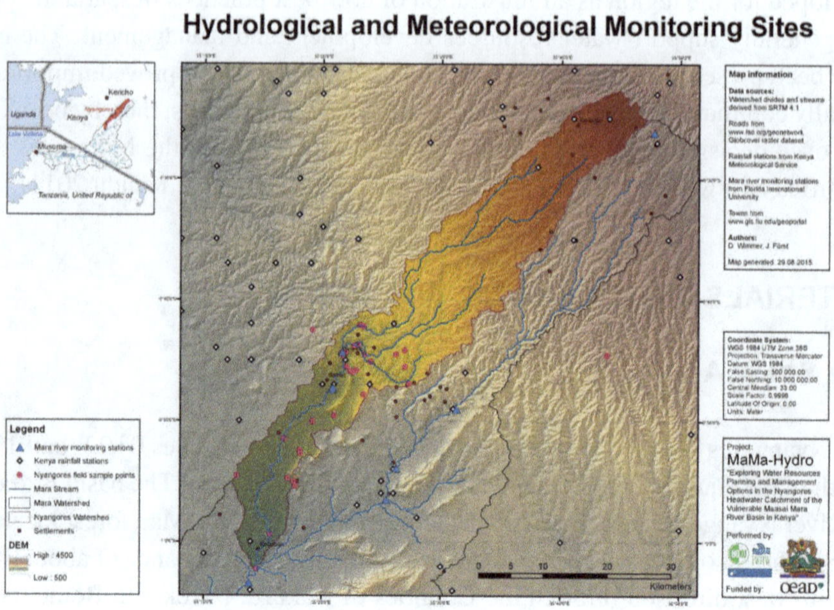

Figure 2. Mean monthly rainfall at the Bomet -Water Supply rainfall station (1967-2008)

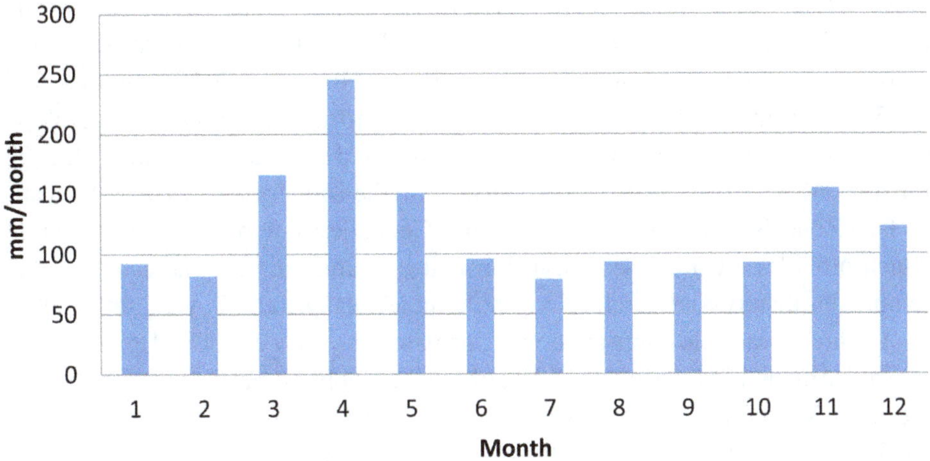

on data for the period 1967-2008. The daily rainfall information corroborates the bimodal nature of rainfall, with the month of February and Julydemonstrating the low flow periods for the region. The data from the station also show that mean annual rainfall for the areais about 1450 mm. The highest annual rainfall of about 1985 mm was recordedin year of 2004. The driest year was 2005, when about 1035 mm of rainfall was recorded. However, these values are only representative of this areas since the basin spans three agro-ecological zones namely the humid (Zone I) on the upper parts, sub-humid (Zone II) in the middle region and semi-humid (Zone III) in the lower parts of the basin (Olang, 2009; Sombroek *et al.* 1982). The long-term average potential evapotranspiration (PET) estimated with the Priestly-Taylor method and a temperature series from Kericho was found to be 1490 mm/year. Thus, the rainfall and PET are in approximate balance in the Nyangores Watershed(Fürst et al. 2015; Juston et al. 2014).

Data Bases and Processing

The spatial datasets for the study were obtained from different global sources, including a fieldwork campaign carried out in the watershed between May and August of 2015. The fieldwork was aimed at estimating the amount and quality of the major water sources and their outlets as a basis of understanding the water availability *versus* demand for water potential allocation studies in future. The collated spatial datasets covered different thematic areas and were obtained from various sources. In cases where more datasets for a particular thematic area were available, for instance soil datasets at various scales and sources, all of the datasets were acquired and processed to illustrate the temporal and spatial variations. Table 1 provides some of the datasets collated for developing a spatial database for the area. With the exception of the data from the fieldwork, the spatial datasets provided in Table were freely accessible from the sources (Fürst et al. 2015; Wimmer et al. 2015). The acquired datasets were organized in format and structure to develop a consistent database that allows ease of navigation and comparison (Hake et al. 2002; Fürst 2004). The datasets were harmonised into the Universal Transverse Mercator (UTM) Zone 36 South based on the World Geodetic System (WGS) 1984 datum, which provides a true to scale transformation of areas around the equator while minimising distortions laterally along the latitude (Olang 2009; Onyando et al. 2005).

To allow invariable interpretation of the maps, a layout was defined to include the title of the maps, a standard legend describing the mapped themes, overview maps of the entire Mara Basin, and a description of the map information, coordinate system and MamaHydro project details. Processing of the spatial data was achieved based on automated procedures available in a Geographical Information System (ArcGIS) software package provided by the Environmental System Research Institute (ESRI). A number of extensions for data conversion, analysis and management including tools for geo-referencing, geo-processing, interpolation, spatial and image analysis were employed in the production of the spatial maps (Mozafari et al. 2011; Kundu et al. 2008). Topographic and spatially based hydrological characteristics of the basin were derived using ArcHydro Tools. Accordingly, the collated analogue maps were geo-referenced, digitised and edited using geoprocessing tools. The collected fieldwork data were also referenced and mapped to establish their spatial and physical attributes (Omonge 2016; Fürst et al. 2015). Using image analysis tool, a consistent and adept classification of the acquired satellite imageries was achieved to discern the spatio-temporal changes of the basin (Ngeno 2016; Fürst et al. 2015). Detailed methodological information of the tools and extensions used in the study can be obtained through the help section of ArcGIS software, which is freely available online through the ESRI website available at www.esri.com.

RESULTS AND DISCUSSION

Water Resources Information

Mapped Water Resources

From the field studies, a total of 52 major water sources in the Nyangores Watershed were identified and mapped with the help of Garmin*Etrex*Global Positioning

Table 1. Description of the spatial datasets collated for the study

Thematic Area	Description of the Dataset	Data Source	Scale/Resolution
Land Cover	Landcover Data 1: Aggregated data for Kenya produced under the Africover program.	Global Landcover Network (GLCN) web portal.	1:200000
	Landcover Data 2: Globcover data for 2009 acquired as a regional map for Kenya	European Space Agency GlobCover web portal.	300x300 m
Soils	Soils Data 1: KenSOTER soil data version 20. provided as the Soil and Terrain database for Kenya (SOTER)	World Soil Information (ISRIC) web portal (with permission).	1:1000000
	Soils Data 2: Soil Atlas for Africa 2014 provided as vector coverage for Africa	Joint Research Centre (JRC) of the European Soil Data Centre (ESDC) web portal (with permission).	1:3000000
Topography	Digital Elevation Model (DEM) 1: One Arc-second DEM from Shuttle Radar Topographic Mission (SRTM)	United States Geological Survey – Earth Resources Observation and Science (USGS-EROS) web portal	30x30 m
	Digital Elevation Model (DEM) 2. Three Arc-second DEM from Shuttle Radar Topographic Mission (SRTM)	Consultative Group for International Agricultural Research - Consortium for Spatial Information (CGIAR-CSI) portal	90x90 m
Geology	Geology Data: Hard copy, Africa-Geological Map for 1962	National Atlas of Kenya	1:3000000
Geomorphology	Landform and Lithology: Data for Kenya produced under the Africover program	United Nation Office for Outer Space Affairs (UN-SPIDER) knowledge portal	1:350000
Hydrology	River Gauging Stations and Data: Locations and daily streamflow data for Bomet station	Water Resources Management Authority (WRMA) in Kericho	
	Rainfall Gauging Stations and Data: Locations and daily rainfall data for region of study	Kenya Meteorological Department (KMD) offices in Nairobi	
	Water Sources and Sinks: Data of the qualities and quantities of major water supply and demand sources.	Fieldtrip carried out in the basin between 2015/2016	
Imageries	Satellite Imagery 1: Data for 1995 from Landsat 4-5 Thematic Mapper (TM) defined by Row 169 and Paths 060 and 061	United States Geological Survey (USGS) Earthexplorer web portal	30x30 m
	Satellite Imagery 2: Data for 2010 from Landsat 7 Enhanced Thematic Mapper (ETM+) defined by Row 169 and Paths 060 and 061	United States Geological Survey (USGS) Earthexplorer web portal	30x30 m
GIS Data	Other spatial Data: Local towns, Roads networks, Local districts, Stream networks, County polygons	Obtained from various global sources as vector data	

System (GPS). The major water sources were identified with the collaboration of the Nyangores Water Resources User Association (WRUA). Overall, 28 springs, 6 boreholes and 1water intake points from the local streams were identified. From the study, it was clear that the upper part of the catchment has fewer and sparsely scattered natural springs. About 50% of the springs are located in the mid part of the catchment, which can be attributed to the close proximity of the area to the forest as well as the hilly nature of the landscape. Figure 3 illustrates the sampled water resources.

Water Quantities and Quality

During the field campaign, the discharge rates from the springs were estimated using a calibrated bucket and a stop watch. From the measurement, it was noted that discharge rates generally varied between 0.09 litres/second and 0.45 litres/second, with a mean value of 0.2 litres/second. Apparently, the lowest discharge rates were observed at lower altitudes within the catchment due to lower precipitation rates combined with the warmer climatic condition leading to higher evapo-transpiration rates in these areas. Besides the water quantity measurements, basic water quality indicators for the springs also estimated included pH, Total Dissolved Solids (TDS) and Electrical Conductivity (EC) as a first rough estimate of the suitability and the quality of the water resources for domestic use. Comprehensive water quality tests would have been useful in this respect, but this would require more sophisticated and expensive laboratory equipment beyond the scope of the assessment. From the results, a pH range of 6.5 to 8.2 was obtained, with the exception of one measurement. This range was considered acceptable (Kilonzo et al. 2014). Based on the electrical conductivity measurements, the Total Dissolved Solids (TDS) were derived. The results obtained indicated a range of 40 to 200 mg/l, with a mean value of 115 mg/l. Figure 4 illustrates the water quality results obtained.

The Landcover Characteristics

Because of the considerable impacts of changes in land use and land cover in the Mara River Basin, there are many studies on this topic. Some studies have also quantified the spatio-temporal effects of detected land cover changes on the hydrology of the region (Ngeno, 2016; Kilonzo et al. 2014). Indigenous information from the local communities, however, indicates that most of the natural forests in the upper parts of the basin were cleared in the year 1930, and deforested parts later replaced with exotic tree species including pines and eucalyptus. With the land division, relocation and settlement plan in 1970, the Mau Forest was again deforested to pave way for further human settlement and subsistence agriculture. Figure 5 below illustrates the

Figure 3. Sampled and mapped major water sinks and sources of the Nyangores Watershed

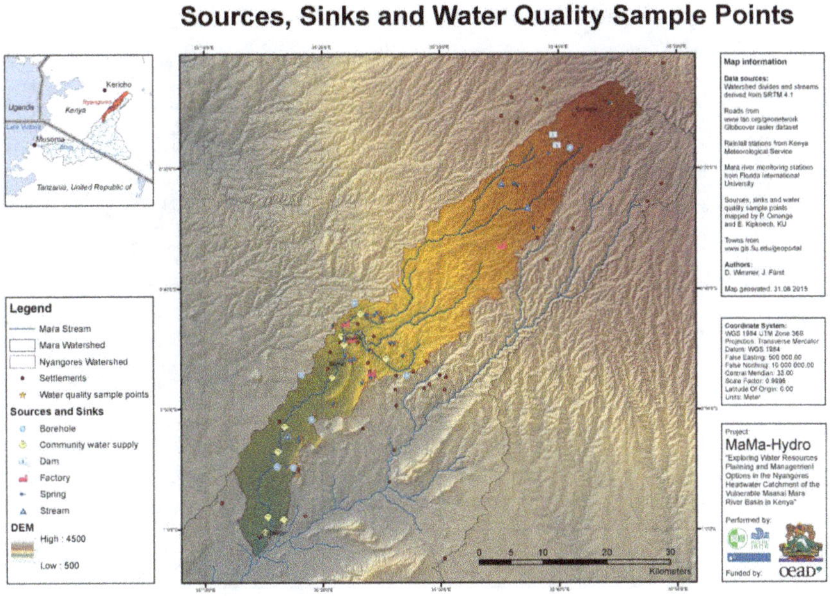

general landcover characteristics of the basin as revealed by the Africover program of the FAO carried out in 2007 at a scale of 1:100,000. Other spatial and temporal information on land cover for the geodatabase were from other global sources and used as a basis for land cover change detection studies for potential hydrological impact assessment of the area (Ngeno 2016; Olang and Fürst 2011). Based on Figure 5, the Forest constituted about 39.1%, Herbaceous crops 37%, Shrublands about 10.3%, Tree savannah 5%, Closed woody vegetation 3.9%, Forest plantations and tree plantations about 2%, Grassland 1.7% and finally Shrub savannah and Shrubland both approximately 0.5%.

Spatial Geodatabases for the Basin

See Figure 5.

The Soil Characteristics

Figure 6 illustrates the soil map of the Nyangore Basin derived from the Soil Atlas of Africa, which was created through a collaboration of several institutions of the European and African Union and the Food and Agriculture Organization (FAO)

Figure 4. Estimates of the Total Dissolved Solids of the springs of the Nyangores Watershed

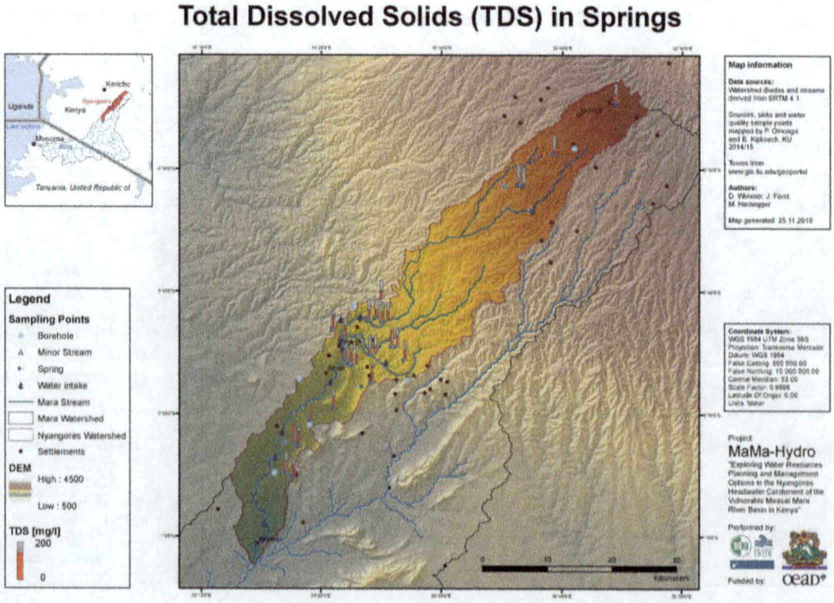

Figure 5. Land cover characteristics obtained from the FAO KenSOTER database

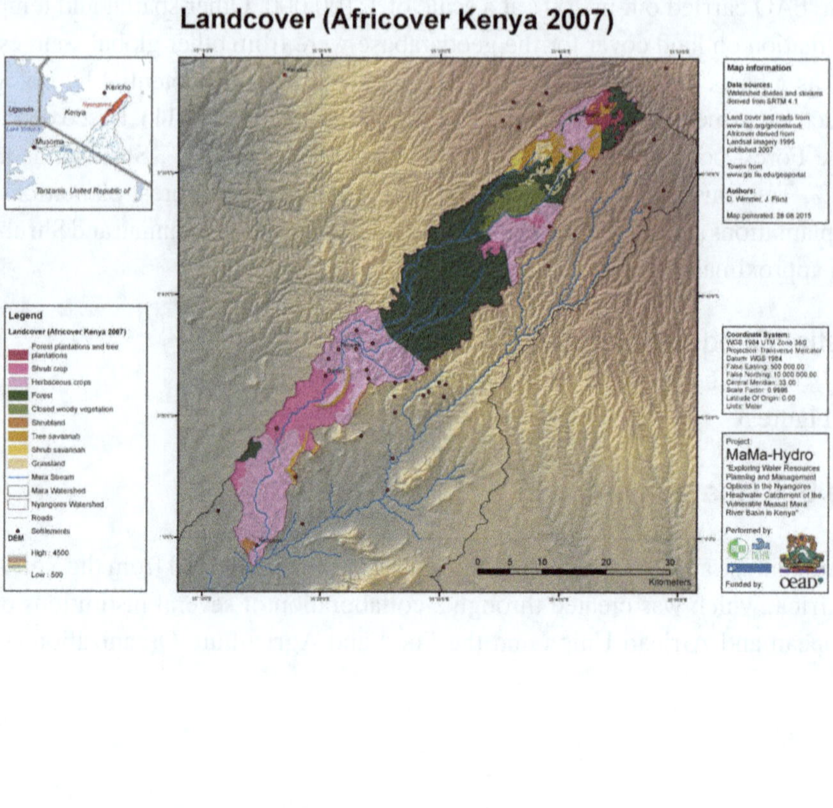

of the United Nation (Joneset al. 2013). The atlas is based on the World Reference Base (WRB) for Soil Resources developed in 1998 in an effort to establish an international reference coded for global soils. In the dataset, every polygon in a shapefile represents a prevailing type of soil that corresponds to well defined and referenced soil group. In addition to this, there exists an atlas with detailed description for the soil groups for various regions of the globe that is available at a cost. The soil properties that influence the ability to generate runoff through the drainage and infiltration processes are defined as Hydrological Soil Groups (HSGs), and are often considered in rainfall-runoff studies (SCS 1986). In water resources management, however, physical and chemical properties of a soil are key factors important for the planning of water resources to avoid compromised qualities due to acidity or alkalinity. In all of these cases, spatially based maps can aid in detailing the extents of the soil properties important for water resources management.

From Figure 6, the Nyangores Watershed is largely dominated by Silandic Andosols in the upper part of the basin. These are young soils developed in volcanic deposits, which are ideal for agriculture. The remaining parts of the basin consist of a mixture of slightly acidic to strongly acidic soils, mainly of loamy soil type in nature. Table 2 provides a summary of the equivalent areas of the major soil types and their characteristics in the Nyangores Watershed. From the table, about 68.2% of the basin constitutes of Andosol type of soils, which hydrologically provides high

Figure 6. The major soil types derived from the Soil Atlas of Africa

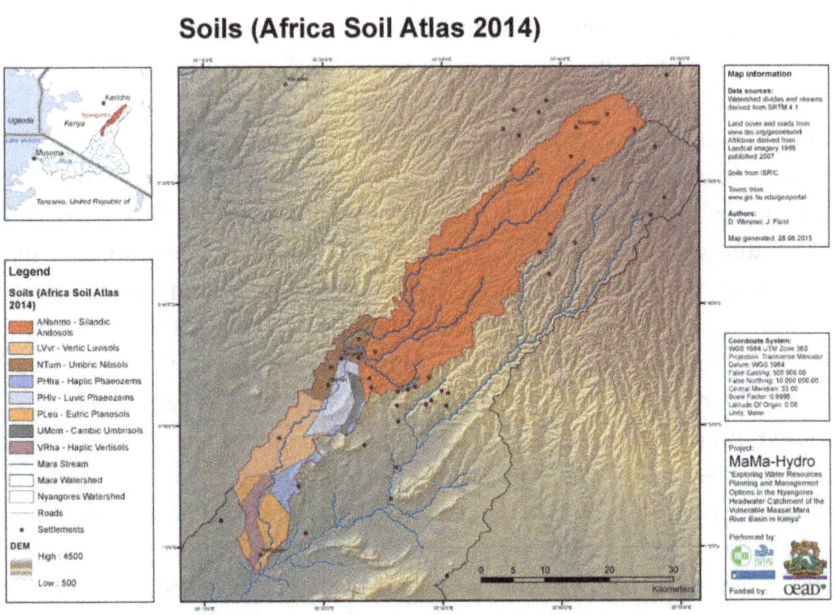

Table 2. Major soil types and their spatial characteristics within the Nyangores Watershed

Soil Type	Description	Area (km^2)	Area (%)
Andosols	Young soil developed in volcanic deposits	636.9	68.3%
Luvisols	Slightly acid soils with a clay-enriched subsoil & high nutrient-holding capacity	79.2	8.5%
Nitisols	Deep red soils with a well-developed, nut-shaped structure, from basalt	48.7	5.2%
Planosols	Poorly structured surface layer abruptly overlying a slowly permeable layer	49.1	5.3%
Phaeozems	Slightly acid soils with a thick, dark-coloured surface layer	71.8	7.7%
Umbrisols	Acidic soil with a dark surface horizon rich in organic matter	15.8	1.7%
Vertisols	Clay-rich soils that develop deep, wide cracks upon drying	31.5	3.4%

infiltration rates and less surface runoffs depending on the land cover and land use practices. The lower parts of the basin constitutes about 8.5% of Luvisols, which largely explains the slightly lower pH values noted in some of the water sources in this area.

CONCLUSION

Managing the water resources requires an understanding of the existing related practices in the region based on comprehensive information on the physical and hydro-meteorological properties of the catchment. From the processed soil maps, it could be noted that the upper part of the basin has low surface runoffs potential based on the characteristics of the soils. However, it should be noted that this also depends on the topography and land cover/land use practices. The maps can also support future water planning in terms of exploitation of the resources in areas with less acidity or alkalinity, mainly in the upstream regions of the basin. The land cover maps generally indicated that area is largely a forested-agricultural catchment. A majority of agricultural activities are predominantly in the upstream and downstream parts of the basin. Already, a preliminary assessment of the land cover changes between 1995 and 2010 indicates that agricultural land increased by about 9% between 1995 and 2010. This observation calls for regulation of the upstream agricultural activities so as to avoid possible degradation of the water quantity and quality of the Nyangores River at the headwaters. The observed discharge rates from the water sources provided a mean value of 0.2 litres/second. The basic water quality parameters measured during the field campaign indicated that the springs conform well to previous estimates from other studies carried out before in the basin. The results show a tendency where higher TDS values were prevalent in the lower parts of the basin, which could be attributed to the accumulation of dissolved solids mainly due to erosion in the lower parts of the catchment.

ACKNOWLEDGMENT

This study was conducted within the framework of the MaMaHydro Project funded by the Commission for Development Research (KEF) at the OeAD-GmbH Austria. The authors would like to acknowledge and appreciate the support from KEF that enabled this study. We would also like to thank our colleagues who supported the work through data collection and analysis. More expressly, we wish to thank the Nyangores Water Resources Users Association (WRUA), through Mr Paul Rono for their support in accessing hydrological and water resource information of basin.

REFERENCES

Ayuyo, I. O., & Sweta, L. (2014). Land cover and land use mapping and change detection of Mau complex in Kenya using geospatial technology. *International Journal of Science and Research*, *3*(3), 767–178.

Baldyga, T. J., Miller, N. S., Driesse, L. K., & Gichaba, N. C. (2007). Assessing land coverchange in Kenya's Mau Forest region using remotelysensed data. *African Journal of Ecology*. doi:10.1111/j.1365-2028.2007.00806.x

County Government of Bomet. (2014). *Medium Term County Fiscal Strategy Paper*. County Government of Bomet.

Dessu, S. B., & Melesse, A. M. (2012). Modelling the rainfall–runoff process of the Mara River basin using the Soil and Water Assessment Tool. *Hydrological Processes*, *26*(26), 4038–4049. doi:10.1002/hyp.9205

Dessu, S. B., & Melesse, A. M. (2013). Impact and uncertainties of climate change on the hydrology of the Mara River basin, Kenya/Tanzania. *Hydrological Processes*. doi:10.1002/hyp.9434

Dessu, S. B., Melesse, A. M., Bhat, M. G., & McClain, M. E. (2014). Assessment of water resources availability and demand in the Mara River Basin. *Catena*, *115*, 104–114. doi:10.1016/j.catena.2013.11.017

Fürst, J. (2004). *GIS in Hydrologie und Wasserwirtschaft*. Heidelberg, Germany: Herbert Wichmann.

Fürst, J., Olang, L., Herrnegger, M., Wimmer, D., Omonge, P., & Kipkoech, E. (2015). *MaMa-Hydro: Exploring water resources planning and management options in the Nyangores headwater catchment of the vulnerable Maasai Mara River Basin in Kenya*. Final Technical Report No. 1. Vienna: Kommission für Entwicklungsforschung (KEF).

Githui, F., Mutua, F., & Bauwens, W. (2009). Estimating the impacts of land-cover change on runoff using the soil and water assessment tool (SWAT): Case study of Nzoia catchment, Kenya. *Hydrological Sciences Journal, 54*(5), 899–908. doi:10.1623/hysj.54.5.899

Hake, G., Grünreich, D., & Meng, L. (2002). *Kartographie: Visualisierung raum-zeitlicher Informationen*. Berlin: Walter de Gruyter.

Jones, A., Breuning-Madsen, H., Brossard, M., Dampha, A., Deckers, J., Dewitte, O., ... Robert, Z. (2013). *Soil Atlas of Africa*. Luxembourg: Publications Office of the European Union.

Juston, J., Jansson, P.-E., & Gustafsson, D. (2014). Rating curve uncertainty and change detection in discharge time series: Case study with 44-year historic data from the Nyangores River, Kenya. *Hydrological Processes, 28*(4), 2509–2523. doi:10.1002/hyp.9786

Kilonzo, F., Masese, F. O., Van Griensven, A., Bauwens, W., Obando, J., & Lens, P. N. L. (2014). Spatial-temporal variability in water quality and macro-invertebrate assemblages in the Upper Mara River basin, Kenya. *Physics and Chemistry of the Earth, 67-69*, 93–104. doi:10.1016/j.pce.2013.10.006

Kiragu, G. M. (2009). *Assessment of suspended sediment loadings and their impact on the environmental flows of upper transboundary Mara River, Kenya* (MSc. Thesis). Jomo Kenyatta University of Agriculture and Technology (JKUAT), Nairobi, Kenya.

Kundu, P. M., China, S. S., & Chemelil, M. C. (2008). Automated extraction of morphologic and hydrologic properties for River Njoro catchment in Eastern Mau, Kenya. *Journal of Science, Technology. Education and Management, 1*, 14–27.

Mango, L. M., Melesse, A. M., McClain, M. E., Gann, D., & Setegn, S. G. (2011). Land use and climate change impacts on the hydrology of the upper Mara River Basin, Kenya: Results of a modelling study to support better resource management. *Hydrology and Earth System Sciences, 15*(7), 2245–2258. doi:10.5194/hess-15-2245-2011

Mati, B. M., Mutie, S., Gadain, H., Home, P., & Mtalo, F. (2008). Impacts of land-use/cover changes on the hydrology of the trans-boundary Mara River, Kenya/ Tanzania. *Lakes and Reservoirs: Research and Management, 13*(2), 169–177. doi:10.1111/j.1440-1770.2008.00367.x

Minaya, V. G., McClain, M. E., Moog, O., Omengo, F., & Singer, G. A. (2013). Scale-dependent effects of rural activities on benthic macroinvertebrates and physico-chemical characteristics in headwater streams of the Mara River, Kenya. *Ecological Indicators, 32*, 116–122. doi:10.1016/j.ecolind.2013.03.011

Mozafari, G. A., Khosravi, Y., Abbasi, E., & Tavakoli, F. (2011). Assessment of Geostatistical Methods for Spatial Analysis of SPI and EDI Drought Indices. *World Applied Sciences Journal, 15*, 474–482.

MRWUA. (2011). *Nyangores River Sub Catchment Management Plan (SCMP)*. Bomet, Kenya: Mara River Water Users Association.

Mutua, B. M., & Klik, A. (2007). Predicting daily streamflow in ungauged rural catchments: The case of Masinga catchment, Kenya. *Hydrological Sciences Journal, 52*(2), 292–304. doi:10.1623/hysj.52.2.292

Ngeno, E. K. (2016). *Assessing the impact of land use changes on the hydrology of the Nyangores sub-catchment, Bomet County, Kenya* (MSc. thesis). Kenyatta University.

Ogutu, J. O., Owen-Smith, N., Piepho, H. P., Kuloba, B., & Edebe, J. (2012). Dynamics of ungulates in relation to climatic and land use changes in an insularized African savanna ecosystem. *Biodiversity and Conservation, 21*(4), 1033–1053. doi:10.100710531-012-0239-9

Olang, L. O. (2009). *Analyses of land cover change impact on flood events using Remote Sensing (RS), GIS and Hydrological Models. A Case Study of the Nyando River Basin in Kenya* (PhD thesis). University of Natural Resources and Applied Life Sciences (BOKU), Vienna, Austria.

Olang, L. O., & Fürst, J. (2010). Effects of land cover change on flood peak discharges and runoff volumes: Model estimates for the Nyando River Basin, Kenya. *Hydrological Processes*. doi:10.1002/hyp.7821

Olang, L. O., & Kundu, P. M. (2011). Deforestation and land degradation of the Mau Forest Complex in Eastern Africa: a review for conservation and restoration purposes. In E. O. Ekundayo (Ed.), *Environmental Monitoring* (pp. 245–262). Rijeka: InTech Publishers.

Olang, L. O., Kundu, P. M., Ouma, G., & Fürst, J. (2012). Impacts of land cover change scenarios on storm runoff generation: A basis for management of the Nyando Basin, Kenya. *Land Degradation & Development*. doi:10.1002/ldr.2140

Oloo, F. O., Olang, L., & Strobl, J. (2015). Spatial modelling of solar energy potential in Kenya. *International Journal of Sustainable Energy Planning*, *6*, 17–30.

Omonge, P. O. (2016). *Simulation of water resource use and allocation in Nyangores sub-catchment of the upper Mara Basin, Bomet County, Kenya* (MSc. thesis). Kenyatta University.

Onyando, J. O., Olang, L. O., & Chemelil, M. C. (2005). Regional Analysis of Conceptual Rainfall-Runoff models for runoff simulation in ungauged catchments of Kenya. *Journal of Civil Engineering Research and Practice*, *2*, 23–37. doi:10.4314/jcerp.v2i1.29135

SCS. (1986). *National engineering handbook*. USDA.

Sombroek, W. G., Braun, H. M. H., & van der Pouw, B. J. A. (1982). *The explanatory soil map and agro-climatic zone map of Kenya*. Kenya Soil Survey Report No. E1.

Wimmer, D., Fürst, J., & Herrnegger, M. (2015). *MaMa-Hydro: Exploring Water Resources Planning and Management options in the Nyangores headwater catchment of the vulnerable Maasai Mara River Basin in Kenya*. Book of Maps: Technical Report No. 2. Vienna: Kommission für Entwicklungsforschung (KEF).

Zlinszky, A., & Timár, G. (2013). Historic maps as a data source for socio-hydrology: A case study of the Lake Balaton wetland system, Hungary. *Hydrology and Earth System Sciences*. doi:10.5194/hess-17-4589-2013

Chapter 9
Multi-Temporal Landsat Remote Sensing for Forest Landscape Fragmentation Analysis in the Yoko Forest, Kisangani, DRC

Jean-Fiston Mikwa Ngamba
University of Kisangani, DRC

Ewango Corneille Ekokinya
World Conservation Society Epulu, DRC

Cush Ngonzo Luwesi
CGIAR Research Program, DRC

Yves-Dady Botula Kahindo
University of Kinshasa, DRC

Muhogwa Jean Marie
University of Kisangani, DRC

Hyppolite Nshimba Seya
University of Kisangani, DRC

ABSTRACT

This chapter assessed the impact of human activities on deforestation and sustainability of water resources and livelihoods in the Congo Basin. It mainly aimed to assess forest degradation in the Yoko Reserve from 1976 to 2015 and investigate the compatibility of Landsat imagery for forest monitoring. Digital image processing for unsupervised classification was done using ENVI software while supervised classification was done by means of ArcGIS 10. Results show that forest landscape faced large-scale human-induced fragmentation over the last 40 years. If these trends continue, they will affect the sustainability of water resources and livelihoods in the Congo Basin of the Democratic Republic of Congo. Hence, policymakers need to look at key drivers and address impacts that may threaten the future of hydrological ecosystems services, including water and land resources in the Congo Basin. Authorities have to apply an integrated management of water, land, and ecosystems.

DOI: 10.4018/978-1-7998-0163-4.ch009

INTRODUCTION

Land cover fragmentation is now recognized as an indicator and a prominent interface between human activities and global environmental change. Human and environment interactions involve conversion of natural landscapes into anthropogenic land use or by changing management practices in human dominated landscapes (Foley et al., 2005). Understanding the role of land use in global environmental change and the long term human environment interaction requires historical reconstruction of past land cover conversions. Sequential information on land cover in most developing countries is quite often missing, outdated or inconsistent (Gregorio and Jansen, 1998; Brink and Eva, 2009). The launch of the Landsat imagery platform in 1972, followed by others such as SPOT and ASTER, has provided satellite remote sensing (RS) capacity to detect LUC change over the past 40 years at most. As suggested by Wasige et al. (2013), "land cover" refers to the biophysical state of the earth surface and immediate subsurface (Turner et al., 2005), and, "land fragmentation" brings out the aspect of human utilization of land (Turner and Meyer, 2004). The combined application of land cover and fragmentation data allows detection of type of change, as well as the way land is changing (Ickowitz et al, 2015).

This study provides fragmentation quantification in land cover. A few environmental studies have been undertaken in the Tshopo District of the Eastern Province of the Democratic Republic of Congo (DRC). This study targets the Yoko forest reserve located in the north-eastern part of the country, in the skirts of the City of Kisangani. The study specifically aimed to establish a detailed assessment of forest degradation over time and space to provide an accounting of environmental changes experienced in this region and the Congo Basin at large. Landsat MSS, TM, ETM and OLI images were used to detect the extent of land degradation in terms of both spatial and temporal patterns as well as magnitude and intensity over the period1976-2015. The following sections present the study area, the type of data and methods used and the results of land cover fragmentation, followed by a discussion and conclusion and recommendations.

STUDY AREA

Yoko Forest Reserve covers a total area of 6,975 ha divided into two blocks, the northern Yoko (3,370 ha) and the southern Yoko (3,605 ha). It is a private property of the Institut Congolais pour la Conservation de la Nature (ICCN) pursuant to Order - Law No. 75-023 of July 1975 establishing a state owned company responsible for managing certain environmental public institutions as amended and supplemented by Ordinance - Act No. 78-190 of 5 May 1988 (Lomba, 2007). The Reserve is located in the administrative sector of Bakumu - Mangongo, to the south of Kisangani. It is geographically bound by the village Banango (kilometer 21; 00 ° 21.439 'N, 025 ° 13.979' E) and village Bagão (Kilometer 57; 00 ° 06.653 'N, 025 ° 17.622 'E) on the Kisangani-Ubundu road, to the left bank of the Congo River (Figure 1). Yoko Forest Reserve is bordered to the east and south by the River Yoko, which is a subsidiary of River Biaro Yoko (Mikwa, 2012).

METHODOLOGY

Acquisition of Remote Sensed Data

Thisstudy's multi-temporal approach is thus clearly illustrated by the 8 Landsat imagery scenes and their coverage in terms of area and period covered, the spectral range and spatial resolution. Medium range satellite images were acquired from the Global Land Cover Facility (GLCF) of the University of Maryland (website: www.glcf.umiacs.umd.edu). They encompass Landsat MSS 1976, Landsat TM 1984, 1990 and 1995, Landsat ETM 2001 and 2005, Landsat TM OF 2010 and Landsat OLI 2015, with sequential improved resolution of 60, 30, 30, 30 and 15

Figure 1. Location of the Yoko Forest Reserve in the map of DRC

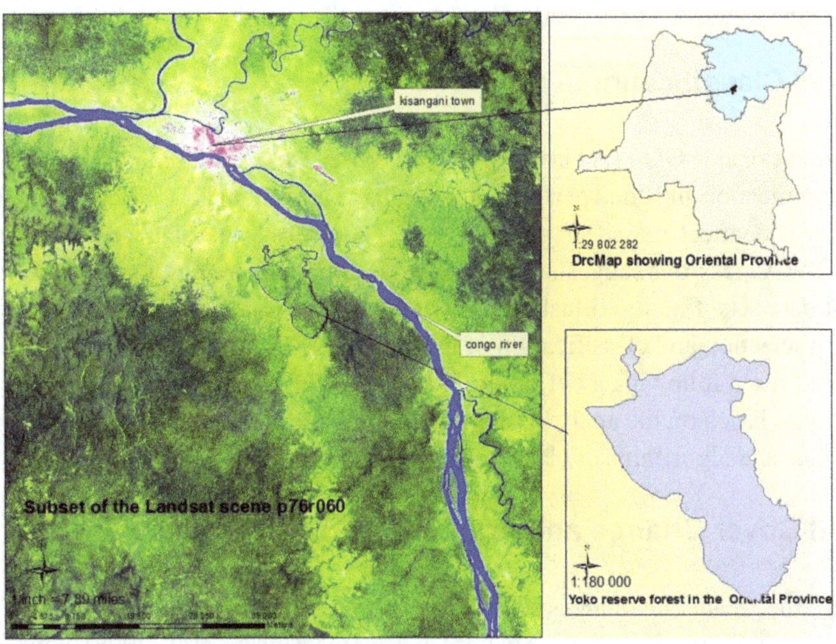

meter for MSS, TM, ETM, TM OF and OLI, respectively. This was an important milestone towards ensuring the best spatial match for the multi-temporal land cover fragmentation analysis.

Pre-Processing of Satellite Data

Landsat images acquired from the GLCF were already georeferenced but showed some spatial shifts from the WGS84 referencing system. Besides, there were significant deviations between images from Landsat ETM+ sensor with those from MSS and TM sensors. Hence, appropriate geometric and radiometric corrections were needed when applying digital image processing techniques to investigate more accurately land cover patterns. Then some atmospheric corrections were applied following the COST technique to ease the shadow effect through mosaicking and inter-annual spectral comparisons of images (Chavez, 1996).In fact, the negative effect of shadowing is the visual quality degradation of the images and is a cause of nuisance during the analysis of these images by modifying spectral response of areas under shadow. Thus, they are often confused with other types of objects which have a similar spectral signature. As result, there was a misclassification of areas covered by the shadow, on one end and on the other, some Landsat MSS images revealed spurious lines. A frequently used technique to correct line-drops consists of replacing the defective pixels with average digital numbers (DNs) of adjacent pixels (Lillesand, 1994). But this method is only efficient in the case of one ortwo pixel wide lines.

Image Classification

The traditional ISODATA classifier was used because of its simplicity in operation, easy application and good performance. Unlike conventional classification methods that use identical classifier structure and feature sets to each land cover type, ISODATA classifier suggests a standard classification strategy that uses different input data sets. Figure 2 illustrates how the satellite images were classified based on the selection and classification of training sites. The training sites were chosen and classified following a definition of thematic classes and their adjustment. The latter was based on the analysis of spectral signatures recorded from training sites, the rules and algorithms of classification selected, their testing and evaluation.

Land Cover Change and Fragmentation Analysis

One of the greatest challenges confronting landscape pattern analysis is quantifying temporal variations in landscape pattern metrics (McGarigal et al., 2008). Landscape metrics are one of the key factors of modern landscape ecological research and in landscape planning. The landscape or spatial metrics, which have been used to quantify spatial patterning of land cover patches and land cover classes in the study area can be defined as quantitative and aggregate measurements showing spatial heterogeneity at a specific scale and resolution (Herold et al., 2003).

This study used similar methods of classification that have been used in analyzing forest cover change in the Brazilian Amazon. Mausel et al. (1993), Brondizio et al. (1994), Moran et al. (1994), Brondizio et al. (1996) distinguished 5 major classes, namely primary forest, secondary forest, fallow, agriculture and bare soil.

This study first created 30 spectral classes using unsupervised classification (ISODATA in ERDAS IMAGINE8.4). Then a combination of unsupervised and supervised procedures led to retaining 4 key landscape unit classes. The latter encompassed: (1) agricultural areas (cropping and livestock rearing areas); (2) initial secondary vegetation; (3) developed secondary vegetation; and (4) forests (large canopy dense forest and well developed secondary vegetation). The supervised classification was based on ground information on vegetation cover collected in January 2014 based on control points and through visual interpretation of high resolution images (IKONOS image of the study area). In a second step, meaningful statistics were derived from these classes to enable a maximum likelihood supervised classification. Then land cover changes over different periods were analyzed and quantified using ArcGis 10and its Spatial Analyst extension for Windows. This

Figure 2. Methodology for the selection and classification of training sites on the image

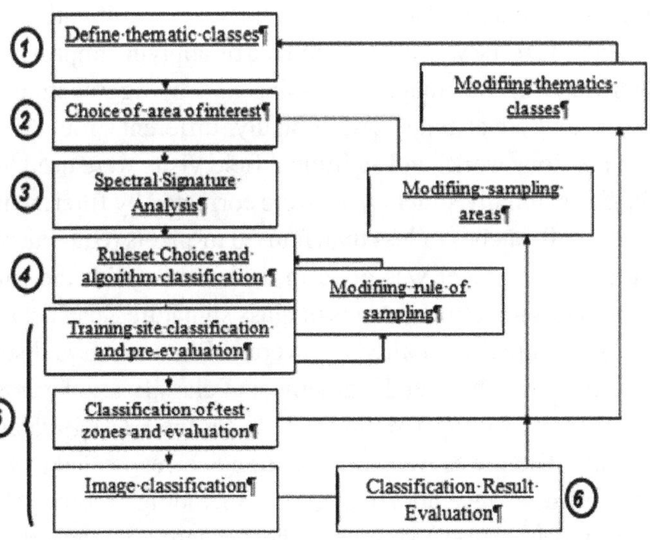

4

enabled quantifying land cover changes and vegetation loss, and the configuration of grid covers for the application of landscape spatial indices. These indices (spatial metrics) were computed using Patch Analyst version 3.5 and version 5, and enabled the comparison of the spatial pattern of vegetation for each time interval in the study area (McGarigal, 2002). The spatial configuration of land cover fragments were quantified and compared. The following key landscape metrics were utilized for comparison purpose: (1) patch area (%);(2) largest patch index (% of the landscape comprised by the largest patch);(3) aggregation index (% likelihood adjacencies between cells of the same patch type), and;(4) number of patches. All of these metrics have been applied in the assessment of spatial attributes of fragmented landscapesin recent vegetation fragmentation studies by McGarigal and Marks (1995), Bogaert (2003), Ndjibu (2007), Barima (2009), Bogaert et al. (2009), and Bamba (2010).

Accuracy Assessment

Classification of land cover types from satellite imagery has never been completely accurate. Landscape pattern analysis presented in this study might have been affected by errors in the classification, which alter its accuracy. Image accuracy tends to improve as the image date becomes more recent. The supervised classification, which was strongly supported by ground-based information, provided a suitable identification of land cover types in each of the satellite scenes processed. Although the image classification based on the oldest scene (MSS 1976) suffered from the disadvantage of not giving room to adequate ground truthing, a cautious interpretation of the results of the spatial analysis considered ground control points that have almost not changed over time. The latter involved high accuracy values obtained from the ground truthing exercise and which enabled a precise identification of land cover classes. Classification errors were also minimized by applying improved topographic correction algorithms in the image processing and by aggregating some of the sub-categories of land cover types. Additionally, different ground-based surveys conducted from previous works and additional field visits were used to support the image classifications. Finally, other errors were corrected by filtering the classified image to remove small patches of less than four 50 m-pixels (equivalent to 0.60 ha).

Image accuracy assessment was done in accordance with the rules of ENVI software for ranges of separability values of class signatures, each classified image achieved 'good' signature separability, the average Bhattacharrya distance ranging between 1.979 and 1.987. The overall agreement of classification was 83.2% for the MSS1976 image, 84.3% for the TM1984, 85.9% for the ETM 1995 and 84.9% for the OLI 2015image. The lowest values of accuracy corresponded to various native forests and other primary vegetation categories, including arborous shrubland and the dense forest, which are very similar in their spectral signatures and stages of succession. The analysis of these classes may easily produce misclassifications between the categories assigned in the training site and those classified by the algorithm. For that reason, the authors of this study decided to group them all in one category of vegetation, namely primary vegetation. To validate the accuracy of the classification procedures as well as the user who achieved the classification, accuracy assessment analysis was completed. Error matrices were developed for both the images classified using unsupervised classification techniques and post-classification operations. User's accuracy of individual classes ranged from 85 to 98% and producer's accuracy ranged from 87 to 99%. Table 1 corroborates with these findings by providing a summary of above analysis in terms of "Average percentage of correctly classified" pixels (Apcc) and the overall Kappa statistic. Results show that the Apcc ranged between 84%and 96% with an overall Kappa ranging from 0.81 to 0.96. These values are significant for any good classification system because they are greater than 75% and 0.75.

RESULTS

Land Cover Change Detection

Changes in the distribution of land use and cover classes in the Yoko forest reserve were quantified as differences between image pixels over the 8 successive periods. The resulting landscape changes from 1976 to 2015 are presented in Table 2 and Table 3. The maps produced after image processing are presented in Figures 3, 4 and 5. The classification expressed different patterns in evolution of classes from 1976 to 2015. The general landscape structure as well as matrices and distribution of spots in classes have all changed. Figure 3 shows high distribution of agricultural

Table 1. Producer and user accuracy of the classification

Year	Image Sensor	Apcc	Overall Kappa
1976	Landsat MSS	87	0.82
1984	Landsat TM	95	0.96
1990	Landsat TM	92	0.93
1994	Landsat TM	93	0.90
2001	Landsat ETM+	86	0.82
2005	Landsat ETM+	84	0.81
2010	Landsat TM	96	0.95
2015	Landsat OLI	94	0.93

Table 2. Land use/cover classes in the Yoko forest reserve (area expressed in hectares and percentage)

	1976		1984		1990		1994		2001		2005		2010		2015	
	Ha	**%**	**Ha**	**%**	**Ha**	**%**	**Ha**	**%**	**ha**	**%**	**Ha**	**%**	**ha**	**%**	**ha**	**%**
Primary forest	6258.9	37.8	7763.5	46.6	8080.1	48.2	8603.2	51.6	8420.9	50.6	7953.5	47.7	5420.5	32.5	4116.5	24.7
Secondary forest	6548.0	39.5	5862.5	35.2	3757.3	22.4	4051.4	24.3	4553.4	27.3	4623.3	27.7	4322.0	25.9	3252.0	19.5
Fallow	1472.4	8.9	1592.2	9.6	2970.1	17.7	1751.7	10.5	1654.2	9.9	2030.8	12.2	3901.2	23.4	4738.2	28.4
Agriculture and bare soil	2300.3	13.9	1444.1	8.7	1953.9	11.7	2255.9	13.5	2024.8	12.2	2054.7	12.3	3018.5	18.1	4556.5	27.34
Total	16580		16662		16761		16662		16653		16662		16662		16662	

Table 3. Sub-period change of areas occupied by different classes in the Yoko forest reserve (in hectares)

	1976	**1984**	**Change (in ha)**	**1990**	**Change (in ha)**	**1994**	**Change (in ha)**	**2001**	**Change (in ha)**	**2005**	**Change (in ha)**	**2010**	**Change (in ha)**	**2015**	**Change (in ha)**
Primary forest	6258.9	7763.5	1504.6	8080.1	316.6	8603.2	523.1	8420.9	-182.3	7953.5	-467.4	5420.5	-2533.0	4116.5	-1304.0
Secondary forest	6548.0	5862.5	-685.5	3757.3	-2105.2	4051.4	294.1	4553.4	502.0	4623.3	69.9	4322.0	-301.3	3252.0	-1070.0
Fallow	1472.4	1592.2	119.8	2970.1	1377.9	1751.7	-1218.4	1654.2	-97.5	2030.8	376.6	3901.2	1870.4	4738.2	836.8
Agriculture and bare soil	2300.3	1444.1	-856.2	1953.9	509.8	2255.9	302.0	2024.8	-231.1	2054.7	29.9	3018.5	963.8	4556.5	1538.0
Total	16580	16662		16761		16662		16653		16662		16662		16662	

Figure 3. Land use/ cover patterns in the Yoko forest reserve from 1976 to 1990

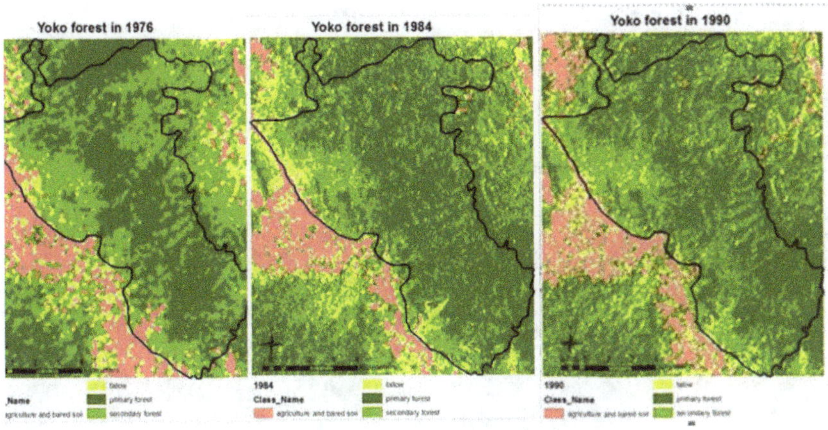

Figure 4. Evolution of the Yoko forest from 1994 to 2005

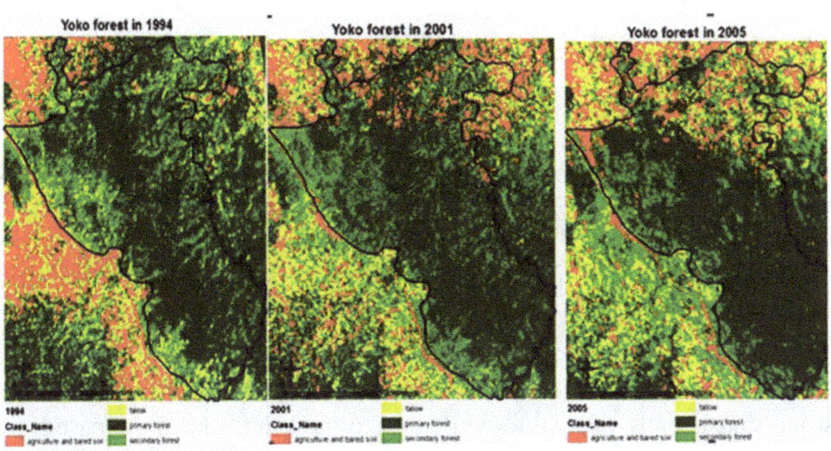

land use activities and bare soil from 1976 to 1990.There are also more gains in primary forests along with some spots of disturbances within the forest reserve. Further disturbances of the in primary forest are detected along the north-eastern side of the forest reserve from 2010 to 2015 (Figure 5). These changes are the results of emerging secondary forest detected from 1994 to 2005 (Figure 4), which changes considerably the landscape from a visually small primary forest to a larger architecture, although it is actually reducing in extent (Table 3). Besides, the area occupied by the primary forest class evolved from 37.8% of the total forest reserve area in 1976 to 51.6% in 1994 and then plummeted from 47% in 2001 to 32% in

Figure 5. Land use/cover in the Yoko forest reserve from 2010 to 2015

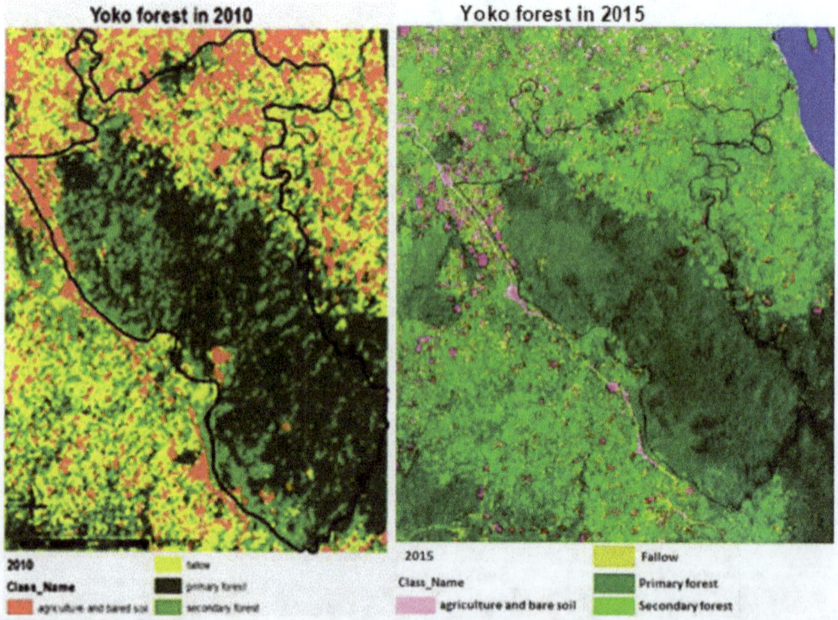

2010, thus resulting in 10 to 40% loss in almost four (4) decades. Regarding the secondary forest class, it occupied 39.5% of the total Yoko reserve area in 1976 but it has lost more than 35% of its area to close at 25% of the forest reserve in 2010. Another notable change is the increase in areas covered by agricultural crops and bare soil and as well as fallow, which represent between 1.9% and 11.4% of the total forest reserve area. About 7% of the primary forest and 15% of secondary forest have been converted to agricultural land between 1976 and 2010.Over the entire period, forest has consistently decreased while agricultural lands have increased in area. Generally, areas covered with primary and secondary forest have been observed to decrease while agriculture and bare soil and fallow have increased.

Large patches of secondary and primary forests have disappeared and have been replaced by smaller spots or often to other classes mainly in the northern part of the reserve. Fallow, agriculture and bare soil have also changed their structure, as the fields under fallow set off across the landscape in 2010. Forests have probably paid the price of farming techniques and habits of local people.

Figure 6 reveals that from 1976 to 2015 the area covered by the forest varies from time to time. Increased forest cover was detected for the period 1976-1995 with a recession of agricultural activities and an increase of fallow, which might be attributed to the flourishing business and trading activities in the then Zaïrean economy. The area of forest loss per forest class varied, with the highest rate of

Figure 6. Temporal evolution of different land use/cover classes in the Yoko forest reserve

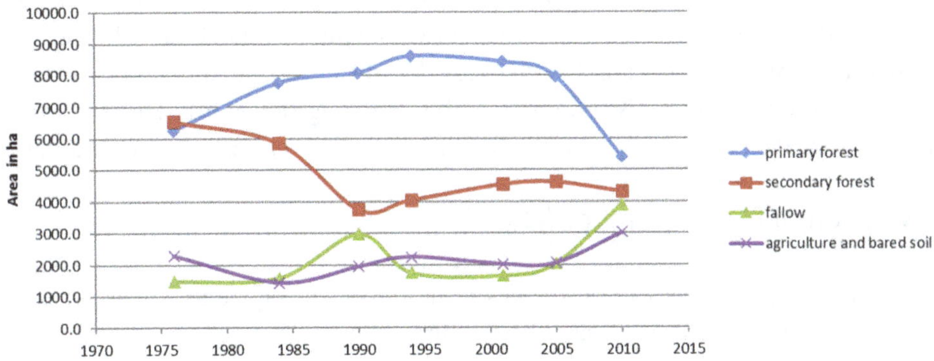

primary forest loss recorded between 2005 and 2010 with the boom of agricultural land use activities, and possibly wood extraction for charcoal, firewood and timber. The lowest rate of primary forest loss is evident between 2001 and 2005 with the pick of secondary forest and slowing agricultural activities and fallow. While forest cover loss in mature forests is lower than in regenerating forests for the last two decades, it has more significant impacts on forest structure and biodiversity. The clearance of the secondary forest is later experienced during the period 2005-2010, as part of a rotational slash-and-burn agriculture practice and fallow regrowth in the Congo Basin (Turubanova et al., 2012).

Hence, between 1976 and2010, both forest classes lost their spatial locational areas in the Yoko forest reserve under intense pressure from human-induced activities. It is apparent that the forest cover has changed markedly from 1976 to 2010 (Table 3), especially for the primary forest and secondary forest classes. This could be attributed to the illegal cutting of trees in the forests for various uses such as charcoal, firewood and timber or clearing for agricultural purposes. On the other hand, fallow and the agricultural and bare soil have increased over the years.

Change Detection for Vegetation Characterization

Evaluation of Fragmentation Using Landscape Metrics

The evolution of class area presented under section 4.1 has clearly shown how the landscape composition looks like in the Yoko forest reserve. The value of the primary forest class was higher than the one of other land-use classes between 1976 and 2015, successively followed by secondary forest, fallow and agriculture and bared soil. The evolution of the patch number will specifically reveal how much of the landscape is comprised in a particular patch type. This is an important measure in a number of ecological applications of habitat fragmentation in quantitative habitat loss. In the study of forest fragmentation, therefore, it is important to know how much of the target patch type (habitat) that exists within the landscape. In addition to the class area and patch number, the edge density and fractal dimension are used to compute class and landscape metrics. Investigating these differences helps to characterize the land-use classes forming the whole Yoko forest reserve landscape.

Evolution of Patch Number (NumP)

NumP is a simple measure of the extent of subdivision or fragmentation of the patch type. In particular, NumP illustrates the diffuse sprawling development and the fragmentation of rural areas. NumP is simply the total number of patches and is a measure of landscape configuration and is closely related to spatial scale of analysis and to the extent of the landscape. Therefore, this metric is more meaningful for comparing the same landscape or landscape with similar size and characteristics over the time. Figure 7 captures the development of the patch numberin the Yoko forest reserve from 1976 to 2010.

The temporal evolution of number of patches shows some significant differences among the land-use classes, with regard to their shape complexity, as reflected by the shape metrics. The same differences were observed in the temporal evolution of classes for each year shows with extreme variation recorded in plots within fallow and secondary forest classes in 1990 and 2001. Heterogeneous human disturbance processes may be the key agents of the varying patch numbers in this landscape over the years.

As illustrated in Table 2 and Figure 7, from 1976 to 2015 the variation of class areas and patch numbers in Kisangani City is featured by increased vegetation cover

Figure 7. Patch numbers of different land use/ cover classes in the Yoko reserve (1976-2010)

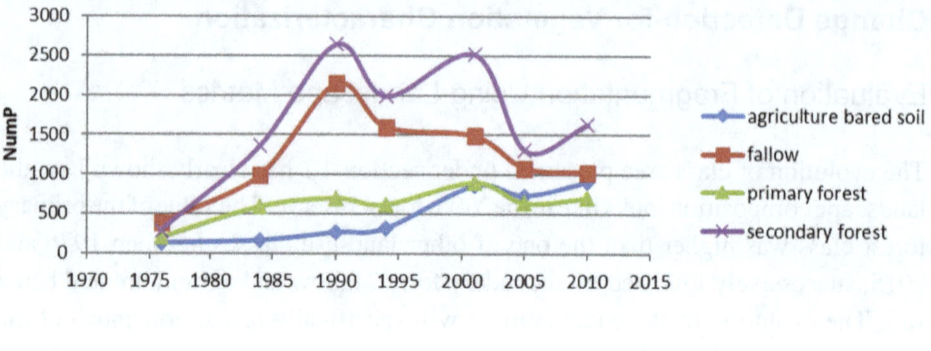

area and decreased bare soil. This evolution is also marked by a decrease of urban patch numbers and bare soil beyond Kisangani Town, which were featured by the apportionment of bare soil for urbanization purpose between 1990 and 2000. Hence, the increase of the total built-up area led to a reduction of the vegetation area and patch numbers. From 2000 to 2011, both trends evolved in the opposite directions with an increase in the number of vegetation patches, and a decrease in the bare soil and built-up area classes beyond Kisangani Town. These results confirm the fact that heterogeneous human disturbance processes may be the key drivers to the varying patch numbers in the landscape over years.

Edge Density

The total edge density index (ED) is based on the ratio between total edges (number of cells at patch boundary) and total area (total cells). It can indicate the fragmentation level of either a landscape or one of its classes. From the table below, the ED fluctuated from 1976 to 2010. Figure 8 indicates that increasing ED values are expected in all classes with picks in 1990 and 2000. From 1976 to 1990 the primary forest class has virtually a slight advantage compared to all other classes, with exception of agriculture and bare soil. From 1990 to 1994 ED values for primary forest and secondary forest classes plummeted but regained a bit of momentum from 1994 to 2001 followed by a decrease between 2001 and 2005.The agriculture and bare soil showed a consistent increase of the ED from 1984 to 2001 followed by a slight decrease from 2001 to 2005. These results confirm the assertion that the Yoko forest reserve landscape changes overtime due to fragmentation and reconstitution of its forest classes.

Figure 8. Edge density evolution in the Yoko forest from 1976 to 2010

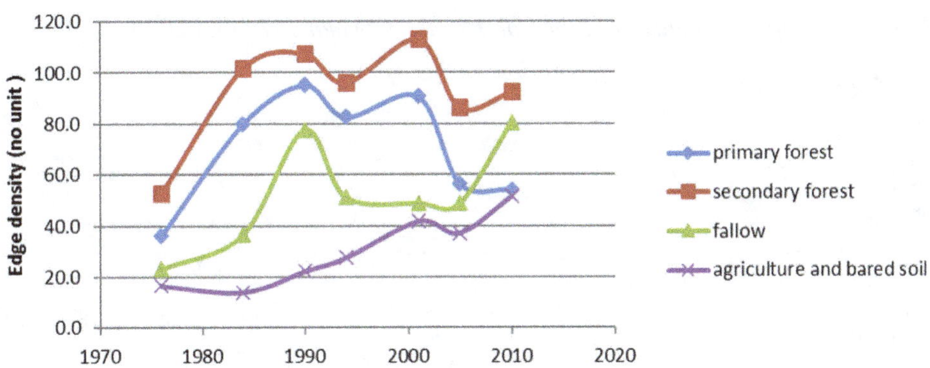

Fractal Dimension

In landscape ecological research, patch shapes are frequently characterized via the fractal dimension (FD) ranging from 1-2 (Krummel et al. 1987). FD values that are close to 2 tend to feature a fragmented class whereas values close to 1 characterize a homogenized class. Figure 9 shows that fallow and secondary forests are more subject to fragmentation than primary forest and agricultural classes, which tend to be homogenous.

From this graph, Yoko forest reserve was not fragmented in 1976; the fragmentation began from 1984 and kept going on up to 2010 with higher fluctuations recorded between 1994 and 2005, and a slight decrease from 2005. Projections show a slight increase of the FD values in the near future in the Yoko reserve. Consequently, agriculture, afforestation coupled with wood exploitation are seemingly key human disturbance processes leading to the varying fractal dimensions of fallow and secondary forest classes in the Congo Basin landscape over years.

DISCUSSIONS

The present study is one of the most extensive land cover fragmentation analysis conducted in the Congo Basin, particularly in the Tshopo area of the Eastern Province of the DR Congo. The results demonstrate how changes in spatial patterns of the land cover may be assessed using multi-temporal satellite data, as achieved recently in other parts of the world. This section discusses methodological approaches used in the study and key findings on land cover classification and fragmentation as well as key drivers to the fragmentation, and which may deeply affect the hydrology of the Congo Basin.

Figure 9. Fractal dimension of the Yoko forest between 1976 and 2010

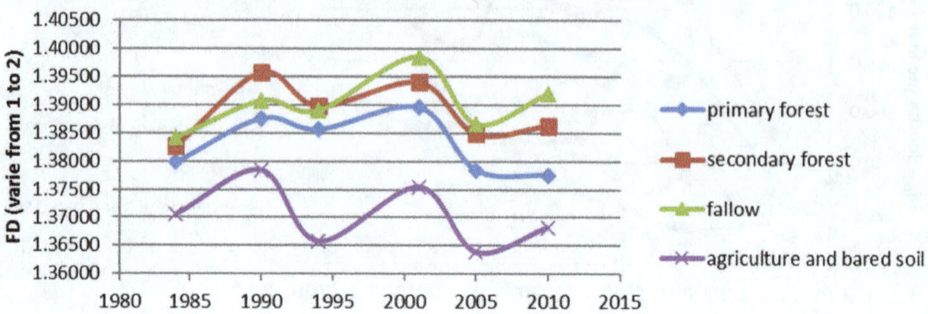

Methodological Approaches

Land Use / Cover Classification

This study used unsupervised ISODATA classification and supervised Land Use / Land Cover (LUC) classification. The first method (ISODATA) was used for change detection in the classes of land use/ cover retained while the second enabled the actual classification using ground truthing and the maximum likelihood algorithm and/or object-oriented change detection analysis. Several authors have used the unsupervised classification for change detection studies (Tou et al., 1974; Bonn et al., 1992; Richards, 1993; Barima, 2009; Daolan Zheng, David O, 1998; Zhanqing et al., 1997, Vancutsem et al., 2009). For instance, Bakr et al.(2010) used ISODATA classification to monitor the change of land use in Egypt, Asia, Latin America and Central Africa.The same methodology was applied by Duveiller et al. (2007) to study deforestation in Central Africa. Desclée et al. (2008) successfully used an unsupervised classification prior to applying an object-oriented classification to check changes of tropical forests. This method was found to be adequate for change detection since it is simple and may provide valuable information on the nature of the changes, which is the most important thing in this case. However, the method is criticized for generating several errors of omission because it tends to underestimate the changes (Béland et al., 2006; Small et al., 2001). For that reason, the present study suggested another method; the Object-oriented change detection analysis, to allow isolation of pixels for which there has been no change and classifies those pixels corresponding to a change only (Jensen, 2004; Desclée, 2007). An alternative would alsobe the application of supervised classification using the maximum likelihood algorithm.

Classification Accuracy Assessment

The classification evaluation was done through a confusion matrix (Godard, 2005). Several studies suggest the use of a confusion matrix to verify the accuracy of a classification (Gao et al., 1998; Zhan et al., 2002; Vancutsem,2008; Desclée,2006). The accuracy of the user and that of the producer were determined according to Campbell(2002) and Story and Congalton (1986) methods and the Kappa (K) coefficient (Congalton, 1999; Hudson, 1987; Rosenfield, 1986). An average accuracy rate of 85% and a value not less than 70% are acceptable, even though USGS recommends a minimum accuracy rate of 85% for the classification of Landsat images (Anderson et al., 1976). Nonetheless, Pontius (2000) argues thata land use/ cover classification can be validated if K is above 75%.Bolstad (1999) provides a minimum overall accuracy target of 85% with no class less than 70% accuracy (Foody, 2002).

IKONOS Visual Validation

This study validated its classification by visual interpretation of large-scale satellite images, namely Ikonos. Turubanova et al.(2012) used Modis and Google Earth images to validate the tree algorithm classification method with a number of additional datasets, including freely available QuickBird images from GoogleEarth. Expert information was also used as reference materials to aid interpretation. Diallo et al.(2009) validation was based on accuracy assessment performed using a random sampling scheme. Various maps, field data, Puer Quickbird image of 2009, were used as references data for validation.

Choice of Landscape Indices

The use of indices of area and perimeter is very common in remote sensing. Bogaert et al. (2005), Bamba et al. (2008), Declercq et al. (2008), Barima (2009) and Bogaert et al. (2009) calculated three sets of measures referring to these initial data of each class that make up the landscape. These metrics included a statistical mean, diversity or heterogeneity measure, and a consistency index.The statistical means encompassed the smallest value, the highest value, the total value, average value and the median.The diversity or heterogeneity measure included the variance, the standard deviation and a coefficient of variation. Finally, the extent, dominance or consistency indices used were mainly Shannon index and Pielou evenness. Likewise, landscapes indices are used for landscape fragmentation analysis to explain the ecological fragmentation of the landscape in response to both endogenous and heterogeneous forces of the landscape fragmentation. In this study, landscape ecology was used to check fragmentation associated with change detection and NDVI.In this case, we considered data bases for each spot, the number of patches, the class area, edge density and fractal dimension of each image class considered by the researchers.

Land Cover Change and Spatial Analysis

The various landscape indices and parameters used to analyze land cover patterns and explain its fragmentation in Kisangani Cityare useful tools for exploring inside variability, cross-site differences and changes over time. The simultaneous use of landscape pattern indices at class and patch levels enables assessing the spatial configuration of vegetation cover and its relationship to other land use and cover types. The increase of the vegetation patch numbers during the first decade of the analysis (1975-1985) with a regression of the aggregation and the large patch index suggest that the area occupied by vegetation has enlarged by the formation of small scattered patches, thus leading to a decrease of the area occupied by bare soil and agriculture. This variation of parameters may be an indication of the regrouping of bare soil and built-up areas, following a concentration of the population in those areas. Besides, the fact that burnt areas decreased strongly suggests a neglect of agricultural activities to the benefit of other socio-economic activities. Indeed this period corresponds to economic growth brought about by governmental investments to stabilize and redeem the DR Congo economy from a period of economic instability that followed the independence of the country. Taking advantage of this economic stability, most of the population ventured into business and trade thus, abandoning agricultural activities.

The next decade (1990-2000) is especially marked by the increase in the large patch index and patch numbers for the bare soil and built-up areas, thus leading to a reduction of the space occupied by vegetation and the burnt areas. Only the aggregation index of vegetation increases during that period while other parameters decrease. This variation shows that vegetation is replaced by agriculture and bare soil, mainly for occupational purpose by small scale squatters. This situation corresponds to a widespread the illegal exploitation of mineral resources and "the African Third World War" that led to the degradation of a few forested zones in Kisangani area during the transitional period.

The 2000-2010 decade is marked by in the number of patches of the vegetation, and the simultaneous decrease in the large patch index for bare soil. This situation illustrates the regeneration of the vegetation and its expansion by small patches, while bare soil are becoming increasingly small, isolated and more consolidated areas.

The evolution of the forest in the last decade (2010-2020)is likely to follow that of various land cover patterns indices and parameters depicted above if only conservation rules are enforced.

Potential Drivers to Fragmentation and Their Projected Impacts

Potentials Drivers to Fragmentation

The presence of large bare soil and burnt areas in the 1970's is very significant to the point that no such land cover pattern as yet been observed in other classifications. The field campaign interviews gave the opportunity to get some insight about what was happening then to identify the drivers, which triggered the large fire episodes in the area, notwithstanding governmental economic measures to stabilize the post-independence economy. Small-scale agriculture, mostly in the form of shifting cultivation, is the major contributor to deforestation in the Congo Basin (Ciza et al., 2015).

Shifting cultivation is an agricultural management strategy that involves the clearing of land and then cultivating and fallowing it in a cyclical fashion (Thrupp et al., 1997). It is the dominant form of agriculture in the Congo Basin. Shifting cultivation is believed to be the cause of erosion problems and extinction of biodiversity through deforestation. However, this type of farming tends to happen in areas with relatively low population density and where there is abundant and available land. If cleared land patches are small and fallow periods are relatively long, fields tend to regenerate to forest once more, although the structure and diversity of such forest is dependent on the fallow length, size of patches and surrounding vegetation, among other things. There are likely to be differences in species composition and some losses in biodiversity, but these are less than in almost any other anthropogenic system (Finegan and Nasi, 2004; Makana and Thomas, 2006).

Nonetheless, with increasing population, pressures on land lead to its shortage and the availability of related resources. Then, fallow periods become too short to allow soils to regenerate sufficient fertility leading to land degradation (Mertz, 2002;Ickowitz, 2006). This issue tends to be a problem in peri-urban areas and areas with a relatively high population density. This situation is ascertained by Iloweka (2004) and Bamba et al. (2008) for Lower Congo, Akkermans et al. (2013) and Bamba et al. (2010) for Kisangani and surrounding areas, Dupain et al. (2008) for Maringa-Lopori-Wamba Landscape, and Desclée et al. (2006) for Virunga National Park and the surroundings. Out of these five (5) case studies, four (4) are related to densely populated areas, which do not represent the low density population of the Congo Basin in general and the DR Congo in particular.

Predicted Impacts on the Congo Forest

Over 12,700 protected areas have been established around the world, accounting for 8.81% of Earth's land surface. Protected areas are key conservation haven for biodiversity and a shelter from harmful human activities(Dupin et al., 2013).By adding geological data layers in the Geographic Information System, remote sensed data are increasingly becoming useful for determining the success of protected area strategy in mitigating human impact on the biodiversity of major river basins of the world, including the Amazon, the Congo, the Ganges, the Mekong and the Nile, to name but a few (Mitchard et al., 2011).

The Yoko forest reserve is well-established in the Congo Basin, as its creation by the Congolese government dates back to 1957. The 1970 forest classification in the Congo Basin brings out a significant observation with the presence of very large entropized patches specifically located around human activity points (Mikwa,2012).

However, the deforestation and fragmentation rates within the primary forest of the Yoko reserve remains almost the same over the last 4 decades studied.

Predicted Impacts on Hydrological Ecosystem Services

The satisfaction of human demand for hydrological ecosystem services (HES) in the Congo Basin largely depend on the provision and the intensity of conservation of natural settings, notwithstanding socio-economic developments that may occur in the near future. In general, the provision of HES by forests is threatened by population growth, agricultural expansion and climate change, which are key drivers to deforestation (Pellikka et al., 2009).Agricultural expansion, water supply for household use, power generation, fisheries, flood mitigation and other ecological produces such as firewood, timber and live shelter are some one of the key benefits people gain from terrestrial ecosystems through their interaction with fresh water (Brauman et al., 2007). Forests here play a key role in determining the quantity and quality of these hydrological ecosystem services through protection of water bodies and water courses. Forests influence the amount of water available from groundwater, surface watercourses and water bodies though interception, evaporation, transpiration and infiltration (Hamilton et al., 2008). Forests actively regulate water quality and ensure timing of water provision by trapping sediments and pollutants from erosion, maintaining or improving soil infiltration and soil water storage capacity (Rockström et al., 2009). Hence, the role of forest conservation in providing ecosystem goods and services, including spiritual, regulatory and support ecosystem services, cannot be overemphasized. Forests with or without management distinctly affect the quantity, quality, location and timing of delivery of water resources, especially in landscapes were vegetation is dominated by intensely cultivated cropland. However, a differentiated view of forestry effect is needed so as to avoid overgeneralization(Calder, 2002; 2005).

Results of this study have shown that secondary forests in the Yoko reserve are subject to recurrent fragmentations and thus the impact of transition systems between primary forest and secondary forest, secondary forest and cropland (agroforestry) need to be addressed to avoid a future disaster in the Congo Basin. Like in many other basins the provision of HES depends not only on the total amount of goods and services but also on the condition of the landscape, mainly the forest cover and underlying soils within the basin. With regards to the natural settings and land use, it is worth noting that forest conservation is crucial for harmonizing the hydrological benefits and production potential of forests with other societal needs for water.

What is needed is a differentiated assessment taking into account the potential benefits of forests but also the risks of maladapted management (Luwesi et al., 2011). The impact of such ecological conservation is provided the mountain forests, which acts as important catchment areas or "water towers" where sources of many rivers spring up to provide water for human consumption, agricultural irrigation, hydro power generation and industrial uses (Fallkenmark, 2003). However, land-cover changes due to deforestation and forest degradation often result in reduction of water supply, deterioration of water quality and loss of forest carbon stocks (Akotsi and Gachanja, 2004). An integrated approach is therefore needed to acknowledge the interactions between water, forest, other land uses and socio-economic factors in complex hydrological ecosystems (GWP, 2015).

The 1992Rio Earth Summit proposed under its Agenda 21, an Integrated Water Resource Management (IWRM) approach to tackle such issues. The latter were suggests a holistic approach for the management of freshwater resources (IWMI, 2007). IWRM represents the "process which promotes the coordinated development and management of water, land and related resources, in order to maximize the resultant economic and social welfare in an equitable manner without compromising the sustainability of vital ecosystems" (GWP 2000).

CONCLUSION

This study is part of the application of techniques of multi-temporal remote sensing and mapping to characterize the dynamics of land use and cover from 1976 to 2015 using post-classification comparisons and landscape metrics. The Congo Basin landscape has changed in the last four decades. Though not very significant, the analysis confirms that the landscape is under human pressure. The fractal dimension shows that the landscape is under anthropogenic pressure. The transition matrix shows a deforestation rate is 0.32% that has led to the degradation of vegetation cover in some part of the Yoko forest reserve. These trends may impact negatively on the biodiversity and hydrology of the Congo Basin. However they are neither finite nor irreversible since ecological conservation is still under application. Besides, other studies may provide different results when using different satellite data types from other sensors. Time has come for the Congo Basin authorities to find new ways to manage their forest resources based on an Integrated Management of Water, Land, and Ecosystems (IMWLE). A REDD+ scheme has been implemented during the last five years, following recommendations from the 1992 Rio conference. This is probably a sustainable way of managing the Congo forest in the long-term under international climate change frameworks and sustainable development regimes agreed upon in Bali (2007), New York (2015) and Paris (2015).

REFERENCES

Akkermans, J., Brenninkmeijer, V., Huibers, M., & Blonk, R. W. B. (2013). Competencies for the contemporary career: Development and preliminary validation of the Career Competencies Questionnaire. *Journal of Career Development*, *40*(3), 245–267. doi:10.1177/0894845312467501

Akotsi, E., & Gachanja, M. (2004). *Changes in Forest Cover in Kenya"s five "water towers" 2000-2003*. Kenya Forests Working Group.

Bakr, N., Weindorf, D. C., Bahnassy, M. H., Marei, S. M., & El-Badawi, M. M. (2010). Monitoring land cover changes in a newly reclaimed area of Egypt using multi-temporal Landsat data. *Journal of AppliedGeography*, *30*, 592–605.

Bamba, I. (2010). *Anthropisation et dynamique spatio-temporelle de paysages forestiers en République Démocratique du Cong (PhD thesis)*. Brussels: Free University of Brussels.

Barima, Y. S. S. (2009). *Dynamique, fragmentation et diversité végétale des paysages forestiers en milieux de transition forêt-savane dans le Département de Tanda (Côte d'Ivoire)* (PhD thesis). Brussels: Free University of Brussels.

Béland, M., Goïta, K., Bonn, F., & Pham, T. T. H. (2006). Assessment of land-cover changes related to shrimp aquaculture using remote sensing data: A case study in the GiaoThuy District, Vietnam. *International Journal of Remote Sensing*, *27*(8), 1491–1510. doi:10.1080/01431160500406888

Bogaert, J., & Hong, S. K. (2009). Landscape ecology: monitoring landscape dynamics using spatial pattern metrics. In S. K. Hong, J. A. Lee, B. S. Ihm, A. Farina, Y. Son, E. S. Kim, & J. C. Choe (Eds.), *Ecological Issues in a Changing World* (pp. 109–131). Dordrecht: Kluwer Academic Publishers.

Bogaert, J., & Mahamane, A. (2005). Ecologie du paysage: Cibler la configuration et l"échelle spatiale. *Annales des Sciences Agronomiques du Bénin*, *7*, 39–68.

Bolstad, T. (1999). Coordinating methodologies for scaling landcover classifications from site-specific to global: Steps toward validating global map products. *Remote Sensing of Environment*, *70*(1), 16–28. doi:10.1016/S0034-4257(99)00055-3

Bonn, F. and Rochon, G. (1992). Précis de télédétection. *Principes et méthodes, 1.*

Brauman, K. A., Daily, G. C., Duarte, T. K., & Mooney, H. A. (2007). The nature and value ecosystem services: An overview highlighting hydrologic services. *Annual Review of Environment and Resources*, *32*(1), 67–98. doi:10.1146/annurev.energy.32.031306.102758

Brink, A. B., & Eva, H. D. (2009). Monitoring 25 Years of Land Cover Change Dynamics in Africa: A Sample based Remote Sensing Approach. *Applied Geography (Sevenoaks, England)*, *29*(4), 501–512. doi:10.1016/j.apgeog.2008.10.004

Campbell, J. B. (2002). *Introduction to remote sensing* (3rd ed.). New York, NY: The Guilford Press.

Chave, Andalo, Brown, Cairns, Chambers, Eamus, … Folster. (2005). *Comparison of various remote sensing data sources in the retrieval of forest stand*. Academic Press.

Ciza, K. (2015). Identification des moteurs de déforestation dans la région d'Isangi. République démocratique du Congo.

Congalton, R. G., & Green, K. (1999). *Assessing the accuracy of remotely sensed data: Principles and practices*. Boca Raton, FL: Lewis Publishers.

Daolan, Z., David, O.W., & Zhanqing, H. (1997). *Rates and patterns of landscape change between 1972 and 1988 in the Schangai Mountain area of China and North Korea*. Academic Press.

Desclée, B., Bogaert, P., & Defourny, P. (2006). Forest change detection by statistical object-based method. *Remote Sensing of Environment*, *102*(1-2), 1–11. doi:10.1016/j.rse.2006.01.013

Desclée, B., deWasseige, C., Bogaert, P., & Defourny, P. (2006). Tropical forest monitoring by object-based detection: Towards an automated method in an operational perspective. *Proceedings of the 1st international conference on object-based image analysis (OBIA 06)*.

Di Gregorio & Jansen. (1998). New Concepts for a Land Cover Classification System. *Land (Basel)*, *1*, 55–65.

Fallkenmark, M. (2003). Freshwater as shared between society and ecosystems: From divided approaches to integrated challenges. *Philosophical Transactions of the Royal Society of London. Series B, Biological Sciences*, *358*(1440), 2037–2049. doi:10.1098/rstb.2003.1386 PMID:14728797

Finegan, B., & Nasi, R. (2004). The biodiversity and conservation potential of shifting cultivation landscapes. In G. Schroth, G. A. B. da Fonseca, C. A. Harvey, C. Gascon, H. L. Vasconcelos, & A. M. N. Izac (Eds.), Agroforestry and biodiversity conservation in tropical landscapes. Island Press.

Foley, J. A., DeFries, R., Asner, G. P., Barford, C., Bo-nan, G., Carpenter, R. S., ... Coe, M. T. (2005). Global Con- sequences of Land Use. *Science*, *309*(5734), 570–574. doi:10.1126cience.1111772 PMID:16040698

Gao, J., & Skillcorn, D. (1998). Capability of SPOT XS data in producing detailed landcover maps at the urban – rural periphery. *International Journal of Remote Sensing*, *19*(15), 2877–2891. doi:10.1080/014311698214325

Godard, V. (2005). Typologie des paysages forestiers du sud du massif de Fontainebleau après la tempête de Décembre 1999. Évaluation des dégâts forestiers à l"aided"un semis de points et d"imageriesatellitale optique. *Revue Internationale de Géomatique*, *15*(3), 281–302. doi:10.3166/rig.15.281-302

GWP, Global Water Partnership. (2015). *Integrated Water Resources Management in Eastern Africa: Coping with "Complex Hydrology". Technical Focus Papers (TFP) No 7*. Stockholm: Elanders.

Ickowitz, Vinceti, Termote, Powell, Kehlenbeck, & Hunter. (2006). *The Contribution of Forests and Trees to Sustainable Diets*. Academic Press.

Iloweka, E.M. (2004). *The deforestation of rural areasin the low Congo province*. Monit Assess.

IWMI, International Water Management Institute. (2007). *Waterfor Food, Waterfor Life - A Comprehensive Assessment of Water Management in Agriculture (Summary Edition by David Molden)*. London: Earthscan.

Jensen, J. R. (2004). *Introductory digital image processing – A remote sensing perspective* (3rd ed.). Upper Saddle River, NJ: Prentice Hall.

Krummel, J. R., O'Neill, R. V., Gardner, R. H., Sugihara, G., De Angelis, D. L., Milne, B. T., ... Graham, R. L. (1988). Indices of landscape pattern. *Landscape Ecology*, (3), 153–162.

Lillesand, T. M., & Kiefer, R. W. (1994). *Remote Sensing and Image Interpretation* (4th ed.). New York: John Wiley & Sons.

Lomba, B. L. (2007). Contribution à l"étude de la phytodiversité de la réserve forestière de la Yoko. (Unpublished Master thesis). Kisangani: University of Kisangani, Faculty of Sciences.

Lu, D., Mausel, P., Brondizio, E., & Moran, E. (2004). Change detection technique. *International Journal of Remote Sensing*, *25*(12), 2365–2407. doi:10.1080/0143116031000139863

Luwesi, C. N., Shisanya, C. A., & Obando, J. A. (2011). Toward a hydro-economic approach for risk assessment and mitigation planning for farming water disasters in semi-arid Kenya. In M. Savino (Ed.), *Risk Management in Environment, Production and Economy* (pp. 27–46). Rijeka: InTech.

Makana, J.-R., & Thomas, S. C. (2005). Effects of light-gaps and litter removal on seedling performance in six African timber tree species. *Biotropica*, *37*(2), 227–237. doi:10.1111/j.1744-7429.2005.00030.x

McGarigal, K., & Marks, B. J. (1994). *Fragstats: Spatial Pattern Analysis Program for Quantifying Structure.* Department of Agriculture, Pacific Northwest Research Station General Technical Report PNW-GTR-351.

Mertz, J., Pons, T., & Moreaux, L. (2002). Photo-induced flip-flop of amphiphilic molecules in lipid bilayer membranes. *Physical Review Letters*, *89*.

Mikwa, N. (2012). *Assessment of forest degradation in the Yoko Reserve forest using remote sensing* (Unpublished Master thesis). Gent: University of Gent.

Mirault, A., Rossi, A., Dupin, N., Bruneval, P., Bel, A., Azarine, A., … Surg, A. T. (2013). Recurrent tamponade and aortic dissection in syphilis. *Stansal Nov*, *96*, 111-113.

Mitchard, Saatchi, Lewis, Feldpausch, Gerard, Woodhouse, & Meir. (2011). *Comment on 'A first map of tropical Africa's above-ground biomass derived from satellite imagery'*. Academic Press.

Moran, E. F., Lambin, E. F., Turner, B. L. II, Geist, H., Agbola, S., Angelsen, A., ... Xu, J. (2001). The Causes of Land-Use and -Cover Change: Moving beyond the Myths. *Global Environmental Change*, *11*(4), 261–269. doi:10.1016/S0959-3780(01)00007-3

Pellikka, P. K. E., Lötjönen, M., Siljander, M., & Lens, L. (2009). Airborne remote sensing of spatiotemporal change (1955-2004) in indigenous and exotic forest cover in the Taita Hills, Kenya. *International Journal of Applied Earth Observation and Geoinformation*, *11*(4), 221–232. doi:10.1016/j.jag.2009.02.002

Richards, J. A. (1993). *Remote sensing digital image analysis. An introduction (2nd ed.)*. Berlin: Springer-Verlag. doi:10.1007/978-3-642-88087-2

Rockström, J., Falkenmark, M., Karlberg, L., Hoff, H., Rost, S., & Gerten, D. (2009). Future water availability for global food production: The potential of green water for increasing resilience to global change. *Water Resources Research*, 45.

Rosenfield, G., & Fitzpatrick-Lins, K. (1986). A coefficient of agreement as a measure of thematic classification accuracy. *Photogrammetric Engineering and Remote Sensing*, *52*(2), 223–227.

Smith, R. C. (2007). Land-use allocation protects the Peruvian Amazon. *Science*, *317*, 1233–1236. PMID:17690260

Story, M., & Congalton, R. G. (1986). Accuracy assessment: A user's perspective. *Photogrammetric Engineering and Remote Sensing*, *52*(3), 397–399.

Thrupp, L. A. (1997). *The diversity and dynamics of shifting cultivation: myths, realities and policy implications*. Washington, DC: World Resources Institute.

Tou, J. T., & Gonzalez, R. C. (1974). Pattern recognition principles. Reading, MA: Addison Wesley.

Tucker, C. J. (1979). Red and photographic infrared linear combinations for monitoring vegetation. *Remote Sensing of Environment*, *8*(2), 127–150. doi:10.1016/0034-4257(79)90013-0

Turner, B. L. II, & Meyer, W. B. (1994). Global land use and land cover change: an overview. In W. B. Meyer & B. L. Turner II, (Eds.), *Changes in Land Use and Land Cover: a global perspective* (pp. 3–10). Cambridge, UK: University of Cambridge.

Turubanova, P.S.A., Hansen, M.C., Adusei, B., Broich, M., Altstatt, A., Mane, L., & Justice, C.O. (2012). *Quantifying forest cover loss in Democratic Republic of the Congo, 2000– 2010, with Landsat ETM+ data*. Remote Sensing of Environment.

Vancutsem, C., Lubamba, J. P. K., Blaes, X., Pekel, J. F., de Wasseige, C., & Defourny, P. (2006). *The Land Cover Map of the Democratic Republic of Congo*. Louvain-la-Neuve: Leuven University Press, UCL-Geomatics.

Wasige, J. E., Groen, T. A., Smaling, E., & Jetten, V. (2013). Monotoring Basin-Scale Land Cover Changes in Kagera Basin of Lake Victoria Using Ancillary Data and Remote Sensing. *International Journal of Applied Earth Observation and Geoinformation*, *21*, 32–42. doi:10.1016/j.jag.2012.08.005

Yacouba, D., Hu, G., & Wen, X. (2009). Assessment Of Land Use Cover Changes Using Ndvi And Dem In Puer And Simao Counties, Yunnan Province, China. *Marsland Press World Rural Observations*, *1*(2), 1–11.

Zhan, Q., Wang, J., Peng, X., Gong, P., & Shi, P. (2002). Urban built-up land change detection with road density and spectral information from multi-temporal Landsat TM data. *International Journal of Remote Sensing*, *23*(15), 3057–3078. doi:10.1080/01431160110104728

Chapter 10
Impact of Climate Change on Water Resources in Eastern Africa

Alfred Opere
University of Nairobi, Kenya

Purity Mueni
University of Nairobi, Kenya

Anne Omwoyo
University of Nairobi, Kenya

Mark Arango
University of Nairobi, Kenya

ABSTRACT

Climate change is causing great impact on water resources in Eastern Africa, and there is need to establish and implement effective adaptation and mitigation measures. According to IPCC, less rainfall during the months that are already dry could increase drought as well as precipitation, and this has great impact on both permanent and seasonal water resources. Increased sea surface temperature as a result of climate change could lead to increased drought cases in Eastern African and entire equatorial region. Climate change will also result in annual flow reduction in various river resources available within the region such as the Nile River. IPCC predicts that rainfall will decrease in the already arid areas of the Horn of Africa and that drought and desertification will become more widespread, and as a result, there will be an increased scarcity of freshwater even as groundwater aquifers are being mined. Wetland areas are also being used to obtain water for humans and livestock and as additional cultivation and grazing land. This chapter reviews the climate change impacts on water resources within the Eastern Africa Region. The climate change impacts on different water resources such as Ewao Ngiro have been highlighted and projection of future climate change on water resources examined. Stream flow for Ewaso Ngiro was found to have a significant increasing trend in 2030s of RCP4.5 and non-significant decreasing trend in stream flow in 2060s for RCP4.5.

DOI: 10.4018/978-1-7998-0163-4.ch010

INTRODUCTION

Climate change will affect the water balance, and particularly the amount of runoff and recharge, which in turn determines the water resources available for human and ecosystem uses. Some parts of the world will experience a reduction in resource availability, while others will see an increase (Urama and Ozor, 2010). Here we assess the effects of climate change on water resources both globally and regionally, comparing these effects with other stresses on the environment.

Interest in global water resources has risen significantly in recent years as increasing population and its concentration within urban areas have increased pressures on the amount and quality of water resources (Kisaka et. al., 2015). The Comprehensive Assessment of the Freshwater Resources of the World estimated in 1997 that approximately one third of the world's population lives in countries experiencing moderate to high water stress, and forecast that by 2025 as much as two thirds of a much larger world population could be under stress conditions simply due to the rise in population.

The contrast between recent observed and simulated trends in a global climate warming has inspired a debate about conceivable events and indications for future climate fluctuation projections. The illustration of the mean state and inconsistency of recent changing climate is indispensable for various purposes in local, regional and global change studies. These incorporate the observation and identification of environmental change and climate model assessment (Orimoloye *et al.*, 2018).

The projection of future climate change involves careful evaluation of models, taking into account uncertainties in observations and consideration of the physical basis of the findings, in order to characterize the credibility of the projections and assess their sensitivity to uncertainties. Different climate models have varying degrees of success in simulating past climate variability and mean state when compared to observations. Verification of regional trends provides further information on the credibility of model projections (IPCC, 2013). With one-third of the population living in drought-prone areas, Africa is considered the most vulnerable continent to climate change (Adhikari *et al.* 2016). This chapter attempts to discuss the temperature projections in the African context within the 4 continental sub regions (Eastern, Western, Southern and Northern Africa) and highlights how those projections will impact on the changes in precipitation.

INDICES OF WATER RESOURCES STRESS

There are many possible indices of water stress, considering different aspects of use and availability. The basic measure used in the Comprehensive Assessment is the ratio of the amount of withdrawals to the renewable resources available in a country, including runoff imported from upstream regions. In the absence of consistent data at more local scales, the indices are calculated using national population and water resource statistics. Baseline estimates of national water resources, both generated internally and imported from upstream, are taken from data prepared for the Comprehensive Assessment, as are figures for current withdrawals. Future withdrawals of water will increase with population. It is assumed that the per capita use will remain constant over the next few decades although this is perhaps unrealistic, because in practice, use would change as economic development changes, water use efficiency increases and irrigation use increases at different rates compared to population. Climate change might also have an effect on the demand for water.

It is generally accepted that water resources become a constraint on development when use exceeds 20% of the total resources potentially available (Falkenmark 1989). In countries prone to severe problems during drought years in absence of secure supplies through investment in supply facilities, development opportunities are constrained. Countries using less than 20% of total potential resources are not immune to water stress and may experience severe problems in some regions and at some times.

Possible Climate Impacts

IPCC predicts that rainfall will decrease in the already arid areas of the Horn of Africa, and that drought and desertification will become more widespread (IPCC 2001). As a result of the increasing scarcity of surface freshwater, groundwater aquifers are being mined and wetland are being used to obtain water for humans and livestock, and as additional cultivation and grazing land. This alters hydrological cycles, leaving the surrounding area more prone to flooding. Catchment degradation results in increased runoff, flash flooding, reduced infiltration, erosion and siltation (Kassian et. al., 2017) and it is a major problem undermining the limited sustainable water resources base in the region. The main causes of catchment degradation are poor farming methods, population pressure and deforestation. Forest excision for resettlement, dependence on wood fuel, illegal logging and encroachment into forests and marginal land enhance catchment degradation.

FUTURE CLIMATE CHANGE RELATIVE TO THE BASELINE PERIOD

In a study to evaluate the climate change projections, the best model dataset was subjected to trend analysis and Gaussian Kernel distribution methods for Tana basin in Kenya. The future climate change scenarios of annual rainfall and temperature distribution for three future climatological periods (2021-2040, 2051-2070, and 2081-2100) relative, for RCP 4.5 and RCP 8.5 to the baseline period 1971-2010 in Upper Tana Catchment were analysed. The trend analysis for the future changes in annual rainfall and mean temperature using the non-overlapping climatic periods 2021-2040, 2051-2070, and 2081-2100 relative to the baseline period 1971-2010 for Nyeri, Embu and Thika is shown in figure 1. The trends in rainfall for RCP 4.5 and RCP 8.5 indicate an increasing trend in the Nyeri and Embu stations and a decreasing trend in Thika station. This has the implication that rainfall is generally increasing gradually over the catchment in both RCPs as shown if figure 1 (a,and b). Figure 1 (c and d) shows that the future temperature is also expected to rise at an alarming rate in both RCPs (Joseph, 2016). However the rate of increase in temperature in RCP 8.5 is higher than that in the RCP 4.5.

In IPCC custom, climate change refers to a change in the state of the climate that can be identified statistically by changes in the mean and/or the variability of its properties and that continues for a period longer than a decade.

Gaussian kernel distributions were used to analyze future changes in annual rainfall and mean temperature for climate in the 30's, 60's, and 90's using a baseline period from 1970-2010 of observed annual total rainfall . In figure 2 (a and b), results showed a general increase of about 30% in annual total precipitation for all the climatic periods for the two RCPs and annual average temperature increase of about 1°C for all the climatic periods for RCP 4.5 and in climate at 30's, 60's for RCP 8.5 figure 2 (c and d). However, an increase of 3 °C was shown in the climate at 90's for RCP 8.5. This change was relative to the baseline period over Nyeri.

Over Embu station, relative to the observed annual rainfall an increase of about 10% for the two RCPs in all the climatic periods was observed as shown in figure 3 (a and b).However, figure 3 (c and d) shows a decrease of about -0.5°C in the annual temperature in the climate at 30's and an increase 1°C in the climate in 60's and climate in 90's for both RCPs.

Thika station on the other hand indicated general decrease of -20% in the annual rainfall for all the climatic periods in RCP 4.5 and RCP 8.5 relative to the baseline period as shown in figure 4 (a and b). In figure 4 (c and d) the annual temperature indicates a decrease of about -0.5°C in the climate at 30's and an increase of 2°C in the climate in the 60's and 90's for the two climate change scenarios.

Extreme whether events like droughts and floods, increased evaporation rates, low rainfall intensity, alteration in the rainfall seasons are some of the events related to climate change. Global warming effects are confirmed by increase in the catchment temperatures. Increasingly unpredictable weather patterns are as result of climate change.

Figure 1. Trend Analysis of Rainfall RCP 4.5 (a) and RCP 8.5 (b) and Mean Temperature for RCP 4.5 (c) and RCP 8.5 (d) for the study area
(Source: Joseph, 2016)

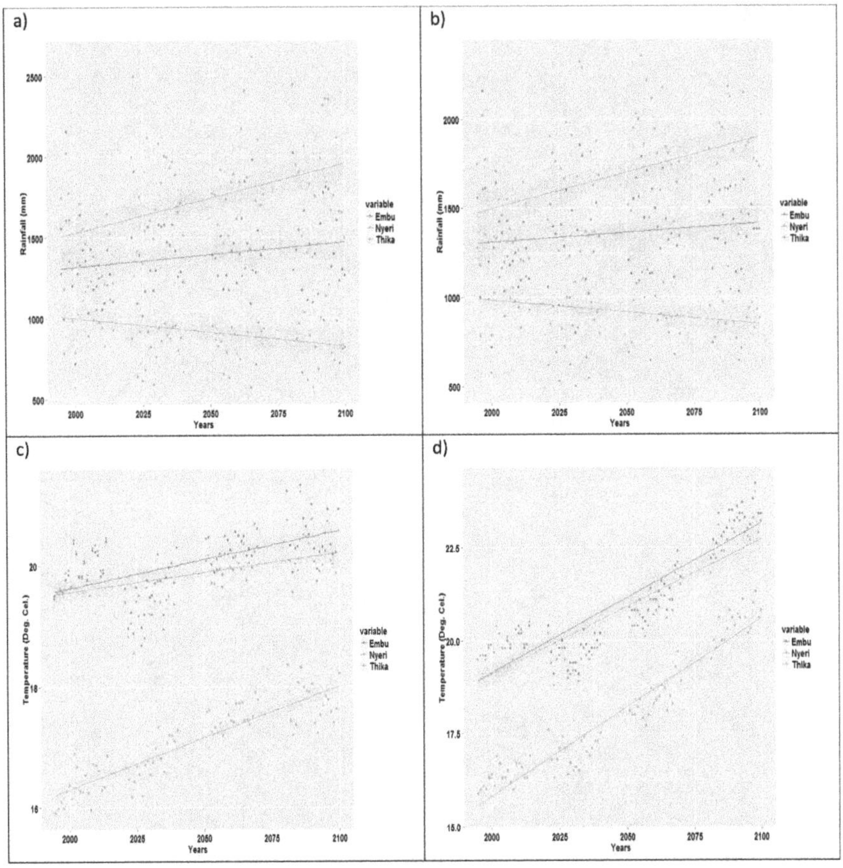

Coefficient of Variation (CV)

This ratio of the standard deviation to the mean shows the extent of variability in relation to the mean of the population. The coefficient is given by;

Formula to find sample mean;

$$\mu = \frac{\sum x}{n} \tag{1}$$

Figure 2. Gaussian distribution of Rainfall for RCP 4.5 (a) and RCP 8.5 (b) and Mean Temperature for RCP 4.5 (c) and RCP 8.5 (d) for Nyeri
(Source: Joseph, 2016)

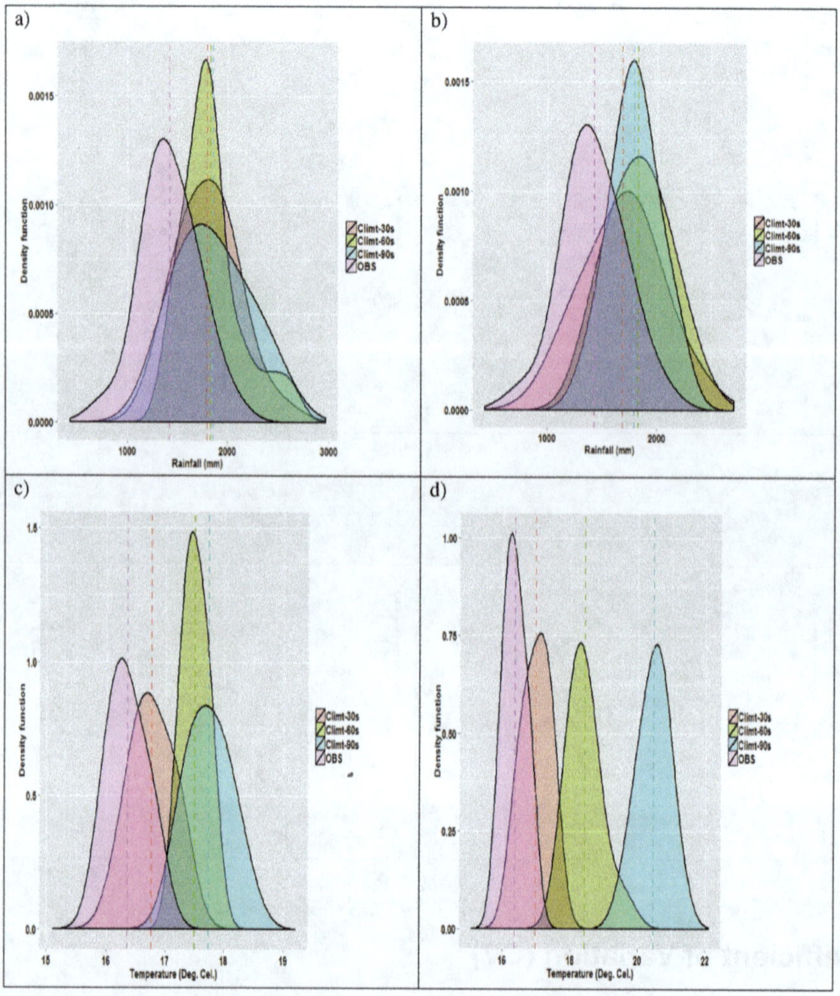

Formula to calculate sample standard deviation;

$$\delta = \sqrt{\frac{\sum\left(x-\mu\right)^2}{n-1}} \quad (2)$$

Formula for coefficient of variation;

Figure 3. Gaussian distribution of Rainfall and RCP 4.5 (a) and RCP 8.5 (b) and Mean Temperature for RCP 4.5 (c) and RCP 8.5 (d) for Embu
(Source: Joseph, 2016)

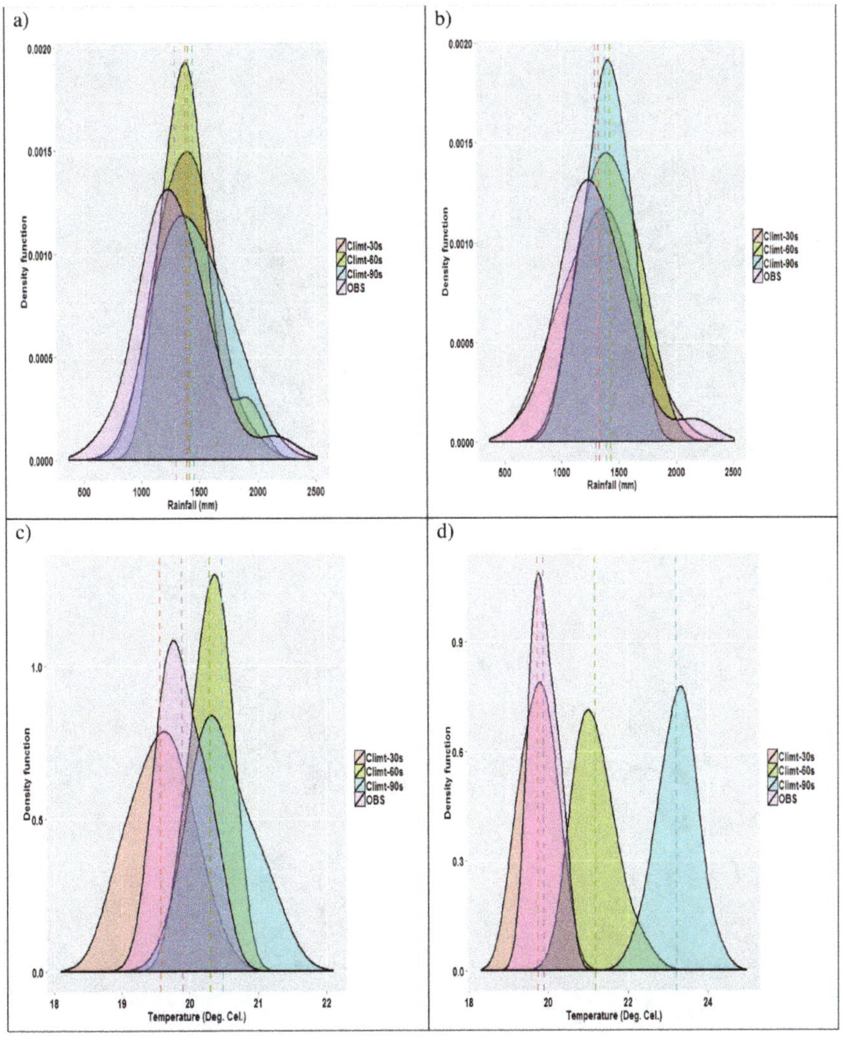

Figure 4. Gaussian distribution of Rainfall RCP 4.5 (a) and RCP 8.5 (b) and Mean Temperature for RCP 4.5 (c) and RCP 8.5 (d) for Thika
(Source: Joseph, 2016)

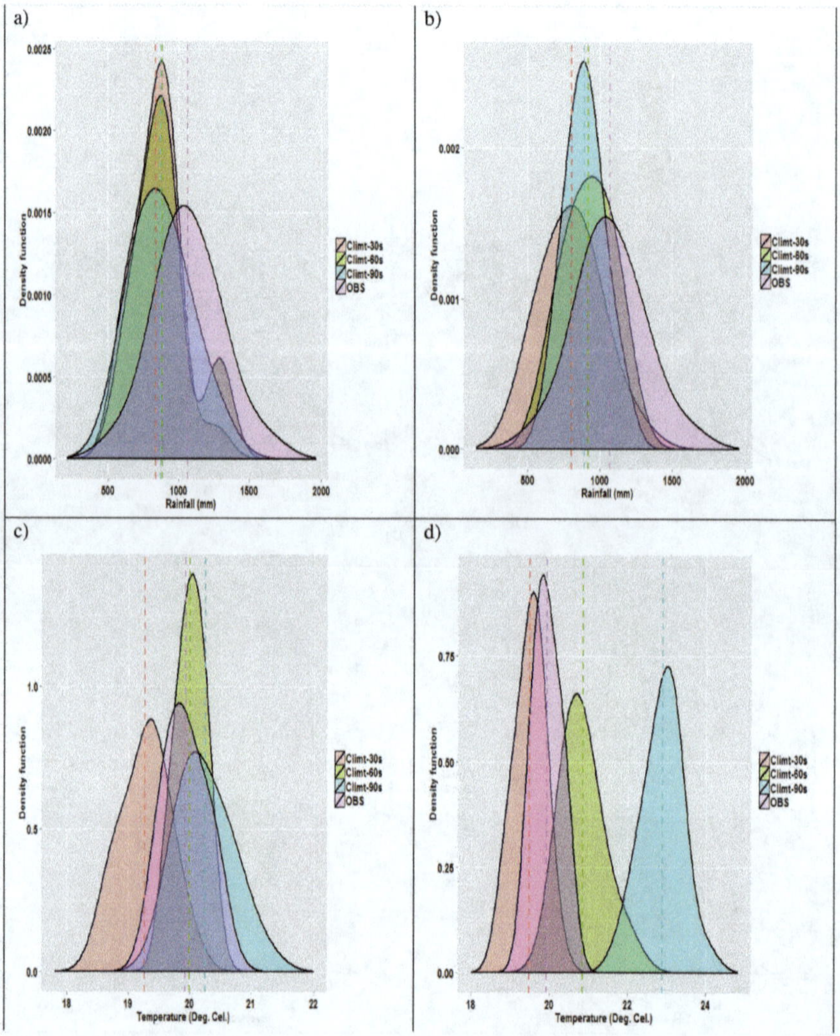

$$CV = \frac{\delta}{\mu} \quad (3)$$

Coefficient of Variability of Mam Rainfall for East Africa

During MAM season, the analysis of Coefficient of variation from 1981 to 2010 from the figure above shows that there was high variability in some parts of Kenyan Rift valley. This symbolized less or lack of rainfall during this period.

COEFFICIENT OF VARIABILITY OF OND RAINFALL FOR EAST AFRICA

During the OND season, the spatial map above shows rainfall variability in different areas from 1981 to 2010. In the North Eastern part of Kenya and some parts of central Tanzania shows high variability during this season(OND), while the western part of Tanzania and Uganda, with Rwanda and Burundi inclusive shows low variability in rainfall. During this season, Kenya experienced more dryness compared to Uganda since high variability covered a wider area.

Figure 5. Coefficient of variability of MAM rainfall for East Africa
(Source: Kisaka et.al., 2015)

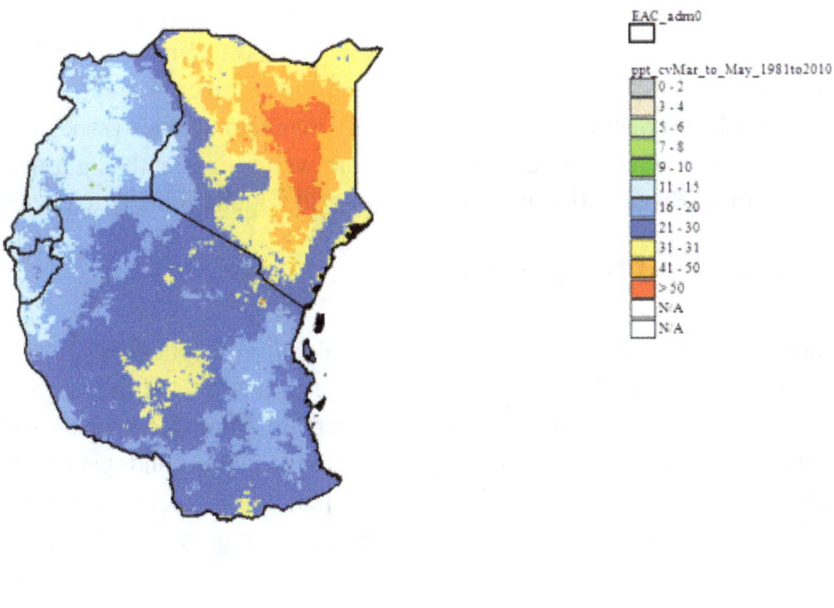

Figure 6. Coefficient of variability of OND rainfall for East Africa
(Source: Kisaka et.al., 2015)

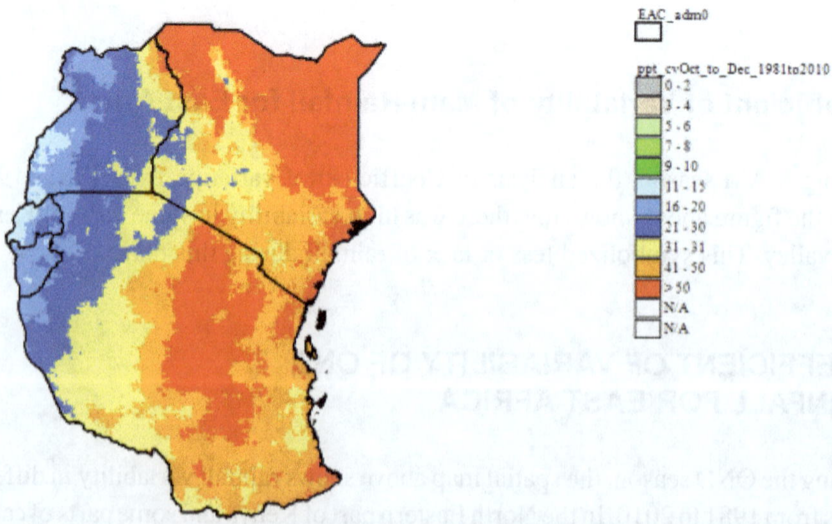

The hydrological cycle can be affected by the future climate change in many ways (Oliver & Oliver, 1994) as briefly described below;

Changes in Precipitation

Generally, on average it is expected that it will be a wetter world. Climate models predict an increase in the global mean of precipitation of 3-15% for a temperature increase of 1.5-3.5˚C (IPCC, 1996). This average masks significant variation in regional precipitation patterns with some regions showing an increase, some a decrease and remarkable inter-annual variability. Increase in precipitation is associated with middle to high latitude continents mostly during winter time with slight decrease in summer time. Precipitation in arid and semi-arid regions is likely to decrease.

Evaporation and Transpiration

Evaporation of water depends largely on the availability of both energy and water. With temperature rise, the available energy for evaporation and the atmospheric demand for water from land and water surfaces increase. More water can be held by a warmer atmosphere as well as by moving air (due to changes in wind speed). Climate models suggest an increase of 3-15% in the evaporation if the CO_2 concentration is doubled. Potential evaporation of humid temperate regions could rise by 40%

(IPCC, 1996). Transpiration from plants is affected by several factors such as plant type, plant cover, root depth, stomata behavior and CO_2 concentration. There are different and conflicting messages on how plants will be affected by climate change. Some studies suggest that plants will decrease their water use if exposed to higher CO_2 levels. Other studies suggest that higher levels of CO_2 will increase the leaf area index (LAI) that counterbalances the improvement in the water use efficiency.

Changes in Soil Moisture

The change in soil moisture is simply calculated as a difference between the input to and output of the soil system. Input to the system is mainly precipitation and output is mainly evaporation, transpiration, surface runoff into rivers and lakes, drainage and deep percolation to groundwater. Soil moisture is vital for crop growth and food production as well as for supporting natural vegetation and determining its type and extent.

Changes in climate in terms of precipitation patterns and evapotranspiration will directly affect soil moisture status, surface runoff and groundwater recharge. In regions with decreasing precipitation, soil moisture may be substantially reduced. It could also be reduced in regions with increasing precipitation as long as the evaporation due to high temperatures is greater than the increase in precipitation. Soil moisture is expected to increase in wintertime in the northern latitudes where precipitation significantly outweighs the increase in evapotranspiration (Gleick, 1998). At the same time, a large-scale drying of the soil surface is expected due to higher temperatures during summer and insufficient precipitation increases or reduction in rainfall. Agriculture production and water demand will be greatly affected by such drying.

Changes in Snowfall and Snowmelt

Snowfall and snowmelt in some regions could be the most important hydrologic impact of climate change.Changes in temperature could lead to important changes in water availability and quality and affect the management of reservoirs and irrigation systems. Temperature increase could increase the ratio of rain to snow in cold months, decrease the length of snow season and increase the rate and intensity of warm season snowmelt. These effects could lead to an increase in average winter runoff and average peak runoff, peak runoff taking place earlier in the year, and spring runoff ending sooner, with faster reduction in soil moisture. Higher temperature causes early snowmelt and will have implications for reservoir storage capacity and operation and for the availability of stored water for domestic and agricultural use later in the year.

Changes in Storm Frequency and Intensity

Climatic extremes and storm events resulting from the changes in the hydrological cycle could have a severe impact on society. It is not clear what global climatic changes may do to the variability of climatic conditions in terms of frequency and intensity of extremes. Little work has been done on the extreme events such as cyclones, hurricanes and more systematic impacts such as those caused by El Nino. However, there are some indications that the variability (internal standard deviation) of the hydrologic cycle increases when the mean precipitation increases and vice versa (Gleick, 1998). In a modeling study, the total area receiving precipitation decreased while the mean precipitation increased. Subsequently, one would expect more intense local storms and runoff as well as other changes in the hydrologic cycle. At present, there is little confidence in the prediction of such changes in variability. However, there are some indications that day-to-day and inter-annual variability of storms in the mid-latitudes will decrease. In addition, the models and empirical considerations suggesting that the frequency, intensity and area of tropical disturbances may increase.

Changes in Runoff, Flood, and Droughts

Climate and regional hydrologic models suggest that some significant changes in the timing and magnitude of runoff are likely to result from quite possible changes in climatic variables. Since society and natural ecosystems are dependent on river flows, any changes would be a good reason for concern. Flood impact may be as serious and widely spread as the adverse effect of droughts. There is indication that flooding is likely to become a serious problem in many temperate regions, which would require some measures not only for droughts and chronic water shortage, but also for flood control and associated damages to dams and reservoirs. If climate becomes more variable, some regions will be subjected to both increases in droughts and increases in floods. River flows have the potential to integrate changes in hydrologic characteristics over large areas and therefore may be used as a good indicator of climate change.

Due to variability in precipitation and temperature the impacts of climate change will affect the water yields in the long term. The river flows in the rainy seasons over the catchment are varied. This has implications on the quantity and quality of water, on the agriculture, in the energy production sectors needs for the aquatic habitat.

HYDROLOGIC SYSTEM AND CLIMATE CHANGE

The water quantity and quality and subsequently the water demand will be affected greatly by the climate change. The water resources availability is expected to affect river flow, which affects navigation, power generation and wetland and ecosystem conservation. The availability of water resources will require change in water policy, in food production and food security and might lead to national and international conflicts as shown in figure 7.

Despite the gaps in data, inadequate climate monitoring, short records and bias in instrumental data, research results show that the changes in the hydrologic cycle are significant and sufficiently different from the past record to be the result of something else other than just a natural variability. The temperature data suggest that the average surface temperature of the earth has increased by 1°C over the past century. The 12 warmest years have occurred this century after 1980 and 1990, 1995 and 1997 were the warmest years in the past 500 years in the Northern Hemisphere (Gleick, 1998). These findings were supported by satellite images, which indicated an increase in length of the growing season by up to 12 days; vegetation bloomed up to eight days earlier in the spring and summer and continued to photosynthesize for four days longer. Precipitation patterns show some increase in precipitation outside the tropics, with a tendency to decline towards the subtropics, especially in the northern tropics of Africa.

Simulated Stream Flow Trend

Trend in simulated stream flow for river Ewaso Ngiro is shown in figure 8 to figure 11. The figures show the projected variation in the stream flow for the Ewaso Ngiro River under different RCPs (RCP4.5 and RCP 8.5)

Stream flow was found to have a significant increasing trend in 2030s of RCP4.5. Change point detection using Pettitt test was found to occur in the year 2043 where an upward shift in the mean was observed (figure 8)

Non-significant decreasing trend in stream flow in 2060s for RCP4.5 was observed and Pettitt test showed a change point in the year 2073 where a downward shift in the mean was observed as shown in figure 9 above.

For RCP8.5, non-significant decreasing trend was observed in 2030s using MK trend test and Pettitt test confirmed the trend where a downward shift in the mean was found to occur in the year 2029 as shown in figure 10.

Non-significant increasing trend in stream flow was observed in 2060s of RCP 8.5. A change point in the year 2065 was observed using Pettitt test where the mean of the annual stream flow shifted upwards as depicted in figure 11. Also, the results of Mann Kendall trend test on stream flow are shown in Table 1.

Figure 7. Impact of climate change on the hydrological cycle
(Ichoku et. al., 2016)

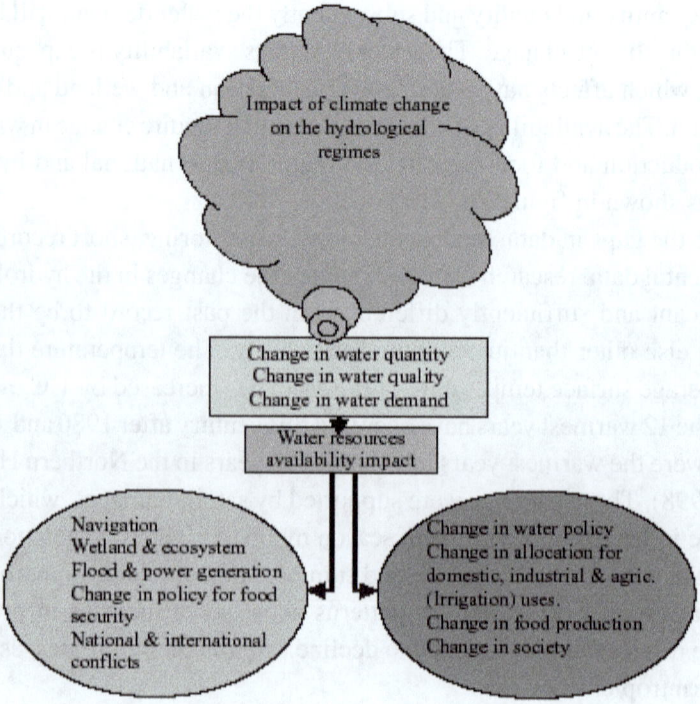

Figure 8. Stream flow in Ewaso Ngiro River in 2030s for RCP4.5
(Source: Omwoyo et. al., 2017)

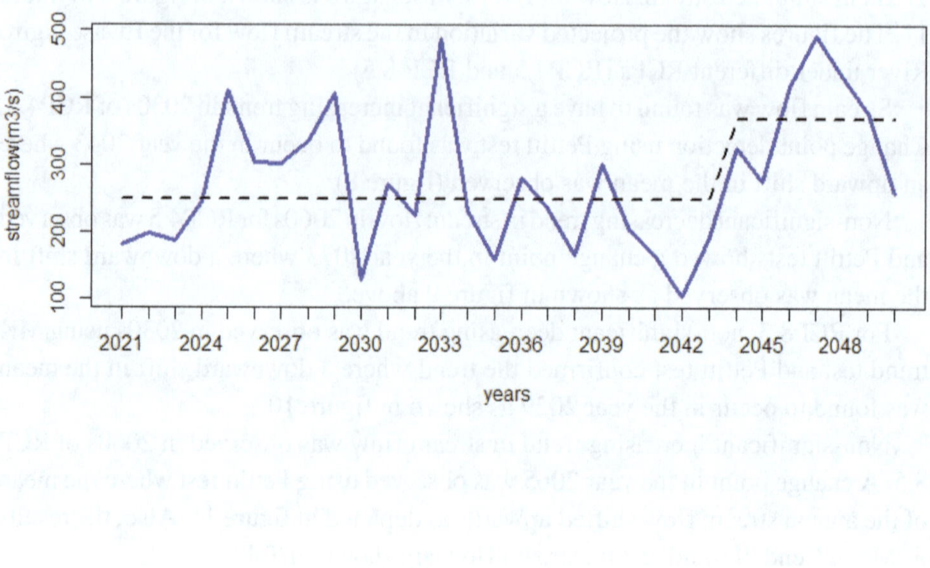

Figure 9. Stream flow in Ewaso Ngiro River in 2060s for RCP4.5
(Source: Omwoyo et. al., 2017)

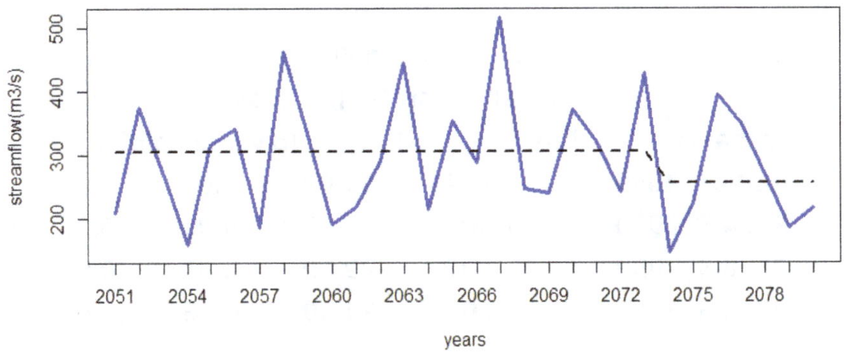

Figure 10. Stream flow in Ewaso Ngiro River in 2030s for RCP8.5
(Source: Omwoyo et. al., 2017)

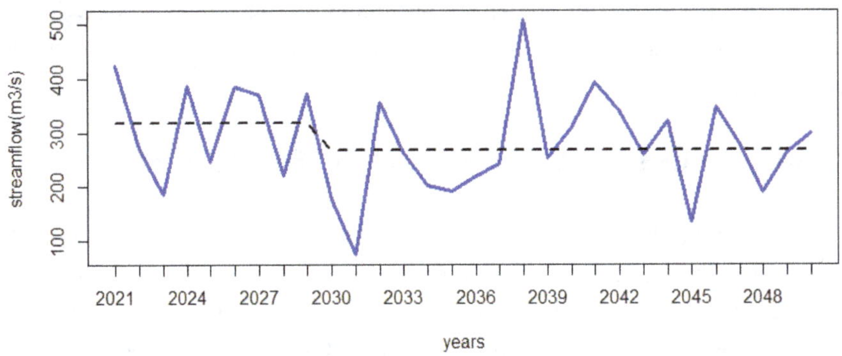

Figure 11. Stream flow in Ewaso Ngiro river in 2060s for RCP8.5
(Source: Omwoyo et. al., 2017)

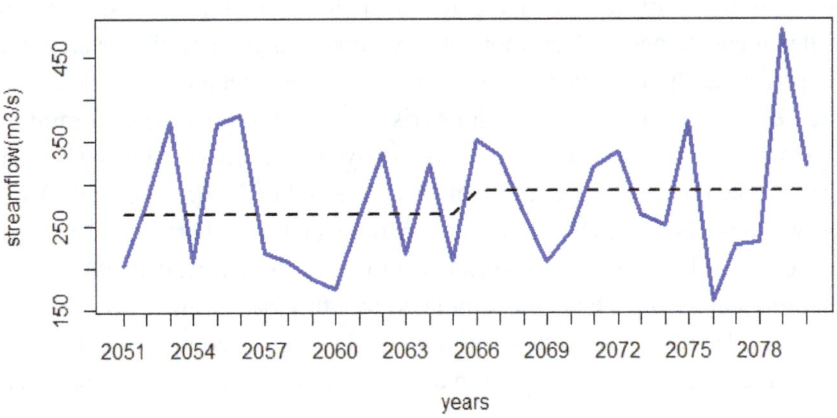

Table 1. Results for Mann Kendall trend test on simulated stream flow

scenario	Kendall's Tau	Mann Kendall statistic (S)	2-sided p value	alpha
RCP4.5_2030s	0.14	59.00	0.00	0.05
RCP4.5_2060s	-0.02	-7.00	0.90	0.05
RCP8.5_2030s	-0.07	-29.00	0.61	0.05
RCP8.5_2060s	0.08	33.00	0.56	0.05

From table 1, the stream flow is expected to significantly increase during 2021-2050 of RCP4.5 and non-significantly increase during the 2060s for RCP8.5, as shown by positive Kendall's statistic S of 59 and 33 respectively. Non-significant decreasing trend in stream flow in 2060s for RCP4.5 and in 2030s for RCP8.5 was also observed.

Percent Changes in Stream Flow in 2030s and 2060s

Results of percent changes in mean monthly stream flow for Ewaso Ngiro River with respect to the baseline are shown in figure 12.

Stream flow during MAM season was found to have a decrease in all scenarios with a greater decrease (-26%) being expected in RCP 8.5 during 2060s and a smaller percent decrease (-10%) during 2060s of RCP 4.5. All the other seasons were found to have an increase in the stream flow with JJA having a greater percentage increase (90-114%) as compared to DJF and SON (figure 12).

A regression of mean annual stream flow percent changes against mean annual rainfall percent changes at Naromoru showed that change in rainfall in all scenarios would significantly influence stream flow within the catchment.

Figure 14 shows marked deviations (rise and fall) in mean annual rainfall and water flow, implying the existence of year-to-year rainfall variability (Kassian et. al., 2017). The highest rainfall amount was observed in 2004/05 (220 mm) and the lowest was observed in 2014/15 (70 mm). Moreover, the highest water flow amount was observed in 1998/99 (484 m3/s) and the lowest was observed in 1999 (12 m3/s).

Figure 15 and 16 presents the simulated monthly basin outflow for both RCPs (i.e., averaged across the ensemble of models). In general, compared to the baseline scenario, the average flow is higher in all the months in all three time periods with

Figure 12. Percent change mean monthly stream flow in the 2030s and 2060s with respect to the baseline for Ewaso Ngiro River
(Source: Omwoyo et. al., 2017)

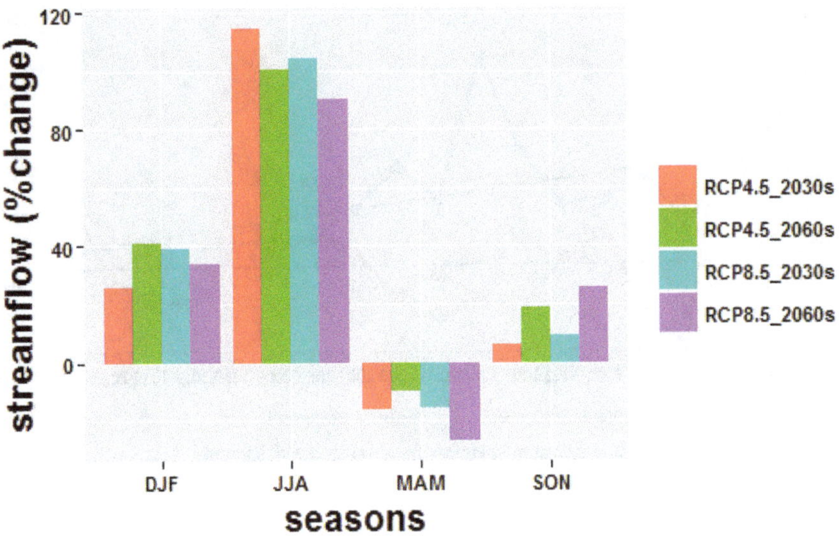

Figure 13. A regression of percent change in rainfall (Naromoru) on percent change stream flow R2=0.9213
(Source: Omwoyo et. al., 2017)

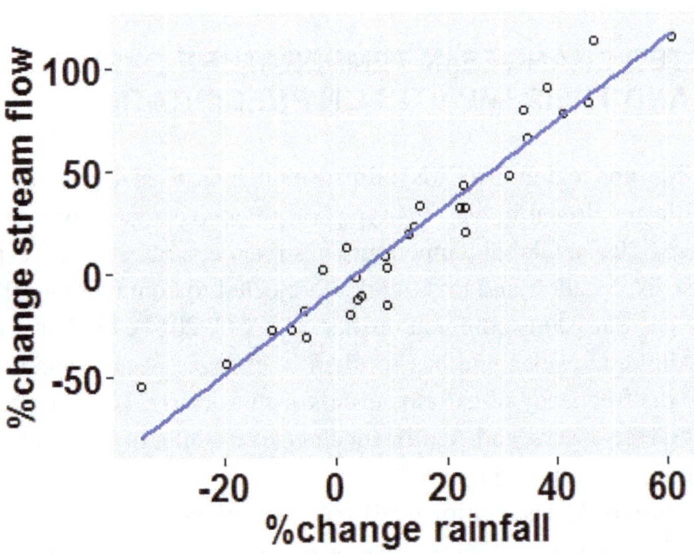

Figure 14. Annual mean flow of Lyandembera River (m3/s) and rainfall from 1997–2014 in Tanzania
(Source: Kassian et. al., 2017)

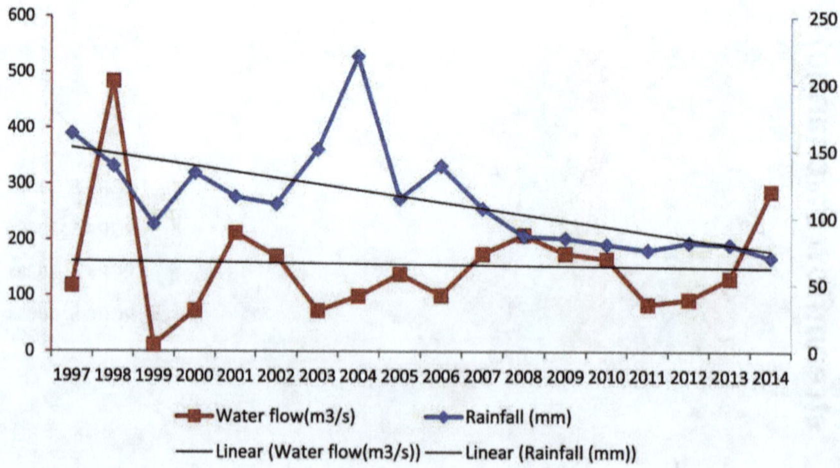

the largest increase in RCP 8.5. Furthermore, the extreme values clearly indicate that, with some exception in the 2020-2049 periods, there is greater variability of flow under RCP 8.5 than RCP 4.5. The results also reflect the gradual shift to the earlier onset of rainfall with peak flows moving to earlier in the year.

PROJECTED CHANGES IN TEMPERATURE OVER AFRICA AND THEIR IMPACTS ON PRECIPITATION

Global (GCMs) and regional (RCMs) climate models are used to produce projections of future climates driven by various types of greenhouse gas emission scenarios (Famien *et al.*, 2017). Global temperature has risen at a rate of 0.12°C per decade over the last six decades, and this trend is projected to continue into the future if greenhouse gas emissions continue unabated (IPCC, 2014). Global warming has changed hydrologic cycles and has resulted in changes in precipitation patterns, increases in the frequency of extreme events such as droughts and floods, melting of glaciers and ice covers, and significant changes in soil moisture and river runoff (Bates *et al.*, 2008, Adhikari *et al.* 2016).

Temperatures in Africa are projected to rise faster than the global average increase during the 21st century. Global average near surface air temperature is projected to move beyond 20th century simulated variability by 2069 under Representative Concentration Pathway 4.5 (RCP4.5) and by 2047 under RCP8.5. However, in the

Figure 15. Monthly variation in outflow from the Tana River basin (averaged across the ensemble of models) for the three time periods for both RCPs
(Source: Aditya et. al., 2017)

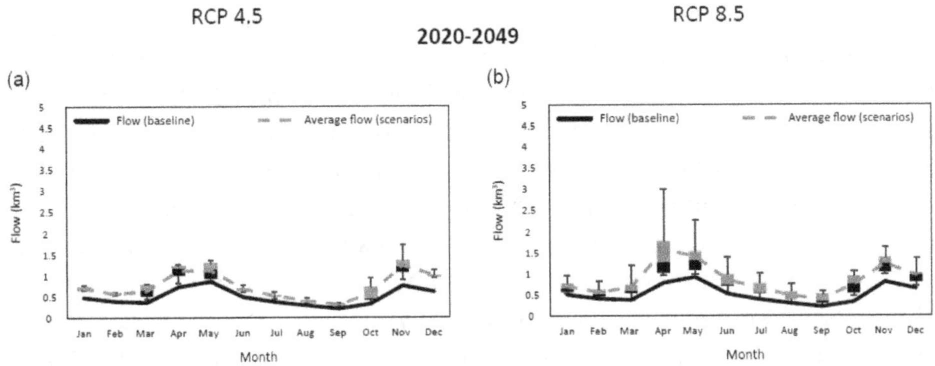

Figure 16. Monthly variation in outflow from the Tana River basin (averaged across the ensemble of models) for the three time periods for both RCPs
(Source: Aditya et. al., 2017)

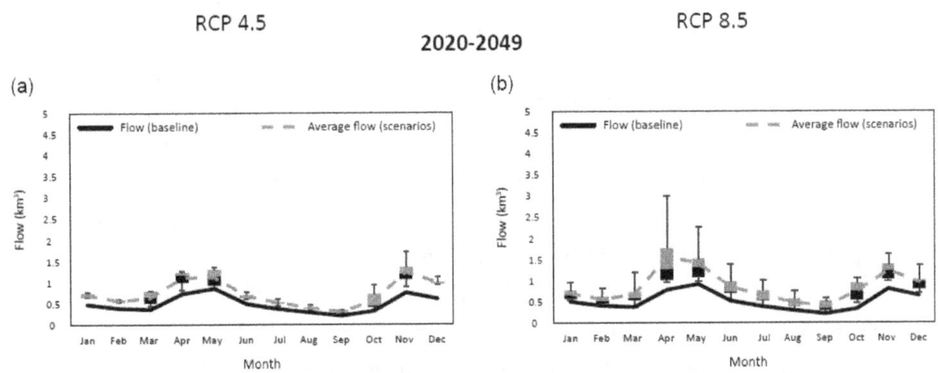

tropics, especially tropical West Africa, these unprecedented climates are projected to occur 1 to 2 decades earlier than the global average because the relatively small natural climate variability in this region generates narrow climate bounds that can be easily surpassed by relatively small climate changes.

Increases in mean annual temperature over all land areas are very likely in the mid and late 21st-century periods for RCP2.6 and RCP8.5. Ensemble mean changes in mean annual temperature exceed 2°C above the late 20th century baseline over most land areas of the continent in the mid-21st century for RCP8.5, and exceed 4°C over most land areas in the late 21st century for RCP8.5. Changes in mean annual temperature for RCP8.5 follow a pattern of larger changes in magnitude over northern and southern Africa, with (relatively) smaller changes in magnitude over central Africa. The ensemble mean changes are less than 2°C above the late 20th century baseline in both the mid and late 21st century for RCP2.6 (Niang *et al.*, 2014).

Moreover, according to IPCC (2007), Africa is one of the most vulnerable regions to weather and climate variability. The most severe effects of global warming will be related not only to a change in the mean climate, but especially to an increase in frequency and intensity of extreme events, such as floods, droughts and heat waves. As an example, Dosio (2017) mentioned that in 2002 north-eastern Nigeria suffered one of the hottest and driest spells in living memory, with recorded temperatures above 50°C and dozens of people dying due to heat wave.In addition, Africa is a large and heterogeneous continent in climatic conditions (Dosio and Panitz, 2015), where climate vary from humid equatorial to seasonally arid and subtropical Mediterranean and exhibits differing degrees of temporal and spatial variability (Hulme *et al.*, 2001).

Case of Eastern Africa

Recently, Adhikari *et al.*, (2016) reviewed the projected change in maximum and minimum temperature under three different climate change scenarios compared to the base period of 1970-1999 (1961-1999 for Rwanda) from an ensemble of multiple GCMs. Those simulations project a 0.9°C to 3.4°C rise in temperature by the 2060s and a 1.3°C to 5.5°C rise by 2090. The projected changes in temperature are higher than the historical trend of 0.5°C in the past century, but most of the temperature projections fall within the global long-term temperature change (0.3°C to 4.8°C) projected by IPCC under various scenarios. The projections further indicate that Rwanda will have the lowest temperature rise.

The same scenarios predict a sea level rise of up to 0.98 m by the end of this century. Rise in sea level will flood or even submerge coastal land, salinize surface water and groundwater resources, increase soil erosion, and reduce the extent of coastal wetlands (Nicholls and Cazenave, 2010). Salinization of freshwater resources becomes more significant during the dry season when river discharge and infiltration rates are low and evaporation rates are high or if the groundwater recharge decreases. Salinization of water resources will imply a reduction in freshwater for agricultural and household use (Bates *et al.*, 2008). The concerned countries are Kenya and Tanzania, the one having coastal areas. For example, in Mombasa, the second largest city of Kenya, a 0.3 m rise in sea level has been projected to submerge 17% of the city area and render large areas unsuitable for agriculture due to water logging and salt stress (Awuor et al., 2008, Adhikari *et al.*, 2016).

East African countries are prone to climate related disasters. Some countries are susceptible to droughts, while others are likely to floods. Kenya is affected by a high frequency of droughts, while Tanzania experiences a higher frequency of floods, and Uganda and Ethiopia are affected by both (Adhikari *et al.* 2016). Due to the large variability in current climatic conditions, the impacts of climate change will not be uniform across the region. Some places are projected to be drier, while others are projected to be wetter; some regions will have positive economic impacts, while others will be adversely affected (Adhikari et al. 2016).

In their review, Adhikari et al., (2016) state that East Africa lies mostly in the equatorial region but is modified by the altitudes in the region, allowing for larger variation in climate. Most of the countries in the region are already water stressed or close to it. Water resources in the region are further threatened by climate change. Projections of temperature rise are consistent in the region, but there are larger uncertainties and variations in precipitation projections (Collier et al., 2008). The IPCC predicts a general increase in precipitation in East Africa, but the increase is predicted to be in the form of increased intensities and shorter durations. Individual studies do not show consistent predictions within the region. Uncertainty in the direction and magnitude of the precipitation variation complicates the estimation of water resources availability in the future.

In conclusion, Adhikari et al. (2016) mention that increase in the evapotranspiration rate along with the increase in temperature will increase the water stress by reducing available soil moisture and will affect the water level in rivers, lakes, and reservoirs. River runoff in East Africa is predicted to increase with the increase in precipitation. Projections of climate change impacts on groundwater resources are uncertain due to the lack of detailed precipitation projections.

In general, deep groundwater resources are more resilient to climate change than surface water. However, there is a concern about the impact of climate change on shallow groundwater resources, which are more sensitive to precipitation variability. Glaciers and ice covers are projected to disappear from the region within one or two decades, but they are not a significant source of water in the region. Therefore, their impact on water resources is predicted to be minimal. Although the extent is small, the rise in sea level is projected to inundate some coastal areas and contaminate freshwater resources, decreasing the availability of freshwater. Small-scale irrigation development would offer resilience to climate change and enhance food security in the region.

Distribution of Rainfall in Eastern Africa

Various studies that have been done over this region have shown that both rainfall and stream flow have shown increasing and decreasing trends. This varies from one region/location to another. Some of these trends have been shown to be significant while others are not. Not all of these changes in trend can be attributed to climate changes. Some are due to changes in land use patterns, increased groundwater abstraction and abstraction for irrigation purposes. Figure 17 shows the mean seasonal distribution of rainfall over Eastern Africa, with the relatively darker/greener areas showing areas of high rainfall potential consequently with higher potential for water resources availability.

Non-Climatic Factors Influencing Water Availability

A broad range of internal and external factors have influenced the availability and quality of water sources; privatization and changing tenure arrangements, poverty and social differentiation, environmental change and degradation, and; population pressure and conflict. In a number of sites in Kenya, water sources that had been accessible to the public had come under private control due to changing tenure arrangements, thus forcing people to pay to regain the right of access or seek alternative sources. Poverty is a major contributing factor to the prevailing poor

Figure 17. Distribution of seasonal rainfall over the Eastern Africa region

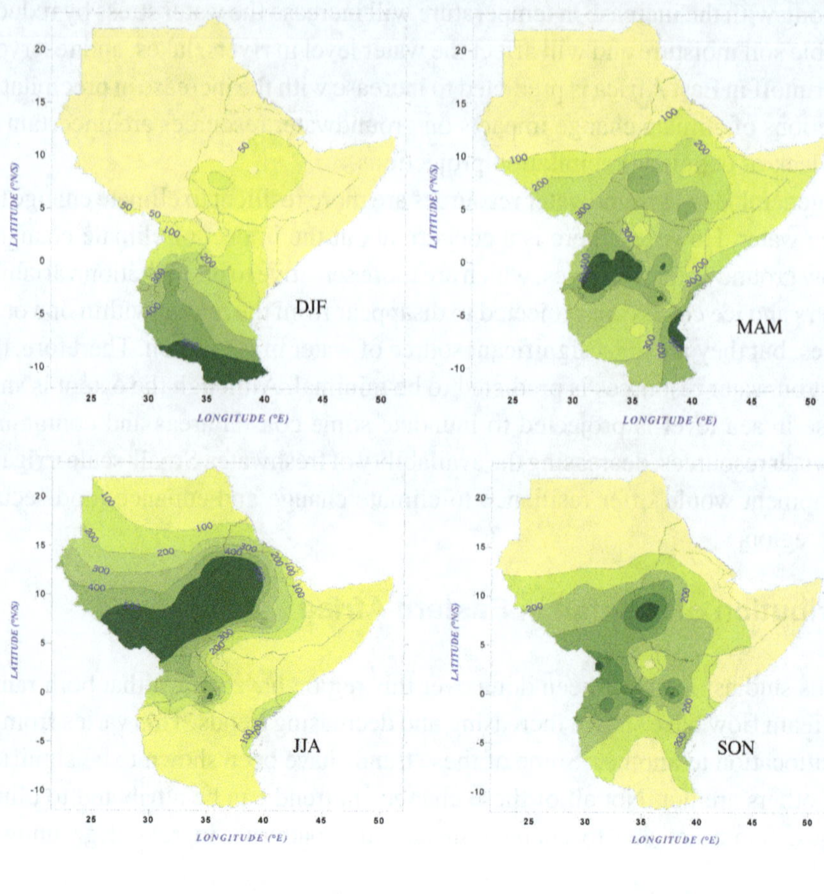

water situation and environmental changes and degradation are evidenced in the draining of wetlands for agriculture.

Impacts of Climate on Ground Water

Groundwater pollution is a particular concern in the more arid countries, which are highly dependent on aquifers for drinking water supplies. The main pollutants are nitrates and phosphates, together with chemical residues from seepage of agricultural runoff, and from inappropriate discharge of industrial and domestic wastewater. In low-lying areas, salinization of groundwater aquifers is a further concern, through the intrusion of saltwater and seepage of irrigation water. Expected sea level rise under climate change scenarios will cause further widespread contamination of aquifers. Therefore, it is critical that action is taken now, in order to ensure the sustained protection and sustainable use of underground water resources.

Higher temperatures will enhance the demand for water for human consumption and agriculture and the highly vulnerable parts of Africa will face increased risks of water shortages. The increasing temperatures will also alter water quality significantly by influencing water temperatures, discharges, runoff rates and timing, and the ability to assimilate pollutants (IPCC, 2014). The magnitude of damage in water quality depends not just on climate change but also on a wide range of other human uses. Effects of degraded water quality include higher thermal stress for cold-water organism, losses of biodiversity, lower content of dissolved oxygen, increases in nutrient cycling and productivity. Flooding will lead to increase in frequency of water-borne diseases where sanitation is inadequate.

CONCLUSION

While water management systems are often flexible, adaptation to new hydrologic conditions may be very costly. Water authorities should begin to re-examine engineering design assumptions, operational rules, system optimization and contingency planning for the present and future planned water management systems under a wider range of climatic conditions than are traditionally used. Water authorities should explore the vulnerability of both structural and non-structural water systems to possible future climate change, not only the past climatic variability. Government at all levels should re-evaluate legal, technical and economic approaches to manage water resources under possible climate changes. Co-operation between the water authorities and scientific communities will help in exchanging information on the state-of-the-art on climate changes and its impact on water resources. The timely flow of information from the scientific communities and water management communities and public would be very important.

Changes in climate in terms of precipitation patterns and evapo-transpiration will directly affect soil moisture status, surface runoff and groundwater recharge. Therefore, in regions with decreasing precipitation, soil moisture may be substantially reduced. On the other hand, soil moisture could also be reduced in regions where evaporation due to high temperatures is greater than the increase in precipitation. A large-scale drying of the soil surface is expected due to higher temperatures during summer and insufficient precipitation increases or reduction in rainfall. Therefore, this implies that agricultural production and water demand will be greatly affected by such drying.

If climate becomes more variable, some regions will be subjected to both increases in droughts and increases in floods. There is indication that flooding is likely to become a serious problem in many temperate regions, which would require some measures not only for droughts and chronic water shortage, but also for flood control and associated damages to dams and reservoirs. The water quantity and quality and subsequently the water demand will be affected greatly by the climate change.

REFERENCES

Adhikari, U., Nejadhashemi, A. P., & Herman, M. R. (2016). *A review of climate change impacts on water resources in East Africa.* American Society of Agricultural and Biological Engineers (ASABE). doi:10.13031/trans.58.10907

Aditya, S., Lal, M., Silva, N. S., & McCartney, M. (2017). Understanding the hydrological impacts of climate change in the Tana River Basin, Kenya. *IWMI Working Paper*, (178).

Alexander, L.V., Zhang, X., Peterson, T.C., Caesar, J., Gleason, B., Klein Tanka, M.G., … Vazquez-Aguirre, J.L. (2006). Global observed changes in daily climate extremes of temperature and precipitation. *J Geophys Res, 111*(D5), 109. doi:10.1029/2005JD006290

Awuor, C. B., Orindi, V. A., & Adwera, A. O. (2008). Climate change and coastal cities: The case of Mombasa, Kenya. *Environment and Urbanization, 20*(1), 231–242. doi:10.1177/0956247808089158

Bates, B. C., Kundzewicz, Z. W., Wu, S., & Palutikof, J. P. (2008). *Climate change and water*. IPCC Technical Paper VI. Geneva, Switzerland: IPCC. Retrieved from http://ipcc.ch/pdf/technicalpapers/climate-change-water-en.pdf

Collier, P., Conway, G., & Venables, T. (2008). Climate change and Africa. *Oxford Review of Economic Policy*, *24*(2), 337–353. doi:10.1093/oxrep/grn019

Dosio, A. (2017). Projections of temperature and heat waves for Africa with an ensemble of CORDEX Regional Climate Models. *Climate Dynamics*, *49*(1-2), 493–519. doi:10.100700382-016-3355-5

Dosio, A., & Panitz, H. J. (2015). Climate change projections for CORDEX-Africa with COSMOCLM regional climate model and differences with the driving global climate models. *Climate Dynamics*. doi:10.100700382-015-2664-4

Droogers, P., Immerzeel, W. W., Terink, W., Hoogeveen, J., Bierkens, M. F. P., van Beek, L. P. H., & Debele, B. (2012). Water resources trends in Middle East and North Africa towards 2050. *Hydrology and Earth System Sciences*, *16*(9), 3101–3114. doi:10.5194/hess-16-3101-2012

Falkenmark, M. (1989). The massive water scarcity now threatening Africa: Why isn't it being addressed? *Ambio*, 112–118.

Famien, A. M., Janicot, S., Ochou, A. D., Vrac, M., Defrance, D., Sultan, B., & Noël, T. (2017). *A bias-corrected CMIP5 dataset for Africa using CDF-t method. A contribution to agricultural impact studies*. Earth Syst. Dynam. Discuss. doi:10.5194/esd-2017-111

Gleick, P. H. (1998). *The World's Water: The Biennial Report on Freshwater Resources 1998–1999*. Island Press.

Gordon, C., Cooper, C., Senior, C., Banks, H., Gregory, J. M., Hagemann, S., ... Wiltshire, A. J. (2013). Climate change impact on available water resources obtained using multiple global climate and hydrology models. *Earth Syst Dynam*, *4*(1), 129–144. doi:10.5194/esd-4-129-2013

Hastenrath, S. (1984). *The Glaciers of Equatorial East Africa*. Dordrecht: Reidel. doi:10.1007/978-94-009-6251-4

Hastenrath, S. (1991). *Climate Dynamics of the Tropics*. Dordrecht: Kluwer. doi:10.1007/978-94-011-3156-8

Hastenrath, S. (1995). Glacier recession on Mount Kenya in the context of the global tropics. *Boletín del Instituto Francés de Estudios Andinos*, *24*, 633–638.

Hastenrath, S. (2001). Variations of East African climate during the past two centuries. *Climatic Change*, *50*(1/2), 209–217. doi:10.1023/A:1010678111442

Hastenrath, S., & Greischar, L. (1997). Glacier recession on Kilimanjaro, East Africa, 1912–89. *Journal of Glaciology*, *43*(145), 455–459. doi:10.1017/S0022143000035048

Hastenrath, S., Rostom, R., & Caukwell, R. A. (1989). Variations of Mount Kenya's glaciers 1963–87. *Erdkunde*, *43*, 202–210.

Hay, J. E., & Mimura, N. (2005). Sea-level rise: Implications for water resources management. *Mitigation and Adaptation Strategies for Global Change*, *10*(4), 717–737. doi:10.100711027-005-7305-5

Hulme, M., Doherty, R., Ngara, T., New, M., & Lister, D. (2001). African climate change: 1900-2100. *Climate Research*, *17*(2), 145–168. doi:10.3354/cr017145

Ichoku, C., Ellison, L. T., Willmot, K. E., Matsui, T., Dezfuli, A. K., Gatebe, C. K., ... Okonkwo, C. (2016). Biomass burning, land-cover change, and the hydrological cycle in Northern sub-Saharan Africa. *Environmental Research Letters*, *11*(9), 095005. doi:10.1088/1748-9326/11/9/095005

IPCC. (1996). *Climate change 1995: the science of climate change. Summary for Policy-makers. Contribution of working group 1 to the Second Assessment Report of the Intergovernmental Panel on Climate Change*. Cambridge, UK: CambridgeUniversity Press.

IPCC. (2000). *IPCC special report: Emission scenarios. Summary for policymakers*. Geneva, Switzerland: IPCC.

IPCC. (2001). IPCC Third Assessment Report of Working Group 1. In *Impacts, Adaptation and Vulnerability: Summary for Policy Makers*. IPCC.

IPCC. (2007). *Climate change 2007: the physical science basis: contribution of Working Group I to the Fourth Assessment Report of the Intergovernmental Panel on Climate Change*. Cambridge, UK: Cambridge University Press.

IPCC. (2013). Annex I: Atlas of Global and Regional Climate Projections. In *Climate Change 2013: The Physical Science Basis. Contribution of Working Group I to the Fifth Assessment Report of the Intergovernmental Panel on Climate Change*. Cambridge, UK: Cambridge University Press.

IPCC. (2014a). Climate change 2014: impacts, adaptation, and vulnerability. In Part B: regional aspects. Contribution of Working Group II to the Fifth Assessment Report of the Intergovernmental Panel on Climate Change. Cambridge University Press.

IPCC. (2014b). Climate change 2014: Synthesis report. In R. K. Pachauri & L. A. Meyer (Eds.), *Contribution of WorkingGroups I, II, and III to the Fifth Assessment Report of the IPCC*. Geneva, Switzerland: IPCC.

Joseph, P. M. (2016). *Climate Change Impacts on Water Resources over The Upper Tana Catchment of Kenya* (Doctoral dissertation). University of Nairobi.

Kaser, G., & Noggler, B. (1991). Observations on Speke Glacier, Rwenzori Range, Uganda. *Journal of Glaciology*, *37*(127), 313–318. doi:10.1017/S0022143000005736

Kaser, G., & Noggler, B. (1996). Glacier fluctuations in the Rwenzori Range (East Africa) during the 20th century — a preliminary report. *Zeitschrift für Gletscherkunde und Glazialgeologie*, *32*, 109–117.

Kaser, G., & Osmaston, H. (2002). *Tropical Glaciers*. Cambridge, UK: Cambridge University Press.

Kassian, L. M., Tenywa, M., Liwenga, E. T., Dyer, K. W., & Bamutaze, Y. (2017). Implication of climate change and variability on stream flow in Iringa region, Tanzania. *Journal of Water and Climate Change*, *8*(2), 336–347. doi:10.2166/wcc.2016.238

Kisaka, M. O., Mucheru-Muna, M., Ngetich, F. K., Mugwe, J. N., Mugendi, D., & Mairura, F. (2015). Rainfall variability, drought characterization, and efficacy of rainfall data reconstruction: Case of Eastern Kenya. *Advances in Meteorology*.

Mariotti, A., Pan, Y., Zeng, N., & Alessandri, A. (2015). Long-term climate change in the Mediterranean region in the midst of decadal variability. *Climate Dynamics*, *44*(5-6), 1437–1456. doi:10.100700382-015-2487-3

Milano, M., Ruelland, D., Fernandez, S., Dezetter, A., Fabre, J., & Servat, E. (2012). Facing global changes in the Mediterranean basin: How could the current water stress evolve by the medium-term? *Comptes Rendus Geoscience*, *344*(9), 432–440. doi:10.1016/j.crte.2012.07.006

M¨olg, T., Georges, C., & Kaser, G. (2003a). The contribution of increased incoming shortwave radiation to the retreat of the Rwenzori Glaciers,East Africa, during the 20th century. *International Journal of Climatology*, *23*(3), 291–303. doi:10.1002/joc.877

Mora, C. F., Frazier, A. G., Longman, R. J., Dacks, R. S., Walton, M. M., Tong, E. J., ... Giambelluca, T. W. (2013). The projected timing of climate departure from recent variability. *Nature*, *502*(7470), 183–187. doi:10.1038/nature12540 PMID:24108050

New, M., Hewitson, B., Stephenson, D.B., Tsiga, A., Kruger, A., Manhique, A., … Lajoie, R. (2006). Evidence of trends in daily climate extremes over southern and West Africa. *J Geophys Res, 111*(D14), 102. doi:10.1029/2005JD006289

Niang, I., Ruppel, O. C., Abdrabo, M. A., Essel, A., Lennard, C., Padgham, J., & Urquhart, P. (2014). Africa. In *Climate Change 2014: Impacts, Adaptation, and Vulnerability. Part B: Regional Aspects. Contribution of Working Group II to the Fifth Assessment Report of the Intergovernmental Panel on Climate Change* (pp. 1199–1265). Cambridge, UK: Cambridge University Press.

Nicholls, R. J., & Cazenave, A. (2010). Sea-level rise and its impact on coastal zones. *Science, 328*(5985), 1517–1520. doi:10.1126cience.1185782 PMID:20558707

Oliver, H. R., & Oliver, S. A. (Eds.). (1994). *The Role of Water and the Hydrological Cycle in Global Change. NATO ASI Series. Series I: Global Environmental Change* (Vol. 31). London: Springer.

Omwoyo, M. A., Muthama, N. J., Opere, A., & Onwonga, R. (2017). Simulating Streamflow in Response to Climate Change in the Upper Ewaso Ngiro Catchment, Kenya. *Journal of Climate Change and Sustainability*, *1*(2), 1229.

Orimoloye, R. I., Mazinyo, P. S., Nel, W., & Iortyom, T. E. (2018). Assessing changes in climate variability observation and simulation of temperature and relative humidity: A case of East London, South Africa. *Research Journal of Environmental Sciences*, *12*(1), 1–13. doi:10.3923/rjes.2018.1.13

Prudhomme, C., Giuntoli, I., Robinson, E. L., Clark, D. B., Arnell, N. W., Dankers, R., ... Wisser, D. (2014). Hydrological droughts in the 21st century, hotspots and uncertainties from a global multimodel ensemble experiment. *Proceedings of the National Academy of Sciences of the United States of America*, *111*(9), 3262–3267. doi:10.1073/pnas.1222473110 PMID:24344266

Schilling, J., Freier, K. P., Hertig, E., & Scheffran, J. (2012). Climate change, vulnerability and adaptation in North Africa with focus on Morocco. *Agriculture, Ecosystems & Environment*, *156*, 12–26. doi:10.1016/j.agee.2012.04.021

Seth, A., Rauscher, S. A., Rojas, M., Giannini, A., & Camargo, S. J. (2011). Enhanced spring convective barrier for monsoons in a warmer world? *Climatic Change*, *104*(2), 403–414. doi:10.100710584-010-9973-8

Shongwe, M. E., van Oldenborgh, G. J., van den Hurk, B., & van Aalst, M. (2011). Projected changes in mean and extreme precipitation in Africa under global warming. Part II: East Africa. *Journal of Climate*, *24*(14), 3718–3733. doi:10.1175/2010JCLI2883.1

Terink, W., Immerzeel, W. W., & Droogers, P. (2013). Climate change projections of precipitation and reference evapotranspiration for the Middle East and Northern Africa until 2050. *International Journal of Climatology*, *33*(14), 3055–3072. doi:10.1002/joc.3650

Thompson, L. G., Mosley-Thompson, E., Davis, M. E., Henderson, K. A., Brecher, H. H., Zagorodnov, V. S., ... Beer, J. (2002). Kilimanjaro ice core records: Evidence of Holocene climate change in tropical Africa. *Science*, *298*(5593), 589–593. doi:10.1126cience.1073198 PMID:12386332

Tramblay, Y., Jarlan, L., Hanich, L., & Somot, S. (2017). Future Scenarios of Surface Water Resources Availabilityin North African Dams. *Water Resources Management*. doi:10.100711269-017-1870-8

Urama, K. C., & Ozor, N. (2010). Impacts of climate change on water resources in Africa: The role of adaptation. *African Technology Policy Studies Network*, *29*, 1–29.

Vorosmarty, C. J., Green, P., Salisbury, J., & Lammers, R. B. (2000). Global Water Resources: Vulnerability from Climate Change and Population Growth. *Science*, 289. PMID:10894773

Yira, Y., Diekkrüger, B., Steup, G., & Bossa, A. Y. (2017). Impact of climate change on hydrological conditions in a tropical West African catchment using an ensemble of climate simulations. *Hydrology and Earth System Sciences*, *21*(4), 2143–2161. doi:10.5194/hess-21-2143-2017

Section 3

Water Resources Management and Policy

Chapter 11
Hydrology and Best Practices for Managing Water Resources in Arid and Semi-Arid Lands in Kenya

Samwel N. Marigi
Institute for Meteorological Training, Kenya

ABSTRACT

This chapter provides a critical analysis and evaluation of the water issues relevant to Kenya's ASALs. This has particularly been centered on the current resource development and management, future resource demand, as well as extent of its vulnerability to climate variability and change. The water development policy interventions have also been evaluated. The analysis has revealed that water resources are being utilized to satisfy a myriad of demands and that in fact a water shortage already exists in these ASALs. A number of factors including population pressure, poor resource use and management, and other socio-economic activities have been noted to increase the vulnerability of the available water to the impacts of climate change. Recommendations for purposes of ensuring the sustainable utilization of the vital water resource have been proposed.

DOI: 10.4018/978-1-7998-0163-4.ch011

INTRODUCTION

The natural resources of the Arid and Semi-arid lands (ASALs) – in particular soil and water- are limited, and are often in a delicate environmental balance. It is also worthwhile noting that easily- developed land in many areas of developing countries in Africa is already exploited. Attention therefore is increasingly turning to ASALs to relieve population pressure and to provide more food for the ever increasing human population. Desert encroachment due to lack of conservation planning, and the dangers of destroying or depleting beyond recovery these productive resources, are evident at the present time and may be disastrous if development is based on short term expediency rather than long-term environmental stability. In general, the distinctive features of ASALs include (1) High levels of incident radiation, (2) generally large diurnal and seasonal temperature variations, (3) low humidity, (3) strong winds with dust and sand storms, (4) sporadic rainfall of high temporal and spatial variability and (5) high sediment transport. What then can the ASALs offer? In agriculture, they can provide a certain well-known range of crops and animals adapted to the range of ASALs climates. In industry, plants and industrial complexes can benefit from the relatively easy availability of extensive areas at comparatively low prices, at least initially, and a warm climate with clear skies, ample sunlight and largely unused solar energy. The most obvious negative features are poor distribution or lack of water. Abundant groundwater may be available in some areas but with small natural recharge and high evaporation, this raises the problems of exploiting a non-renewable resource. Other problems in development of ASALs may arise from the sudden introduction of modern technology to societies unprepared for it. For instance traditional water extraction from the ground has been based on manual methods using limited power. The rate of withdrawal of groundwater and natural recharge was therefore achieved without substantial use of groundwater storage and field efficiency was relatively high. Electrical or engine- driven pumps have greatly increased the capacity of the wells. The absence of effective controls on drilling and pumping has resulted in groundwater harvesting with the consequences of decreasing yields, deteriorating water quality, and wastage of water. On the other hand, climate change which has set in is even complicating if not worsening the very existence of an already vulnerable and fragile ecosystem. The difference between climate variability and climatic change is a delicate point that arouses much debate. However, in the real world, what matters is the fact that the climate is indeed never static and that a study of the past is still the best basis on which to plan for future. This is because the prediction of the climate is too uncertain. In fact even a small "change" of the climate, such as an increase in the rainfall variability about the mean even if the mean remains unchanged can be sufficient to tip the balance against- or for- say the cultivation of a certain crop in an area where rainfall is marginal. Tough irrigation provides a form of risk insurance, the effects of even a small climatic "change" may be significant too. Surface run-off in rivers and recharge to groundwater represent a small residual between two much larger terms in the hydrologic equation, rainfall and evaporation. It therefore follows that a small change in one of these factors will have a significantly greater effect on water availability in the ASALs ecosystems.

Water Resource Management in the ASALs

Water is a basic need to all socio-economic developments and for maintaining healthy ecosystems. However, water availability is increasingly being limited by an ever-increasing demand, profligate use, and increased incidences of pollution. There are several indicators of water resource stress, including the amount of water available per person and the ratio of the volume of water withdrawn to the volume of water potentially available. When withdrawals are between 20% and 40% of the total renewable resources, water stress is acknowledged to be a limiting factor on development. Withdrawals of 40% or more represent high stress. In 1990, approximately one third of the world's population lived in countries using more than 20% of their water resources, and by 2025 around 60% of a rather larger total, would be living in such stressed countries. Water stress is widely acknowledged to be a problem if a country or region has less than 1700 cubic metres /capita/year(source) Simple numerical indices, however, give only partial indications of water resources pressures in a country or region, as the consequences of "water stress" depend on how the water is managed. About 80% of Kenya is classified as arid to semi-arid (Gok 2015). The discussion in this chapter therefore, though specific for Kenya with regard to water issues represents issues relevant to many ASALs of the world.

Limited Water Resources

Kenya is classified as a chronically water-scarce country. The country's natural endowment of freshwater is limited by an annual renewable freshwater supply of only 647 cubic meters per capita. Globally, a country is categorized as "water-stressed" if its annual renewable freshwater supplies are between 1,000 and 1,700 cubic meters per capita and "water-scarce" if its renewable freshwater supplies are less than 1,000 cubic meters per capita (World Bank 2000). Only 8.3 percent of the countries in the world are classified as water-scarce, while 9.8 percent of the countries are considered water-stressed. By comparison, Kenya's neighbors, Uganda and Tanzania have annual per capita renewable freshwater supplies of 2,940 and 2,696 cubic meters per capita per year respectively.

The availability of water is often a key factor in determining the patterns of human settlement and the value of land for agricultural and livestock production. Within arid and semi-arid lands (ASALs), the food security of pastoral and farm households improves considerably during the wet years. Improved grazing fundamentals in several pastoral areas has resulted in favorable livestock body conditions, increased calving rates, and improved milk output—together bringing market improvements in pastoralist food security.However, the limited endowment of water resources places an added financial burden on the population of Kenya compared to other countries. This burden is even greater because of the country's vulnerability to rainfall variability. In these circumstances, water ought to be treated and managed as a scarce resource with real economic, social, ecological, and political values.

Underdevelopment of Water Resources

Although Kenya has one of the lowest water endowments in the world, it still possesses adequate total stocks of surface and groundwater, although these resources do not always coincide geographically with demand. However, the country has not invested adequately in the water storage capacity needed to exploit these stocks, as called for by the 1992 National Water Master Plan and the subsequent AfterCare Study Report of 1998a. In fact, even the 1992 infrastructure has not been properly maintained in the face of damage from the 1997/98 El Niño floods, siltation of storage facilities, and general depreciation. The consequence is that, given both population growth and the depreciating asset base, the maximum per capita storage of surface water for water supply has continued to decline (Figure 1).

This very low level of per capita storage reduces the reliability of supply and increases the vulnerability of supply to minor variations in rainfall. The uses of surface and groundwater resources differ. Development of groundwater resources is preferable in areas where no or very little surface water is available, such as the arid and semi-arid lands (ASALs). In these regions, groundwater is almost the only resource available for dryland agriculture and grazing during long periods of the year, and so is the cornerstone for socioeconomic development.In other regions, such as Nairobi and other cities and towns, development of surface water resources is preferable because of the volumes required and the limited recharge rates of groundwater. Development of surface water resources requires construction of storage areas to capture both storm flows and regular river base flows. Development of groundwater calls for resource assessments, determination of the safe yield levels, monitoring of resource quality, and investment in appropriate water extraction technologies.

Figure 1. Per capita water availability in Kenya
(WARMA, 2009)

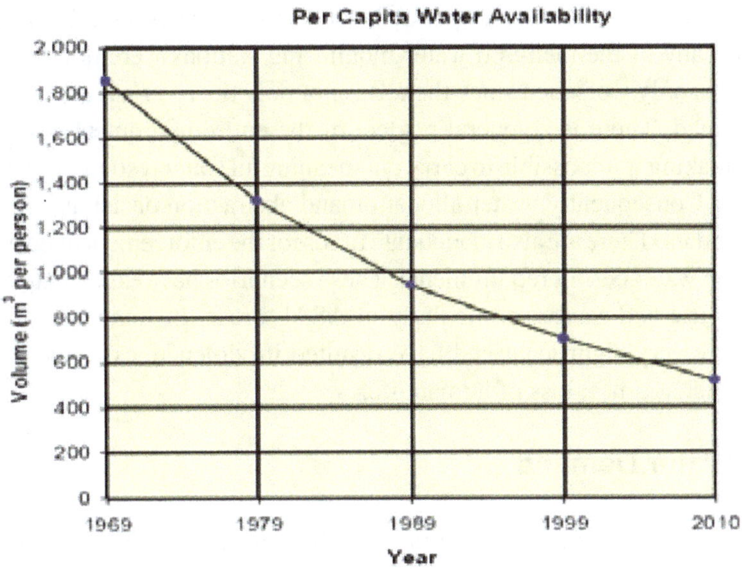

Management of a Limited Water Resource Base

Kenya is classified by the U.N. as a chronically water-scarce country (Republic of Kenya 1992). The country's natural endowment of freshwater is highly limited, with an annual renewable freshwater supply of about 647 cubic meters per capita, significantly below the 1,000 m^3 per capita set as the marker for water scarcity. The current level of development of water resources in Kenya is very low. Only 15 percent of the safe yield of renewable freshwater resources has been developed currently.The low level of development means that water supply storage per capita has declined dramatically from 11.4 m^3 in 1969 to about 4.3 m^3 in 1999—simply because of population growth.

Deteriorating Management of Water Resources

Government investment dominates surface water development and management. However, the private sector (for example, the hotel industry) invested heavily in groundwater development during the 1999/2000 drought. Given the decline in government investment and limited funds available, there are opportunities for more private sector investment in water resources development and management. for example, in the water resources assessment area.Flow gauging is vital for sustainable and equitable management, but has been chronically underfunded, with equipment falling into disrepair and flows not always being recorded even when monitoring stations are operational. It was pointed out in the 1992 National Water Master Plan that many of the then-900 water-monitoring stations were not working. The situation has only worsened since then. It is not only the river gauging stations that are debilitated. There is a general neglect of the entire hydrometeorological data network, making it impossible to carry out meaningful water resources planning and operations. Consequently, water allocation and abstraction decisions are based on inadequate data. There are also inadequate funds for the enforcement of decisions and to follow up water permit requirements. These scenarios have exacerbated conflicts over water use and access, particularly in ASAL areas. In some cases, water use conflicts are very significant and have resulted in violence, extensive damage to property and, at times, loss of human life.

Future Water Demand

Based on the national development target in Kenya Vision 2030 and the country's socioeconomic frameworks, the water demand required for water resources development and management planning have been estimated by JICA team for the domestic, industrial, irrigation, livestock, wildlife and inland fisheries water use. The present and future water demands estimated and projected are given in the table1 below, but the future water demands are projected including irrigation water demand for new development area of 1.2 million ha targeted in Kenya Vision 2030 without benefit of any water balance study.

Table 1. Present and Future Water Demand by Subsector (JICA, 2013)

Subsector	2010 (a)	2030 (b)	(b)/(a) (%)	2050 (c	(c)/(a) (%)
Domestic	1,186	2,561	216	3,657	308
Industrial	125	280	224	613	490
Irrigation	1,602	18,048	1,127	18,048	1,127
Livestock	255	497	195	710	278
Wildlife	8	8	100	8	100
Fisheries	42	74	176	105	250
Total	3,218	21,468	667	23,141	719

(Unit: MCM/year)

Table 1 shows the seriously tight situation of water balance at present due to large water use as demonstrated by high water demand/water resource ratio. In the target year 2030, water demand will most likely increase in all catchment areas, and water balance is expected to be even tighter in all areas owing to both population increases and increasing demand to meet the country's national development goals. Climate change, with its attendant impacts on water resources and the environment, is likely to further complicate matters and therefore needs to be factored at every stage of decision making.

EXTENT OF KENYA'S WATER VULNERABILITY TO CLIMATE VARIABILITY AND CHANGE

The vulnerability of a water resource system to climate change is a function of a number of physical features and social characteristics. The physical features, which are associated with maximum vulnerability of the water resources in a country, include:

1. A current hydrological and climatic regime that is marginal for agriculture and livestock;
2. Highly seasonal hydrology due either to seasonal precipitation or dependence on snowmelt;
3. High rates of sedimentation leading to reduction of reservoir storage;
4. Topography and land-use practices that promote soil erosion and flash flooding conditions;

On the other hand, the social characteristics that would increase vulnerability of water resources include:

- Poverty and low-income levels;
- Inadequate water control infrastructures;
- Inadequate maintenance and deterioration of existing infrastructure;
- Inadequate human capital skills for system planning and management;
- Lack of appropriate and empowered institutions;
- Absence of appropriate land-use planning and management;
- High population densities and other factors that inhibit population mobility;
- Increasing demand for water because of rapid population growth; and
- Unwillingness to live with some risks as a trade-off against more goods and services.

The degree of vulnerability varies from place to place because of the way the physical and social factors of vulnerability inherently combine in a given area to produce a unique degree of vulnerability over that region. In general, the surface water resources in the country are and will continue to be most vulnerable in the arid and semi-arid areas, while ground water resources in the coastal regions will be most vulnerable in the future. Kenya's water vulnerability therefore arises from a combination of the country's very limited natural endowment of water, the high variability with which annual rainfall occurs, the heavy dependence of the economy on water resources, and inadequate preparedness for regularly recurring climate shocks to the economy. The agriculture sector—which accounts for about 25 percent of GDP, 60 percent of foreign exchange earnings, and 70 percent of total employment—is particularly sensitive to climate changes. In 1999, for example, the country realized only 69 percent of the expected maize harvest (a total of 2.25 million tons were harvested) mainly because of the failure of the rains, particularly in the Eastern, Central, and Northern Provinces (UNEP/GOK 2000). There was a further 12 percent decline in the maize harvest (to 2 million bags) in 2000, resulting in the importation of 409,000 tons of maize valued at Ksh 4,700 million ($59 million) for relief and commercial purposes (Republic of Kenya 2001b). The failure to develop surface and groundwater resources has exacerbated the country's vulnerability. Consequently, there is very little stored water per capita, so when severe droughts occur, water storage areas are rapidly drawn down. Those dependent on surface water resources very quickly experience water shortages.

The country's vulnerability is exacerbated by the extensive degradation of the nation's water resources and weak water resource management. In the last decade, the country has severely reduced its expenditure on water resources operations and maintenance. It is widely accepted that good catchment and aquifer management influences the quality and reliability of water resources (UNEP/GOK 2000). Thus, heavy rains cause erosion from the cleared forests and poorly-maintained agricultural lands, leading to accelerated siltation and loss of storage capacity in the country's water storage dams and pans. Noticeable population increases have occurred in the ASALs as a result of migration. This has affected the ecosystems of these regions, making them more vulnerable to disasters such as drought and environmental degradation (Glantz 2001). Many dams and reservoirs silt up years before their design life. Of the estimated 3,200 dams and pans nationwide, between 80 percent and 90 percent have lost at least 50 percent of their expected economic life from siltation (source). This depletion in volumetric capacity arises from both siltation and lack of investment in new infrastructure. A variety of strategies can be used to optimize water resources and reduce vulnerability, including the protection of groundwater recharge areas, siting and operation of boreholes, protection against groundwater pollution, and the conjunctive use of surface and groundwater.

A classical example is that the mean annual discharge of the Maragua River declined by 13 percent over 38 years and the Molo River declined by 26 percent over 27 years. While there were probably a number of factors involved, excessive abstractions almost certainly played a major role. In another well known example, the median decadal flow of the EwasoNyiro North at Archer's Post during the month of February had dropped from 9 m^3/sec in the 1960s to about 4.5 m^3/sec in the 1970s, 1.2 m^3/sec in the 1980s, and 0.9 m^3/sec in the 1990s (Gichuki eta l, 1998). In 1984, 1986, 1991, 1994, 1997, and 2000, the river dried up completely for a 60 kilometer stretch upstream just above Buffalo Springs, an event that the communities along the river and recent studies both note to be unusual (Republic of Kenya 2001f).

Similarly, there was an evident increase of water abstraction from the river during both flood and drought periods (Figure 2). The EwasoNyiro basin cuts through Archer's Post, Chanler'sFalls, MalkoBulfayo, Merti, Sericho, and Habaswein. The river flow reached Habaswein (which is 250 km from Archer's Post) during the early 1900s. Between the 1950s and 1970s, the river flow reached Sericho (200 km from Archer's Post) even during the driest period, but dried up between Sericho and Habaswein. In the 1980s and 1990s, the river frequently dried up between Archer's Post and Merti, which is 123 km from Archer's Post. More than 65 percent of water abstractions on this river are considered unauthorized. The effect of illegal abstractions on downstream water users has serious implications. Although this report concentrates on the impacts and costs arising from droughts and floods, periods of excess rainfall also bring long-term benefits by recharging aquifers and filling storages. The natural recharge of aquifers can be augmented during floods by either injecting floodwaters into the aquifer or holding the floodwaters into areas for long periods to allow extended natural recharge to occur. These techniques, called Aquifer Storage and Recovery (ASR), are being tested and implemented around the world. However, in spite of its wet-dry cycles, Kenya does not have any trials of ASR techniques under way

Despite the established pattern of droughts and floods in Kenya, the response has been wanting. A coordinated response to water and sanitation issues during droughts has now been developed with institutional structures and arrangements to mitigate the negative impacts of climate variability or to take advantage of the positive elements of climate variability—for example, recharging groundwater reserves during El Niño events. However, inadequate human and financial resources limit the country's responsiveness to rainfall variability. The country is in the process of preparing a National Disaster Management Strategy that "recalls and learns from past events, in order to anticipate and reduce the impact of disasters" (Republic of Kenya 2001d).

All the above are indicators of the government's efforts to address water resources development and management. Other recent efforts include the new Water Act, the formation of the Ministry of Water and Irrigation, and the Water

Figure 2. Water abstraction rate in Upper Ewaso Nyiro North River
(Mogaka, etal,2006)

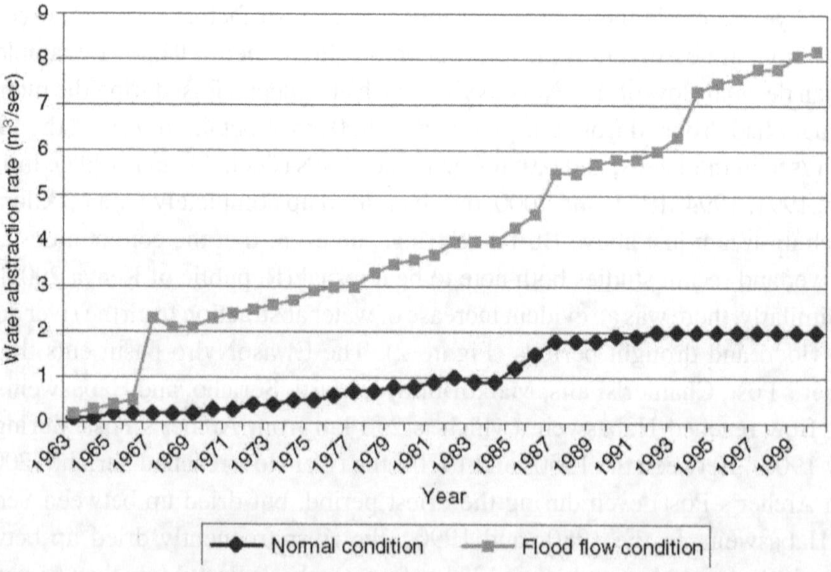

Resources Management Authority. Some regional development authorities (for example, EwasoNyiro North Development Authority, and Lake Basin Development Authority) have put in place strategies for effective water resources development and management, including catchment conservation. However, the government budget is inadequate to support effective implementation of the NWP, the Water Act, and the RDA's plans and strategies.

BEST WATER MANAGEMENT PRACTICES

Best practices generally relate to adaptation in the water usage demand in the face of climate variability and change as well as population pressure and developmental issues. A wide range of adaptation techniques have been developed and applied in the water resources sector in different parts of the world (FAO, 1981). Broad distinctions can be made between "supply-side" adaptive techniques (changing structures, operating rules and institutional arrangements) and "demand side" techniques (which change the demand for water or protection against risk and include institutional changes too). Examples of supply-side adaptations include increasing flood defences, building weirs and locks to manage water levels for navigation, and modifying or extending infrastructure to collect and distribute water to consumers.

Demand-side techniques include water demand management (such as encouraging water-efficient irrigation and also through water pricing). Distinctions can also be made between anticipatory and reactive actions. The former is taken in advance of some change, while the latter is a response to a change. Reactive actions include both short-term operational adaptations, such as the temporary exploitation of new sources and longer-term measures. A major flood or drought, for example, often triggers a change in water management. However, whilst many reactive options do exist in Kenya, anticipatory options have not been implemented satisfactorily in many situations. The ability to adapt to climate variability and climate change (anticipatory options) is affected by a range of institutional, technological and cultural features at the international, national, regional and local levels in addition to specific dimensions of the change being experienced. Among the most important features are the following:

1. The capacity of water-related institutions to act, skilled personnel, the capability and authority to consider a wide range of alternatives (including but not limited to supply-side and demand-side interventions) in adapting to changed conditions, the capability and authority to use multi-objective planning and evaluation procedures in the assessment of policy alternatives, procedures for conflict resolution, and the incentives to undertake serious post analysis of policies and projects to learn what has and has not really worked in the past.
2. The legal framework for water administration that always constrains, for better and for worse, the options that is open to water management. Naturally, laws change as needs change, but the changes is slow and greatly lag behind changing needs. In Kenya, the legal framework for water management is moving in a direction towards increasing environmental protection. Such a direction on the one hand poses further constraints on options to address climate change, but if the move reflects an increasing concern with sustainable water management, then the opportunities for considering adaptation to climate change are increased.
3. The wealth of nations in terms of natural resources and ecosystems, human-made capital (especially in the form of water control systems), and human capital (including trained personnel) that determines what nations can "afford to commit" to adaptation. This should include the ability and willingness to transfer wealth among population groups and regions within a country and among nations. This is the major constraint on adaptation to climate change in poor countries such as Kenya.
4. The state of technology and the framework for the dissemination (or monopolisation) of technology, especially in the fields of bioengineering of drought resistant and salt resistant varieties of plants and techniques for the desalination of seawater.

5. The mobility of human populations to change residential and work locations in response to severe climate events or climate change.
6. The speedof climate change is crucial in determining the capabilities of societies to adapt and change water management practices. Speed of change and the cumulative extent of change affect the impacts on society in non-linear fashions.
7. The complexity of management arrangements may also be a factor in response. In principle, the fewer the number of agencies involved in water management, the easier it will be to implement adaptation strategies. If there are many stakeholders to involve - perhaps with conflicting requirements, management goals and perceptions, and each with some management control over part of the water system - then it may be more difficult to adapt to changing circumstances. There is evidence that in some mature infrastructure systems there may be substantial opportunities for increasing the resilience of water resource systems through institutional changes as well.

The management of existing water supply systems is constrained by many factors especially,

- Shortage of funds for development, operation and maintenance of water supplies and management of water resources.
- Institutional weaknesses, especially the scarcity of qualified manpower and inadequate skills among users to properly operate and maintain water supplies.
- Institutional weaknesses to deal with climate extremes.
- Water resources unavailability due to its uneven distribution both in time and space.
- Poor choice of technology in water supply and sewerage development, and inconsistent project selection criteria which has resulted in adoption of technologies and delivery mechanisms which are not well suited for the sector development.
- Inadequate co-ordination of the various actors in the sector.
- Inadequate inter-linkages with other water related sectors.

The water-sector policy as outlined in the Session Paper No. 1 of 1999 does not adequately incorporate issues that are important for adaptation to climate change. The following are some of the factors, which need to be considered and probably added into the existing water policy for ease of adaptation to climate change.

1. Changing operating rules for existing structures and systems.

2. Poverty eradication measures.
3. Increasing protection against drought and flood hazards through: Continuous maintenance of infrastructure;
    a. Introduction of an effective flood and drought disaster management strategies;
    b. Limitations on sand harvesting in river channels;
    c. Introduction of a settlement and land tenure policy in coastal zones 1, 2 and 3; near river banks; around major reservoirs and forests and other state reserves; and
    d. Investing in alternative power sources.
4. Introduction and implementation of policies to encourage the use of groundwater through:
    a. Subsidies for community efforts to sink boreholes;
    b. Financing of research into cheaper bore-sinking technologies; and
    c. Enhancing groundwater protection.
5. Development of cheap technologies for desalination of seawater for use in the coastal areas especially for ships, sanitation (hotels, domestic) and industries.
6. Review the land tenure policies to encourage large-scale land developments and discourage land subdivisions which often lead to migration of people and the creation of sedentary settlements and lifestyles based on poor land use and management practices which in turn adversely affect the hydrological systems.
7. Develop and implement strategies which would ensures that small water supplies projects (including low-cost rainwater harvesting) are distributed uniformly over wide areas and therefore spreads out the settlements against the creation of large water supply projects, which encourage dense settlements around such projects.
8. The capacity of water-related institutions must have the capability and authority to consider a wide range of alternatives that include not only the supply and demand interventions but also the adaptation to climate changed conditions.
9. Sustainable development and expansion of urban centres.
10. Public awareness to prepare communities to learn how to live in a sustainable manner with any uncontrolled risks arising from climate change.

In recognition of the limitations posed by the absence of the above factors in the water policy, the government has initiated efforts to revise the existing water policy. A new draft water policy includes options encompassing the above factors. The government has also initiated efforts to develop and institute a disaster policy, which includes, among others, disasters, which are either directly or indirectly related to the water resources in the country.

The implementation of the new water and disaster policies in Kenya will greatly ease the burden of the water managers to plan adaptation strategies to reduce the vulnerability of the water resources in the country as a result of climate change. However, whether these steps towards adaptation take place or not may be heavily influenced by the occurrence of extreme events in the future. Such events are often the catalyst for change, and may serve two roles. First, they may expose failings in the implementation strategies of both the water and disaster policies in relation to the water management system. The failings often reflect weaknesses in development control, planning guidance, public education and fiscal incentives. Secondly, they may raise the perception amongst decision-makers of the possibility and dangers of climate change - even if they cannot be directly attributed to climate change.

The government has also identified other factors, which are important in reducing vulnerability of water resources to climate change, and is putting in place the relevant measures to minimise the limitations due to these factors. These other factors include:

- **Data for Monitoring:** Adaptive water management needs reliable data for use in making decisions, calibrating models and developing projections for the future. These data should cover not just hydrological characteristics, but also indicators of water use.
- **Understanding Patterns of Variability:** An understanding of patterns of variability - and in particular the stability of a "baseline" climate - is important for medium-term water management. It is increasingly recognised that even in the absence of climate change, the recent past may not be a reliable guide to the hydrological resource base of the near future.

The impact of climate change in the water sector is a function of the biophysical change in water quantity and composition, the use to which the water is put, and the way in which those uses are managed. The implications of climate change on water resources must therefore be seen in the context of the many other pressures on water resources and their management. These pressures, and the management responses to them, are evolving rapidly, and the water management system (infrastructural, legal and institutional) in the future may be very different from that pertaining today. Considerable efforts are under way in many international agencies and organisations to improve the way water is used and managed, and these actions will have very significant consequences not only for economies, access to safe water and the environment, but also for the impacts of climate change. Adaptation to climate change in the water sector must be seen in the context of these other changes and, of course, climate change must be considered as a factor in the development of improved management techniques. These other sectors are: ecosystems, coastal and marine zones, settlements, financial services, health, energy, and water and conflict.

WATER DEVELOPMENT POLICY INITIATIVES

The vision for the Kenya Water Sector is that of achieving sustainable development and management of the country's water resource as a basis for poverty reduction and promotion of socioeconomic development (Republic of Kenya 1999). Key national water policy objectives are to:

- Conserve and protect available water resources,
- Apportion water resources in a sustainable, rational, and economical way,
- Supply adequate and quality water to meet various needs,
- Ensure safe disposal of wastewater to safeguard ecological and environmental processes,
- Establish an efficient and effective institutional, policy, and legal framework to achieve systematic development and management of water resources, and
- Develop a sound and sustainable financing mechanism for effective water resources management, supply, and sanitation systems

In recent years, the Government of Kenya has been developing a water supply and sanitation sector strategy and an integrated water resources management strategy as part of the overall country's efforts to support economic growth and reforms in the water sector. The 1992 National Water Master Plan, which called for major investments in infrastructure and administrative capacity, was accepted in principle by the government and regarded as an important guide to the country's water resource development and management. The 1999 National Policy on Water Management and Development is now being implemented through both legislative changes and the development of a comprehensive Country Strategy on Integrated Water Resources Management.

The Regional Development Authorities also have developed their regional development strategies, which recognize the importance of water resources in supporting regional economic activities.All these documents recognize the need for adequate water resources to meet both domestic and industrial water needs. In both the Eighth and Ninth National Development Plans and now the Vision 2030 (1997–2001; 2002–08, Vision 2030), the government acknowledges the scale of water resource threats. It cites the degradation of water catchment areas as one of the main challenges confronting sustainable water resource management efforts (Republic of Kenya 1997). Other relevant and recent initiatives that underscore the inadequacy of water resources development, as well as critical constraints to development and the importance of sound water management in the national economy, include the 9th National Development Plan, Kenya Rural Development Strategy Paper, the National Health Sector Strategic Plan 1999–2004, the Sessional

Paper on Environment and Development and; the Poverty Reduction Strategy Paper. Kenya has signed and ratified a number of international and regional conventions, treaties, and agreements with significant components on sustainable water resources conservation, development, and management. The government has, for example, ratified the Convention on Biological Diversity (CBD), whose key objectives include sustainable conservation of biodiversity (including important water catchment areas), as well as the United Nations Convention to Combat Desertification (UNCCD). Achieving the objectives of the latter involves long-term integrated strategies that focus on improved productivity of land and water resources

CONCLUSION AND RECOMMENDATIONS

Kenya has a wide spectrum of the water resources base, which consists of both surface and groundwater resources (fossil and rechargeable). The terrestrial-surface-water resources systems are strongly influenced by rainfall. Most projections of future climate change in Kenya indicate that rainfall decreases in the arid and semi-arid areas. All these imply that, terrestrial surface water resources are very vulnerable to the impacts of climate change. Further, the surface water resources in the country have been utilized differently and at various degrees of exploitation. In the arid and semi-arid areas, estimates show that more than 40% of the existing surface water resources are being used for different purposes, but mainly in horticulture. This implies that the current surface water resources are stressed and in fact a water shortage exists in the arid and semi-arid areas. This is compounded by factors related to population pressure and migration, poor resource use and management and other socio-economic factors mainly poverty and inappropriate land tenure practices. The factors increase the vulnerability of water to the impacts of climate change. The government is currently reviewing not only the Water Act but also the water and disaster policies. This review when completed and implemented will form important tools for reducing the vulnerability of the surface water resources.

On the other hand, the groundwater resources in Kenya also vary spatially, because of the complex geology in the different parts of the country. Nevertheless, this resource is sufficiently abundant in many parts of the country. However, groundwater resources are currently under-utilised. Perhaps, this is due to the fact that the medium to large-scale groundwater projects requires considerable investments. Thus, most communities, groups and individuals prefer to develop and utilise surface water resources for domestic, irrigation and livestock farming requirements to groundwater. This is unfortunate because groundwater resources are less sensitive to climate variability than surface water resources. It is anticipated that the impacts of climate change on the groundwater resources will be more gradual than in the case with surface water resources. This is a good attribute of the ground water resources in terms of adaptation to the impacts, which may be caused by climate change. The government is therefore encouraging individuals, groups and communities, through development of appropriate policies and provision of financial assistance to utilise groundwater more substantively but simultaneously, wherever possible, with surface water resources. However, great care should be exercised in the exploitation of the groundwater resources in the arid and semi-arid areas where the recharge processes may be negatively impacted on by climate change and also in coastal areas where salt intrusion, compounded by sea-level rise could be a serious set-back. The government is also reviewing the settlement and land tenure policies to discourage migration and settlements in environmentally fragile areas.

The study makes the following recommendations

Development of Adequate Water Storage

Conjunctive use of surface water and groundwater banking, water reuse, and water demand management are all mechanisms for improving the efficiencyof water use, and should be considered along with investment in new sources of water. Good catchment management must always accompany any developments if these assets are to be fully utilized. Groundwater is an especially important resource for large portions of Kenya. Key areas of improvement here include understanding of the geology and hydrology of specific areas with high potential prior to drilling; regular monitoring and evaluation of water levels, water discharge, and water quality; establishment and enforcement of standards; and development of groundwater programs that are based on known recharge rates.

Raising Awareness of Water Resources Management Challenges

Efforts to develop and launch a National Water Campaign must be supported to highlight the problems facing the country, to elevate water resources management as akey national priority, and to raise awareness about individual and collective responsibilities as stewards of the nation's water resources. A longer-term goal of the National Water Campaign will be to initiate actions to shared experiences and knowledge and inform water users about practical solutions, and promote water-wise behaviours. The campaign should be targeted to water users such as farmers,

Industrial companies, and urban residents, and should seek to include the private sector and NGOs in supporting the reforms. The younger members of society could be sensitized on the importance of preserving the nation's scarce water resources by building on existing programs such as the UNICEF water and sanitation "WASH" program in schools.

Financing Water Resources Management

If water resources management reforms are to be successful, it will be essential that water is treated as a social and economic good and financial practices are realigned accordingly. Sound and fair financial management, with a goal of cost-recovery and financial sustainability is needed to improve the efficiency of services, provide additional resources for reinvestment, encourage demand management, promote catchment management, and pollution prevention and control. This should, however, be undertaken in a manner that protects the needs of the poorest citizens

Catchment Management

According to the Kenya Government Sector plan for Environment, Water and Sanitation (2008), the Formulation of guidelines and code of practice for groundwater investigation, drilling and test pumping, construction of dams and boreholes with long-life span, protection of water catchment areas and ground water artificial recharge in five drainage basins is aimed at improving water catchment management.

Consequently, in pursuant to Section 15 of the Water Act 2002, the Water Resources Management Authority, in consultation with the public, is responsible for the formulation of Catchment Management Strategies (CMS) for the management, use, development, conservation, protection and control of water resources within each of the catchment areas in Kenya. The CMS is a tool that describes the framework for the management of the water resources and related land resources in the catchment and it outlines how the concept of Integrated Water Resources Management (IWRM) can be implemented at the catchment level. This is because catchment degradation is one of the most serious water resources threats facing Kenya. It is a lose-lose situation for all concerned. Not only does soil erosion and excessive use of agro-chemicals represent a loss for the landowner, but it also causes large costs to downstream water users. There is good evidence from Kenya, as well as the rest of the world, that most landowners want to manage their properties sustainably if they can generate an adequate income and have access to the technologies and knowledge required. Catchment management groups consisting of upstream and downstream stakeholders, assisted by government agency staff, are envisaged.

Good catchment management is so central to many of the issues including siltation of dams, nutrient pollution of Lakes and rivers through both river-borne and atmospheric pathways, and potential loss of high-quality water supply from cleared forests. However, other reforms, such as in land tenure, may also be needed to ensure that responsibility is clear for land management decisions.

Enforcement of Regulations

In many cases the appropriate legislation and regulations are in place, but water resources degradation occurs because of a lack of enforcement. This is partly because of a lack of funding (for example, it is difficult for regional staff to travel to check for compliance with regulations), partly because of a lack of information and data. Stakeholders have little incentive to comply with the regulations when the procedures for granting permits and licenses are slow and cumbersome. More effective controlof abstractions and pollution will safeguard water quality and sustainability. Perhaps the most serious example is the large number of illegal water abstractions on the arid and semi-arid zone rivers which have led to severe conflicts between upstream and downstream water users.

Development of an Early Warning System

It is widely accepted that Kenya was unprepared for the impacts of both the recent El Niño and La Niña events (UNEP/GOK 2000; Glantz 2001). However, there is good institutional capacity in the country to anticipate and respond to these extreme events. The Meteorological Department provides seasonal weather forecasts which are used by the relevant institutions in an early warning system. However, farmers are sceptical of these forecasts because they are often non-specific and hard to understand (UNEP/GOK 2000). More can be done to make these forecasts accessible and believable at the local level. When warnings are received, typically six months in advance, the government can make preparations for the floods and droughts so as to minimize their impacts. Lessons can be learned from the preparations of other countries during the 1997/98 El Niño event (Glantz 2001). Studies in the developed world show that benefits from ENSO forecasts can range upto $2.60 per ha (Hill and others 1998). Even though there is good evidence that these benefits cannot be realized without infrastructure and knowledge, unfortunately which is inadequate in the developing world, there is also evidence that large numbers of lives and wholesale destruction of property can be prevented in the developing world with good warning (VanAalst and others 2000). In addition, the actions of government agencies can be better coordinated during both floods and droughts so that the impacts, particularly from droughts, can be anticipated and dealt with across the different sectors of government. Flood management strategies and plans can help reduce the dangers to people living in floodplains and identify risks associated with infrastructure in floodplains. Flood risks can also be reduced with a functioning flood forecasting and warning system and watershed management programs.

Establishment of a Monitoring System

Water resources assessment, particularly monitoring the extent and quality of the nation's water resources is extremely important. Without an adequate knowledge of climate, water use, and sources of pollution, it is impossible to make wise decisions on water allocation, enforce pollution controls, plan and design water resources infrastructure, and target investments in better catchment management. Monitoring also provides the means to detect water quality and flow problems at an early stage, thereby allowing management interventions to occur before the problems grow to the size of some of those documented here. Water resource assessment involves not just the physical infrastructure for rainfall and Riverflow gauging and sampling and ground water level monitoring, the laboratory analysisof the samples taken, and the recording of the results in a widely accessible computer database. It also should involve investment into studies that link and establish rainfall – runoff relationships and the physical, chemical, and ecological processes linking the cause of degradation or pollution and the effects.

A well-designed hydrological and water quality monitoring network and systematic water resources assessment is essential and some monitoring stations have been in existence for a while. However many have now fallen into disrepair and require replacement, while others needto be rehabilitated. A comprehensive hydro-meteorological system is essential for making and enforcing water allocation decisions at the catchment level and could be funded, at least partially, through water user charges.

Building Capacity for Effective Water Resources Management

Knowledge and information alone are not enough to bring about changes. Enabling and empowering individuals to use that knowledge will require training and capacity building. Institutions involved in water resources management will have to emphasize interdisciplinary approaches to water resources management, and generate and disseminate knowledge on good practices. Training in water law, conflict resolution and mediation, water resources economics, environmental planning and management, and financial administration maybe required so that the community and water users possess the skills and knowledge to be confident and effective in their roles with regard to water resources management.

REFERENCES

FAO. (1981). *Arid zone Hydrology for Agricultural Development.* Irrigational Drainage Paper.

Gichuki, F., Liniger, H., MacMillan, L. C., Schwilch, G., & Gikonyo, J. K. (1998). Scarce Water: Exploring Resource Availability, Use and Improved Management. *Eastern and Southern Africa. The Geographical Journal, 8*, 15–27.

Glantz, M. (Ed.). (2001). *Once Burned, Twice Shy: Lessons Learned from the 1997-98 El Niño.* Tokyo: The United Nations University.

Government of Kenya (GoK). (2008). *Sector Plan for Environment, Water and Sanitation, 2008-2012*. Author.

Government of Kenya (GoK). (2015). *Second National Communication to the UNFCCC*. Author.

Hill, S. (1998). The Value of El Niño Forecast Methods for the U.S. Winter Wheat Producer: Do They Differ? *Proceedings of the American Agricultural Economic Association (AAEA)*.

JICA. (2013). *Project on the Development of National Water Mater Plan 2030 for Kenya. Nippon Koei Co.* Ltd.

Mogaka, H., Gichere, S., Davis, R., & Hirji, R. (2006). *Climate variability and water resources degradation in Kenya: Improving water resources development and management*. World Bank paper No.69.

Republic of Kenya. (1992). *The Study on the National Water Master Plan*. Nairobi Climate Variability and Water Resources Degradation in Kenya.

Republic of Kenya. (1997). *National Development Plan*. Author.

Republic of Kenya. (1998a). *The Aftercare Study on the National Water Master Plan*. Tokyo: Nippon Koei Co., Ltd.

Republic of Kenya. (2001a). Economic Survey 2001. Nairobi: Author.

Republic of Kenya. (2001b). *National Disaster Management Policy*. Nairobi: Draft.

Republic of Kenya. (2001c). *EwasoNyiro North River Catchment Conservation and Water Resources Management Study*. Tokyo: Nippon Koei Co. Ltd.

Republic of Kenya. (2002). *National Development Plan 2002–2008*. Author.

UNEP/Government of Kenya. (2000). *Devastating Drought in Kenya: Environmental Impacts and Responses*. Nairobi: United Nations Environment Programme.

Van Aalst, M. K., Fankhauser, S., Kane, S. M., & Sponberg, K. (2000). *Climate Informationand Forecasting for Development. Paper No. 79*. Washington, DC: Climate Change Series, The World Bank Environment Department.

Water Resources Management Authority. (2009). *Integrated Water Resources Management and Water Efficiency Plan for Kenya*. Author.

World Bank. (2000). World Development Report 2000/2001: Attacking Poverty. Washington, DC: Author.

Chapter 12
Kenya Success Story in Water Resources Management:
Participatory Capacity Building in Integrated Watershed Management

Joy Apiyo Obando
Kenyatta University, Kenya

Anthony Ogutu Opiyo
South Eastern Kenya University, Kenya

Cush Ngonzo Luwesi
University of Kwango, DRC

Chris Shisanya
Kenyatta University, Kenya

Nele Förch
Water Sector Reform Program, Kenya

Gerd Förch
Independent Researcher, Kenya

ABSTRACT

Management of water resources is at the heart of political discourse to raise awareness among local stakeholders for support in policy formulation and implementation of water sector development plans. The concept of integrated water resources management (IWRM) has been largely disseminated by the Global Water Partnership. Though theoretically appealing and sound, the process of implementation of participatory water resources management still has potential to yield results at local level. One reason is that the top-down approach used is too broad to be implemented and neither does it facilitate better understanding of the needs of each sector involved in the inter-sectoral collaboration to foster planning and benefit sharing of water resources. It is in favor of such practical action for water sector planning and development at small-scale catchment level that the concept of "light" IWRM or integrated watershed management (IWM) was developed to reduce various threats and severe water constraints affecting local stakeholders.

DOI: 10.4018/978-1-7998-0163-4.ch012

INTRODUCTION

Water is the elixir of life and a principle indicator of sustainable development (Lal 2015). With the growing global population there are often outcries about the status of major river basins and lakes, which in turn poses serious water provision and availability issues leading to new development challenges. The management of water resources is hence at the heart of implementation of Sustainable Development Goals (SDGs) (UN 2016). In Sub Sahara Africa most countries could not adequately supply clean water to meet achieve the Millennium Development Goals (MDGs) on access to water and sanitation by 2015. Benjamin Franklin, one of the founding fathers of the United States and a renowned polymath of the 18th century was then right to say: "*Many people think that water comes from the tap in the same way milk comes from the cow… When the well is dry, we learn the worth of water*" (quote from Cyber-nook.com 2011). Savenije et al. (2014) relate this saying to water scarcity in the tropics. "In most climate zones", they record, "freshwater availability fluctuates with the seasons and is scarce during some months each year. Given the vital nature of water for humans, all societies located in such climate zones developed ways to arrange and secure access to water for domestic and productive uses. Those societies that survived over time found ways to use water in a sustainable manner, or at least allowed the water resource to regenerate itself and did not destroy the natural cycle".

There are various reasons to the ineffective water policies in Sub Sahara Africa. Some of them include weak institutional frameworks that lead to sectoral and fragmented approaches; inadequate water infrastructural development with competing interests, uses and conflicts; very little or no concern about the transboundary nature of water resources in public policies; and poor and unsustainable management schemes. Since most uses of water resources are interdependent, the problem of coordination of both supply and demands becomes very critical in most countries where the long history of water policy follows a unilateral ministerial sector based approach.

Most analysts and professionals would argue that water management has been sectoral and reductionist for too long, and that there is a need to better coordinate management of different components of the resource (e.g. groundwater and surface water), between various sectors and stakeholders, across links in the water chain (the pathway from drinking water supply to wastewater treatment) and across administrative boundaries. In that sense, Integrated Water (Resource) Management has been widely hailed among those working in water as a welcome aim or vision. Nonetheless, efforts are being made to identify and implement water management solutions through collaboration at national and regional levels to increase self-reliance, reduce conflicts and achieve socio-economic and environmental sustainability for improved livelihoods. Such an integrated approach is very crucial, since water as a resource cannot be served or transported in a vacuum, just like radio or television signals cannot be sensed without a device. Hence, the need to consider both the catchment area and its natural resources, water included, to understand the complexity of "Integrated Water Resources management" (IWRM).

IWRM AND IWM CONCEPTS

The development of the concept of IWRM is generally explained by an increasing recognition of the linkages between natural water systems and human development in order to strike a balance between fulfilling human needs and sustaining ecosystems rather than merely exploiting water to the full to maximize production (Rockström et al. 2009). In fact, prior to the 1950s food crises, water research was solely the field of hydrologists and engineers. The projected 1970s multifaceted economic crises brought about "water resources development" concept, then "water resources management" (WRM) in the 1980s and more popularly "integrated water resources management" (IWRM) in the 1990s with a handful of newly related paradigms such as "integrated watershed management (IWM)", "food-energy-water nexus (FEWN)", "water smart production (WSP)", "green water saving and management (GWSM)", "global water accounting (GWA)", "virtual water trade (VWT)", etc. (Savenije et al. 2014).

The history of Integrated Water Resources management (IWRM) can be traced back to the Tennessee Valley Authority in the 1930s (Molle 2008), but the more recent incarnation of the principles of IWRM is associated with the 1977 Mar del Plata UN Water Conference, and the 1992 Dublin Principles adopted during the 'Earth Summit' in Rio de Janeiro (ICWE 1992). These principles provide a framework for managing water both as key "resource" (that is the content of the water catchment) and a unique "ecosystem" (that is water catchment as the structural container of water and other related resources). Hence, IWRM seeks to operationalize these principles, that water is recognized as "a public good with both social and economic values" and its management done in a broad and holistic perspective with the appropriate involvement of users at all levels" (Lenton and Muller 2009). Due to the high degree of sector intersection around water, all IWRM approaches stress the need to improve efficiency in water use (the economic rationale), promote equity in access to water (the social or developmental rationale) and to achieve sustainability (the environmental rationale) (Box 1).

Ideally, IWRM approaches promote the coordinated development and management of water, land, and related resources, in order to maximize the resultant economic and social welfare in an equitable manner without compromising the sustainability of vital ecosystems as a major goal while this concept has been operationalized.

Box 1. Degree of inter-sectoral integration

Low mono-sectoral view: Focus on water as resource (with attention to externalities	**Partial sector integration**: Focus on water in connection with another resource sector (with attention to externalities)	**Higher sector integration**: Focus on water and multiple other resources (with attention to externalities)
➢ Integrated management of water quantity and water quality aspects in particular water courses ➢ Ensuring the conjunctive use of groundwater and surface water resources	➢ Land and water management in irrigation ➢ Wastewater reuse in agriculture ➢ Control of minimum flows in rivers for ecosystem protection and agricultural water supply	➢ Integrated resources management in a watershed ➢ River basin management ➢ Various control measures for ecosystem valuation and protection
European Union-Water Framework Directive (IHE Definition of IWRM Water management concept)	**CGIAR Water Management Concept** (IWRM for agricultural ecosystem management)	**GWP IWRM Definition** (IWRM for Integrated Water Resources Management) **IUCN Concept** (Ecosystem based approach)

Knowledge from various disciplines insights from different stakeholders to plan and implement efficient, equitable and sustainable water solutions. At the local level, IWRM approaches have been implemented in the country specific cases depending principally on the nature of water, services and occurring problem. Many countries in the region have implemented the Johannesburg Plan of 2005 World Summit, which involves creating IWRM and water services efficiency as an opportunity to take a coherent approach to improving how they develop, manage and use water resources to further sustainable development goals and meet development challenges. Some countries have chosen to de-concentrate water resources management services through zoning and others through decentralization as a strategy to focus and foster IWRM.

According to GWP (2015), stakeholder participation in water planning and management is enshrined in all water legislation across the East African region and is seen as the next step along the pathway to introducing the IWRM approach. All countries follow a catchment based planning approach, embark on stakeholder participation, and incorporate gender issues in water resources management. Indeed, participation can under certain conditions; help towards the aim of ensuring wider access to water, by allowing an individual or communal service to expand into neighborhoods that previously had none (Jaglin 2002). However, top-down approaches prevail when establishing the legal and regulatory set-up, where-as catchment based strategies involving local stakeholders and water resources users for sustainable management of watershed resources have been implemented on large scale in Kenya only.

In general, many countries within the region have established IWRM structures including legal as well as institutional; and continue formulating management plans at catchment level, while incorporating water management into current national development strategies. Regardless of the initial approach, strategies and plans are striving to go beyond the actions needed to solve current problems in water and move towards strategic Integrated Water Management. Implementation at lowest levels is in general hampered by lack of capacity and reliable data on water resources and their utilization.

Having all these considerations in mind, GWP (2000) provides a definition of IWRM:

Integrated Water Resources Management (IWRM) is a process that promotes the coordinated development and management of water, land and related resources, in order to maximize the resultant economic and social welfare in an equitable manner without compromising the sustainability of vital ecosystems.

It shall be recognized that integrated approaches to water resource management has regained a lot of attraction because of the food-energy-water nexus (FEWN) approach, which is an alternative or complementary approach to IWRM championed by the UN (FAO), the International Water Management Institute (IWMI) and other institutions focusing on different perspectives of the nexus between water and food or water and energy (Hoff 2011). Further philosophical divide has emerged between those who place yield gains as paramount in the context of population growth and, Sustainable Intensification (SI) and those who advocate a more agro-ecologically focused approach based on Sustainable Land Management (SLM) (Cassman 1999) or a more holistic approaches to ecosystem services (Dore et al. 2011).

Considering the holistic approach introduced by the European Union (EU) a whole set of specific definitions, objectives and constraints emerge from various aspects of the water management EC directive 2000/60 (Meire et al. 2008). The latter include aspects of water quality management, policies, economic aspects, ecology, price, and sustainable development). Whether driven by knowledge intensive management or the humanitarian need to alleviate hunger, a key difference between the three approaches is always the starting point for water resources management. When considering the inter-relationships between water, land, food and energy GWP focuses on the catchment area, while the EU and UN-FAO and IWMI base their consideration on water as a natural resource and water for agriculture, respectively (van der Bliek et al. 2014).

Thus, IWRM seeks first to ensure the *sustainability* of water resources; IWRM recognizes the *interdependency* between water resources and ecosystems as well as living creatures therein. It also integrates all types of water users and key stakeholders in water management and development for *equitable*distribution of the resource. Finally, it shall manage water resources as an economic asset by integrating all costs incurred by competing uses and interdependent interests to achieve *Efficiency*. With all these different objectives, IWRM can be challenging and unyielding for implementation on the ground. Concerns relating to the way the IWRM concept has been interpreted and implemented led researchers to invest in development of the concept of "light" IWRM (Moriarty et al. 2004) often referred to as "Integrated Watershed Management" (IWM) in North America (Förch, et al. (2005, 2006, 2007, 2008).

Implementing the concept of IWRM requires multidisciplinary efforts that encompass the social and natural sciences and engineering fields to generate integrated scientific methods and approaches and make good use of the new ICT tools to rapidly tackle some of the more complex and expanding water, land and ecosystems issues. Research in this area has seen considerable changes and developments over recent years. Disciplinary research into issues such as hydrological modelling, the engineering of irrigation systems, land and soil erosion and the distribution and species composition of ecosystems has continued with an emphasis on increasing productivity and sustainability while recent developments have seen increasing demands for minimizing social, economic and environmental trade-offs (Savenije et al. 2014). Whilst an attractive idea, this key principle of Integrated Water Resource Management (IWRM) has been widely adopted and applied at international and regional scales but with very little or no palpable results in most developing countries. In addition, different countries within the same region overcome many challenges and meet the water needs of people respective of their situation and development priorities in their countries, without considering the transboundary nature of their water catchments. Paradoxically, the IWRM concept has also been criticized for being: 1) Too broad to have any real meaning and 2) So narrow that it focuses mainly on water and ignores important linkages between land and water management (Merrey et al. 2005). In addition, some critics have argued that as a concept IWRM is naïve and idealistic (Biswas, 2004). Others have maintained that IWRM takes insufficient account of the politics that are at the core of most (if not all) important water-related decisions (Jensen, 2013). Molle (2008) concludes that it is often pretended that all these different objectives of IWRM can be maximized simultaneously, when in reality there will nearly always be trade-offs and, at best, only a balance can be achieved. Whilst the FEW nexus approach has significant merit, it has also attracted criticism for being unnecessarily limiting and prescriptive, for example, by not explicitly highlighting inter-linkages with climate change, poverty and pro-poor development (NSF 2016) [1].

However, there are quite a number of examples of it working in practice, especially at a river basin level (Meire et al. 2008). In order to maximize the contribution of IWRM to sustainable development and minimize conflicts among stakeholders, a "light" format of IWRM, often referred to as "Integrated Watershed Management" (IWM) in North America has been suggested and tested with success in India and East Africa (Moriarty et al. 2004).

Integrated Watershed Management (IWM) is the process of managing human activities and natural resources on a *watershed* basis. This approach allows us to protect important water resources, while at the same time addressing critical issues such as the current and future impacts of rapid growth and climate change. Watershed management implies a prudent use of the resources such as land, soil, water, biodiversity and biomass in a watershed for optimum productivity with minimal disturbance to the environment. Bamutaze et al. (2014) define "Integrated Watershed Management" (IWM) as follows:

Integrated Watershed Management is a holistic and integrated approach for sustainable management of a watershed area (...) Integrated Watershed Management is a process of rational decision making in successive steps. Systematically the available management options are compared, and a Watershed Management Plan is developed that is mainly a rural development concept.

Integrated management, as the authors of this paper understand it, aims to address trade-offs and minimize the negative impacts that might be created by the actions of one particular sub-sector, stakeholder or time, on others. It seeks to avoid inefficiencies and conflicts that are a feature of less-integrated approaches. Besides, in contrast to prescriptive top-down IWRM (Shah and van Koppen. 2006), IWM or Light IWRM aims to be problem-focused, opportunistic and adaptive/iterative when applying core IWRM principles, especially at the water-users level. The intended outcome of applying IWM is a system of managing water resources and water services delivery that has developed incrementally over many years and, as a result, is better adapted or tailored to the political economy of a given area.

The concept of IWM puts emphasis the principle of public participation to ensure inclusiveness and create interactions between the public and private sectors with the civil society, people of different genders and generations, urban locations and rural areas to achieve a sustainable management of water resources, as put forward by the Dublin Principles 2 and 3.

Principle 2: *Water development and management should be based on a participatory approach, involving users, planners and policy-makers at all levels.*

Principle 3: *Women play a central part in the provision, management and safeguarding of water.*

IWRM values promote gender mainstreaming and intergenerational equity to build strong partnerships in catchment management. A good case is given by the provisions made in the constitution of the water users' associations, which provides a fair share of 50% of female cadres in the management of the associations in a male dominant cultural setting. This recognition of the key role of women in the custody of water resources has led to the improvement of many catchment areas. Increasingly IWM approach is being implemented across the region by Alemu and Kidane (2014) for instance.

IWM thus implies a move away from traditional sub-sector foci that address domestic water supply, wastewater, irrigation, industry and the environment separately (often within different agencies or government departments) to a more holistic approach. All case studies in Kenya can be used to address problems both within the regions and specific water user communities across a wide scope of stakeholders in water resources development and management. Importantly, they can be used to address problems at the very local level with aggregated effects at the national, regional and even global levels in view of the preparation of IWRM specific country strategies and plans. There still remain the sense and need for inter-sectoral coordination and a solution to implement the approach at larger landscapes than a small catchment area shaped. Hence the need for national planning actions that involve bottom-up action plans.

The ideals of both IWM put emphasis on the holistic approach as captured in the following SDGs:

Goal 2: *End hunger, achieve food security and improved nutrition and promote sustainable agriculture.*

Goal 6: *Ensure availability and sustainable management of water and sanitation for all*

Goal 13: *Take urgent action to combat climate change and its impacts*

Goal 14: *Conserve and sustainably use the oceans, seas and marine resources for sustainable development*

Goal 15: *Protect, restore and promote sustainable use of terrestrial ecosystems, sustainably manage forests, combat desertification, and halt and reverse land degradation and halt biodiversity loss*

The value of integrated watershed management (IWM) focuses on identifying land or environmental resources such as water and socio-economic potential; for development and allowing the community and decision-makers to select management options with least negative impacts. At the strategic level, IWM calls for use of readily available scientific and other information such as political, cultural, economic and legal; relevant to the strategy or policy initiative at hand and that can be used to develop a risk management plan. The risk management plan describes the ways and means of coping with the predicted risks and managing these to acceptable levels.

Aims of the Study

This chapter provides a case study on the role played through the participatory process within IWM with the planning right from the lowest level at the sub-catchment level and eventually contributing to catchment management. The Integrated Watershed Management Network (IWMNet) led by Universität Siegen (Germany) and hosted by Kenyatta University (Kenya) contributed to this participatory processes, thereby catalyzing the sub catchment management plans (SCMPs) for management the water resources. Already, WRMA was mandated with formation of Water Resource User Associations (WRUAs) as stipulated in Section 15(5) of the Water Act 2002 (GoK 2002),

(...) the catchment management strategy shall encourage and facilitate the establishment and operation of WRUAs as fora for conflict resolution and co-operative management of water resources in catchment areas.

The chapter analyzes the opportunities and challenges resulting from the participatory planning, management, monitoring and evaluation process through "DAAD Alumni Summer Schools" held in Meru (Kenya) from 2006 to 2008 are also presented. A survey involving 68 (out of 100) participants of the above summer schools and a documentary review on the developments of the water sector in Kenya guided the study.

With the capacity acquired during the implementation of the Water Act 2002, water sector actors are more equipped to effect the revisions and amendments required for compliance to the devolved water governance system enshrined in the Constitution of Kenya 2010 (GoK 2010). The enactment and implementation of the new water rules by local stakeholders at county level as well will be guided by good practices and lessons learned from the IWM capacity building process. Prior to this review, let us revisit the concept of Integrated Watershed Management.

Application of Integrated Watershed Management in Kenya

In the Kenya, the enactment of the Water Act 2002 (GoK 2002) saw the start of the implementation of the water sector reforms (Box 2). These reforms were propelled

Box 2. Kenya response to water resource management in 2002

Achieving water sustainability requires integrated approaches towards catchment management (Förch et al., 2005; Cap-Net et al., 2007). The latter involves a strong cooperation between upstream and downstream stakeholders. Ngigi and Macharia (2007) report that this has never been a priority for the political class leading the water sector in Kenya from 1963 (independence) to 1997. In effect, the reforms initiated in 1965 targeted the improvement of water quality and quantity through adequate financing mechanisms (GoK, 1965). This was reiterated by the "Water for all by 2000" motto upheld by the 1974 National Water Master Plan (NWMP), which led to establishment of a national water development corporation in 1988. The first guidelines for community participation appear in 1997, when the government invited the private sector to take part to a decentralised form of water governance. These guidelines were formally released in 1999 as the National Policy on Water Resource Management and Development (GoK, 1999). They were enacted as law under the Water Act 2002 (GoK, 2002).

Figure 1 illustrates well the separation of roles and mandates introduced by the Water Act 2002:
(i) water resource management moved away from the supply of water and sewerage services;
(ii) water policy-making departed from its daily administration;
(iii) creation of a Water Services Regulatory Board (WASREB) alongside a Water Resources Management Authority (WRMA);
(iv) permitting of dozens of Water Service Providers (WSP) in major cities of the country and recognition of hundreds of Water Resources Users' Associations (WRUAs); and
(v) creation of Water Appeal Boards (WABs) for dispute arbitration.

The promulgation of a new national constitution in Kenya in 2010 led to a devolved system of water management, which draft policy and act are still pending for enactment (GoK, 2010; Nyongesa, 2013). In the year 2012 a new bill was introduced to parliament to align the provisions of the Water Act 2002 to the devolution enshrined in the Constitution of Kenya 2010 (GoK, 2010; 2014). This gave way to the development of the National Water Master Plan 2030 (JICA and GoK, 2013).

SOURCE: Luwesi, CN, Shisanya, CA and Obando, JA. (2012). *Warming and Greening: The Dilemma facing Green Water Economy under Changing Hydro-Climatic Conditions in Muooni Catchment (Machakos, Kenya).* Saarbrüken: Lambert Academic Publishing, pp. 20-22.

by the mounting pressure of water to support the development activities of a rapidly increasing population coupled with degradation of the water catchment areas (GoK 2013).

There were persistent problems in water resources management as well as service provision caused by centralized governance, the lack of investments and a weak enforcement of existing water laws. The Water Act provided for the formation of a Water Resources Management Authority (WRMA) in 2003 as the lead agency for national water resources management, which is responsible for the management of lakes, aquifers and rivers, among other functions. Thus the water resources management system was changed from administrative basis to the five main catchment areas (Figure 2). After the enactment of the Water Act 2002, the GoK has been implementing water sector reforms and several other developments have since taken place that contribute to the management of water resources. Indeed, within the devolved government, chapter 5 of the Constitution of Kenya (GoK 2010) stresses the need for catchment area conservation encompassing biodiversity and land, soil, forest and water resources. Furthermore, more recently, the National Water Master Plan (GoK 2013) and the Masterplan for the conservation and sustainable management

Figure 1. Legal framework under the Water Act 2002
(Source: GoK, 2002)

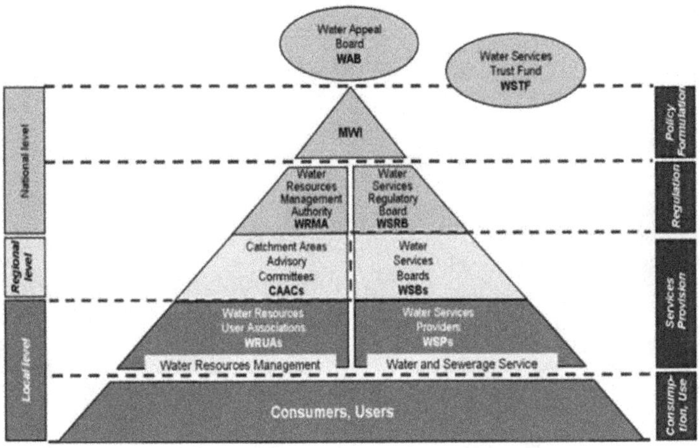

of water catchment areas in Kenya (GoK 2012) have also been informed by the work spearheaded by the participatory planning for sub-catchments. The Kenya Vision 2030 (GoK 2007) can only be achieved with proper implementation framework and planning of water resources so as to manage them sustainably within the scope of affordable water demands in tandem with the respect of natural ecosystems' thresholds while conserving the catchment.

LESSONS LEARNED FROM IWM CAPACITY BUILDING IN KENYA

Opportunities Arising From Participatory IWM Learning Process

Results show that through North-South cooperation, the Center for International Capacity Development (CICD) of the Universität Siegen established a network for Capacity Building on Integrated Watershed Management (IWMNet) in Eastern and Southern Africa. The latter encompassed several universities, including Kenyatta University (Kenya), Makerere University (Uganda) and the University of Dar es Salaam (Tanzania), to name but a few. This network has enhanced the teaching and research of IWRM and IWM programmes at postgraduate level using different media and options, including course works (in situ and online), fieldworks and summer schools.

Figure 2. Five water catchment areas of Kenya
(Source: WRMA 2010)

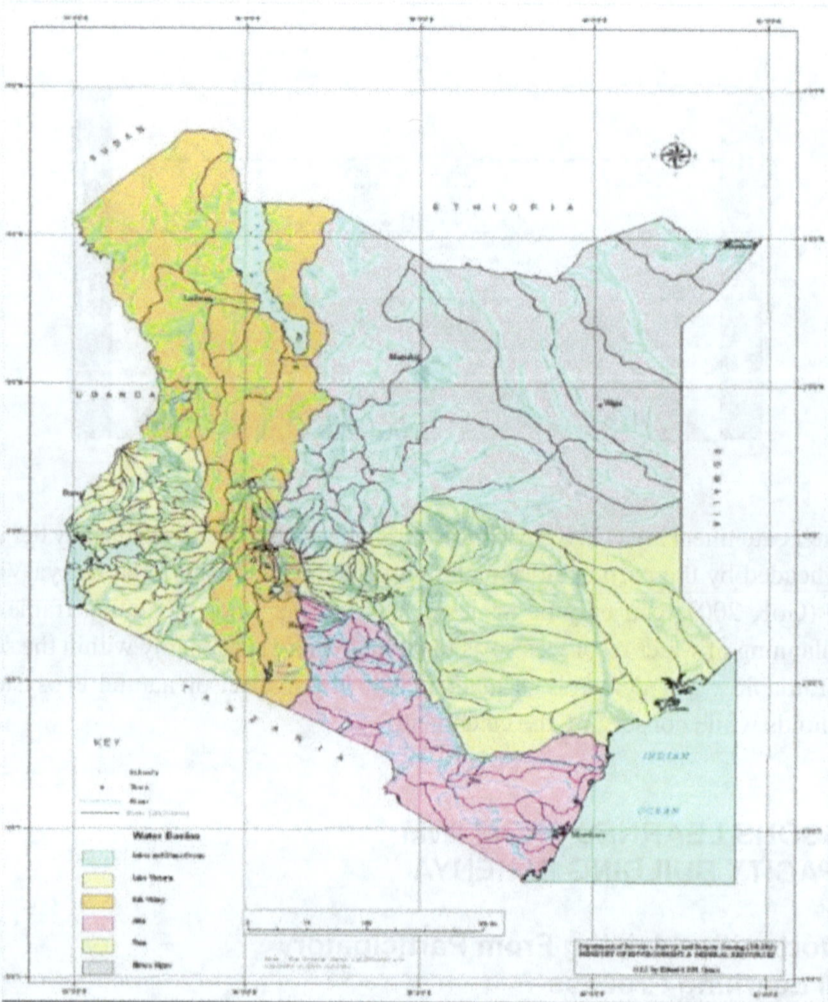

The summer schools aimed at building their capacity to gather information from the ground, to analyze the problems, to plan and solve them, based on their own understanding. They used a theory of change underlying the role of German Universities' alumni in facilitating the transfer of professional expertise and know-how in IWM to local stakeholders. This included academic knowledge and competence transfer, and the use of holistic and interdisciplinary approaches toward problem assessment and resolution during participatory fieldworks. Hence, the capacity building process focused on developing participatory sub-catchment management plans (SCMP) to empower the Water Resource Users' Associations (WRUAs) to formulate IWM problems, plan for adequate solutions and implement them as they monitor the achievements. Thus, the WRUA would play the most important role in the IWM process, both as key stakeholders and beneficiaries (Box 3).

The SCMP serves as framework for the management of the water and related land resources in the catchment and it outlines how the concept of integrated watershed management (IWM) can be implemented at the sub catchment level.

Box 3. Water rules enforcement in the Tana Basin, Kenya

1.1 Bwathonaro Water Resources Users' Association
Background Information

Bwathonaro Sub-Catchment is one of the first sub-catchments within Kenya in general, and the Tana River Catchment Area in particular, where the IWM capacity building programme was implemented through the summer school models. Geographically, the Bwathonaro catchment is bounded by latitude 37° 53′00″ E and 38° 05′ 00″ E and 0° 12′ 00″ N and 0° 21′ 00″ N. The catchment drains the Kithetu, Maua, Sanga, Buathini, Ikombo and Kiraro Hills that are part of the Nyambene volcanic system. The drainage is through the Bwathonaro River system in the South-East direction into the Meru National Park. The upper catchment above the Meru National Park covers an area of 149 km^2.

In order to ensure effective management of water resources and facilitate conflict resolution within the catchment area, Bwathonaro Water Resources Users' Association (BWARUA) developed a Sub-Catchment Management Plan (SCMP) with the support of WRMA, the Universität Siegen (Germany) and Kenyatta University (Kenya) in 2006. The latter implemented, monitored and evaluated in a participatory way during a period of three years following its inception in 2005. Positive results as well as challenges have been outlined below to give an overview of BWARUA's experience on enforcement of water rules in Kenya.

BWARUA's Achievements

BWARUA's Visionwas to "assure availability of water resources for the wellbeing of all people in the Bwathonaro Catchment". Therefore, its Missiondealt with 'protecting the catchment, to ensure availability of good water quality and sustainable management of the environment'. From 2006-2009, BWARUA's Goal was to "plan and work towards an environmentally and socio-economically healthy watershed that benefits the communities living in the Bwathonaro watershed".

Committed to its vision, mission and goal, BWARUA assigned to itself the following targets, which it achieved:
(i) Improved water quality at source through control and management of sources of pollution;
(ii) Reduced conflicts over water arising from illegal water abstractions and over-abstractions for irrigation through enforcement of the Water Act 2002 rules on water abstraction
(iii) Promotion of efficient irrigation practices;
(iv) Conservation and management of the water resources in the catchment through protection and rehabilitation of water sources such as springs and wetlands;
(v) Promotion of 'agro-forestry' in the catchment;
(vi) Awareness creation on the danger of planting water unfriendly trees at the expense of local indigenous trees;
(vii) Minimization of soil erosion through soil conservation measures;
(viii) Reduced cases of human-wildlife conflicts through maintenance of reserve flows entering the Meru National Park; and
(ix) Enhanced cooperation with relevant stakeholders.

The WRUA made several sensitization meetings to inform the public on the dangers of illegal activities in the Bwathonaro river system, including its tributaries. Water users were therefore informed on the process of acquiring water abstraction permits. Through the WRMA office in Meru, all the water projects within the catchment were given pre-conditions in their water permit, notably to install master meter and water controlling devices. Moreover, communities possessing several water projects within the Bwathonaro catchment were sensitized on the payment of water use charges during public meetings. This led to a reduction of illegal water abstractions and over-abstractions in the catchment along with increased water revenues collected by WRMA.

Further, the need to protect springs for recharge was noted, and by 2010 Gitwe, Kathithi, Makutano Nkanga and Matiandui springs had been protected. The Athindi wetland likewise was fenced and the vegetation regenerated enhancing the general environment. An emphasis was put on catchment conservation and protection during sensitization and awareness meetings through public meetings. Community members were informed on the benefits accrued from good conservation and protection measures put in place. Riparian land owners agreed to conserve within the riparian zone, thus leading to improved natural water recharge after protection and conservation of the riparian land within catchment area.

continued on following page

Box 3. Continued

Challenges Related to Water Stress and Conflicts
Bwathonaro is a water scarce catchment, and therefore one of the BWARUA objectives was to prevent and resolve any conflicts related to water resource use. Conflict resolution was done in conjunction with WRMA and/ or the local provincial administration. To prevent conflicts, water projects requesting water use permits within the catchment had to make their applications to WRMA office. The latter would be forwarded to BWARUA for validation and seconding, to ascertain whether the application would likely increase the risks of water stress and conflicts, or impact on water quality. That is how the WRUA has been able to manage the challenge related to water stress and conflicts arising from illegal water abstractions and over-abstractions.

1.2 Ngaciuma-Kinyaritha Water Resources Users Association
1.2.1 Background Information
Ngaciuma-Kinyaritha is a sub-catchment of about 167 Km^2 located at longitudes 37.5° E and 37.75° E, and latitudes 0.04° N and 0.15° N within Imenti North Constituency of Meru County of Kenya. It has a population estimated to 65,000 people in 2010. Almost all the rivers originate from Mount Kenya Forest and generally flow eastwards to join Kathita River. Like for BWARUA, Ngakinya Water Resource Users' Association (NGAKINYA WRUA) was among the first WRUAs to be formed in the Tana Catchment Area to implement a process that involved stakeholders' participation in the management of Ngaciuma and Kinyaritha sub-catchments. To achieve this participatory water resources management a plan was developed in 2007 with the support of WRMA, the Universität Siegen (Germany) and Kenyatta University (Kenya). This plan was smoothly executed over the 3 years following its inception. The first monitoring and evaluation activity in Ngaciuma-Kinyaritha was conducted in November 2008, and covered activities of 8 months, the implementation being interrupted by the post-election violence that consumed the country from December 2007 to March 2008. Moreover, a water demand management plan was developed during the DAAD Alumni Summer School 2008, and it was to be implemented in the short-term (0-2 years) and the medium-term (2-5 years).

1.2.2 NGAKINYA WRUA's Achievements
NGAKINYA WRUA was founded with a Vision of'ensuring the availability of adequate amount and good quality water resources for the well being of all the people and the environment in the sub-catchment'. Consequently, its Mission consisted of 'involving the community and stakeholders to protect, regulate, equitably allocate and conserve water resources' with the Goal of 'working toward socio-economic and environmental sustainability of the sub-catchment'.
The following targets have been successfully achieved for the period 2007-2010:
(i) Control of illegal water abstractions and over abstractions;
(ii) Promotion of efficient and beneficial use of every water drop abstracted;
(iii) Dissemination of efficient soil and water conservation practices;
(iv) Improved management and development of water resources for sustainable use; and
(v) Establishment of a frank cooperation between different stakeholders involved in the management and conservation of the Ngaciuma and Kinyaritha riparian lands.
In 2009, the WRUA enumerated 351 water abstractors in Ngaciuma and Kinyaritha sub-catchments. Agriculturalists were among the most important water abstractors (79.4%) while householders, institutions and Industrialists accounted for 15.8 and 4.8% of the total water abstractions, respectively. In the agricultural sector, water use for irrigation represents the highest water demand. Water use for domestic purposes is higher in urban areas (54%) rather than rural areas (46%). The water resources in the catchment are mainly from springs that feed Ngaciuma and Kinyaritha streams. Other water sources are insignificant and include boreholes and roof catchments for rainwater harvesting. However, these water resources are inefficiently managed owing to poor irrigation practices, limited use of water harvesting methods and high level of illegal abstraction. To curb this trend, Ngakinya WRUA advices WRMA in the allocation and authorization of water permits for any type of work and application falling within its catchment area.

1.2.3 Challenges Related to Water Stress and Conflicts
During dry periods, water use conflicts escalate, owing to the depletion of the available water resources. To minimize this tension, and in accordance with the Water Act 2002, the WRUA with support from WRMA successfully enhanced the enforcement of water rules to ensure that every water users is catered for within the limits of policies that improve water use efficiency. To that regard, the principles of have been much appreciated by local stakeholders as being fundamental for sustainable management of their scarce water resources in a participatory and gender sensitive approach. By them recognizing the economic value of water and committing themselves financially to WRMA efforts, all stakeholders were keen ensuring that resources generated through revenue collection were reinvested in catchment conservation to enhance water flows and good water quality.

SOURCE: WRMA. (2010). *Enforcement of Water Use Charges and Water Quality Thresholds in Kenya.* WRMA Evaluation Training Workshop. Meru: Three Steers Hotel, 24-28 January 2010, pp. 43-47.

Figure 3 shows how the other participants to the summer schools played different roles to exchange ideas on integrated methods of managing watersheds. These participants came from different backgrounds encompassing geographers, engineers, foresters, agriculturalists, social scientists, economists, other scientific fields and local stakeholders. However, three key actors were pointed out in the participatory learning process, namely DAAD Alumni (including professors, professionals and students), WRMA officials (representatives of the Water Resources Management Authority) and WRUA members (mainly local stakeholders and other professionals).

First, DAAD alumni were invited to train other participants in watershed planning, management, monitoring and evaluation processes, as well as in methodological approaches pertaining to IWM. WRMA officials were called in order to provide a keen interpretation of the law and of governmental policies. And WRUA members were the major beneficiaries of the summer schools. They were also regarded as major representatives for community interests, awareness creation on the water sector reform, and summer schools participants' guidance on key issues occurring on the ground. Students and other stakeholders were invited to ensure the sustainability of the summer schools' impact through further awareness creation, scientific research, etc. Then, a progressive change of attitudes, behaviors and working methods would

Figure 3. Roles played by different categories during the summer schools
(Source: Förch and Ngonzo, 2009)

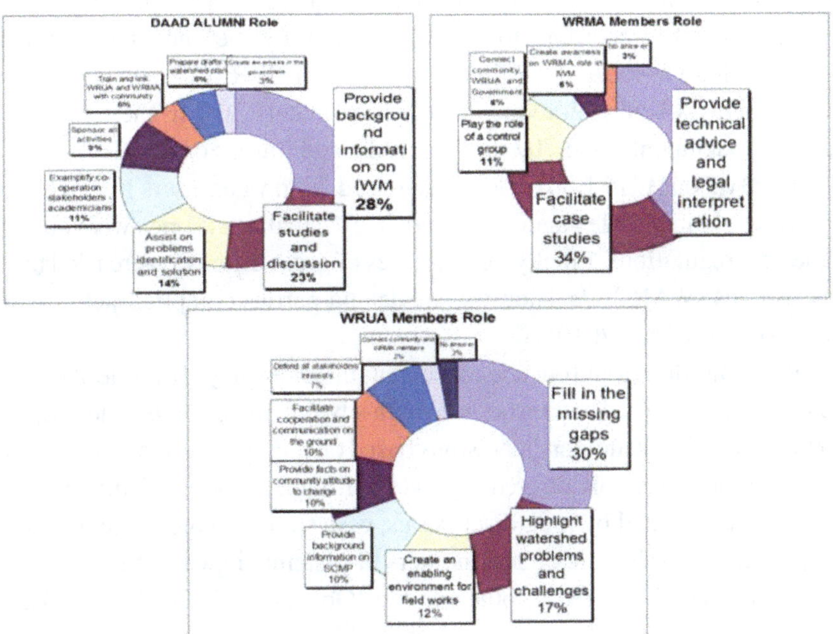

follow in the management of water resources. WRMA officials were able to explain more adequately principal regulations on water resources and their implementation in the watershed. Once in the field, they adopted participatory approaches to implement water rules with regard to water use charges and effluent discharge control plans. This enabled them monitoring illegal water abstractions and protect the catchments with the support of the WRUAs working within the six large basins of the country, namely: Athi, Ewaso-Ngiro, Lake Victoria North, Lake Victoria South, Rift-Valley, and Tana.

The relevance of interdisciplinary approaches was overemphasized as a channel for enabling scientists from different fields to work together with local stakeholders in IWM. Community members and other key stakeholders were trained by Alumni and outside consultants on concepts and field techniques pertaining to IWM. They had at the same time the opportunity to share their experiences in the field during workshops and fieldworks. WRUA members had a chance to monitor and evaluate watershed management by assessing their achievements and measuring the success of their former plans. They also set measurable indicators for the work they had to perform in the following periods (financial years). At the end of the process, they made recommendations for optimizing the implementation of their plans on the basis of their findings in the field. Indeed, Ondieki (2013), indicates that emphasis should be place on collaborative efforts and use of sustainable best practices would require input of various stakeholders including WRMA, Basin Authorities and National Environmental Management Authority (NEMA). In less than two years, this mismatch of participants brought the message a long way to 292 sub-catchments across the country. According to WRMA (2010), by the end of the year 2009, 30% among these WRUAs had already matured in IWRM (Box 2). Also, illegal water abstractions had been reduced by approximately 30% in the upper parts of most of the catchments, and 70% in the middle and lower zones. Also, about KES 126,104,300 (about US$ 1,801,490) were raised from water users in 2009. Finally, about 21.9% of large scale water users and 78.1% small scale ones were complying with the new regulations. Finally, only 7 cases of gross offenses were filed to Water Appeal Boards (WABs), decision was made and parties complied with the ruling of the WABs (WRMA, 2010).

Participating Alumni and outside consultants therefore played the role of facilitators in the learning process. They restrained themselves from checking and judging the outcomes of local stakeholders' decisions based on their personal opinions. The fact that these summer schools offered a good environment for acquiring and sharing knowledge was attested by long discussions, case studies and explanations evoked by the participants. The role of stakeholders in sustainable watershed management has further been underscored by other studies (Obando et al., 2015; Ondieki, 2013,

Shisanya et al., 2010). Further assessments of the capacity building process in IWM include, Forch and Luwesi, (2011, 2010, 2009); and Luwesi et al., (2010)

Finally, the summer schools offered them an opportunity to water sector professionals and agencies to network, cooperate and exchange their experiences. For instance, in an the EU Water Facility funded the IWMNet project, but it has become a very strong network for IWRM capacity building in East Africa, thus enabling local institutions to receive funding from other donors, notably AFD, AfDB, CIDA, DANIDA, DfID, KfW, JICA, SIDA, USAID, the World Bank, and others. After the closure of the IWMNet/ EU Project, the Center for International Capacity Development (CICD) of the Universität Siegen (Germany) established partnerships with the six universities of eastern and southern Africa and water ministries in Kenya, Tanzania and Uganda (including the WRMA and WRUAs). There is a strong partnership between these institutions and their development partners. This has been illustrated by periodic meetings held in each country to harmonize the concepts and policies, and increase accountability among partners. These have opened a route to WRMA officers; WRUA and community members to change their attitudes toward the water sector reforms, and thus use innovative and efficient methods for managing their natural resources.

Challenges Arising From Participatory IWRM Learning Approaches

The participation of professionals from several research fields alongside people from diverse social backgroundsoften inhibited the freedom of speech of many among local stakeholders. DAAD Alumni may have dominated the discussions using their strong theoretical backgrounds. This was substantiated by interminable debates during which representatives of WRUAs and WRMA, and other key beneficiaries of the summer schools were out of track. Also, their participation to the planning, monitoring and evaluation processes could have been factual. It may have been more interesting for Alumni to go straight to the business by showing the practicality of the theories on the ground through problems identification and resolution. This would have avoided miscommunications among participants. Though their input was also well appreciated by all the participants during case studies, WRMA officials displayed a defensive attitude, so as to "protect" the interpretation of the law and of governmental policies. By so doing, they concealed some weaknesses that needed to be addressed during the summer schools. Also, being the main providers of funds to WRUAs, their defensive attitudes would have threatened WRUA members, discouraging them to interact and express themselves freely on those weaknesses.

This might have been one of the reasons why some participants felt that the summer schools were not such environments enabling free cooperation and easy interaction among participants, despite the fact that a majority among them rejected such a statement. Concerning WRUA members, their guidance in the field, reports from the previous summer schools, and the lack of clear records on achievements and financial matters are key highlights of their involvement in the participatory learning process. WRUA members usually targeted some sites where there were apparent achievements from the planned activities, even though these activities may have been achieved by other social groups. Such self-defensive attitude may have distorted the results of the monitoring and evaluation process. Therefore, more exposure is needed in the future, to allow thorough investigations and acquaintances with the WRUA management tools, methods and realizations.

Good Practices and Lessons Learned

The lessons learned from the evaluation process can be categorized at five levels as summarized in Box 4.

WRMA officials have particularly learned how to measure the actual impact of their Catchment Management Strategy (CMS) on the ground using continuous assessment tools of the environment and, monitoring and evaluation of the management process. Therefore, local communities have been empowered with capabilities to manage effectively their vulnerable water resources and related resources using their financial and material means as well as their human resources. These local communities were offered opportunities to network with all stakeholders living in the watershed, to cooperate and to share relevant information on the new water laws, policies and regulations. Interrelations between different stakeholders were known

Box 4. Lessons learned

1. Stakeholders' sensitization is the starting point for a progressive change of attitudes toward a reform.
2. Sustainable management of natural resources requires stakeholders' participation in planning, implementation, monitoring and evaluation.
3. A good evaluation of the success of the water sector reform is done where both biophysical and socio-economic data are available.
4. The cost of water shall be understood as a key to water security and as a response to the Sustainable Development Goals.
5. Stakeholder involvement in the management of the water sector shall be perceived as a value added to good governance and sustainable development.

through a logical design defining the functions of some and roles of others, and the type of relation or cooperation existing among them. They were encouraged to support the enforcement of the IWM principles in their respective watersheds. An emphasis was put on rules related to equitable allocation and use of the resources among homogenous hydrological units and different groups of stakeholders and interests to sustain water balance and gender equity. An agreement was to be made with all stakeholders on the future quantity and quality of water resources to be conserved. For that purpose, WRMA was to set measurable standards of water protection and storage, good land husbandry and conservation. Local stakeholders were expected to use their technical know-how for identifying the issues on the stake, setting agenda and developing plans for addressing the issues of water scarcity and conflicts on resource use. The summer schools provided them with an enabling environment for training on water policies implementation, strategic and operational management, institutional coordination of efforts made by local stakeholders, and monitoring and evaluation of watershed management. Political and technical strategic issues were articulated into management tools and merging organizational tools such as water sector reform legislations, policies, programmes, tactical projects, budget lines and schedules, functional and relational designs.

CONCLUSION

Solving issues of water stress and scarcity requires the formulation of common, integrated, sustainable approaches for managing the water system from a multidisciplinary perspective, and the definition of new professional skills, requirements which become even more evident where transboundary areas are concerned. In order to take decisions in an equitable, sustainable, and ethical fashion, water resource management does in fact call for extensive knowledge of the complex relationships between citizens and their water system. Integrated Water Resource Management (IWRM) is an approach designed to achieve sustainable development. It offers a way to manage competing demands for water and the linkages across sectors. The IWRM core principle can be described as the recognition that water is a public good with both social and economic values. A major challenge in IWRM is managing different interests of different sectors and users that might conflict. The ecosystem requires a basic amount of water to remain its integrity and to replenish (Dutch Ministry of Foreign Affairs, 2014). Though Kenya can be credited for having succeeded to initiate and implement a participatory water governance system, despite various financing and investment challenges, it has another challenge of implementing the water rules to be enacted soon according the devolution system enshrined in the Constitution of Kenya 2010. Besides, the country has to get prepared to tackle water deficits arising from the increasing water demands while the resources are likely to deplete by 2100. Hence, the country shall capitalize on the good practices and lessons learned from the IWM capacity building process. The activities prioritized include; awareness creation on water scarcity and environmental trends, water allocation and conflict resolution, pollution control, catchment rehabilitation and conservation, WRUA financing, monitoring and evaluation.

REFERENCES

Alemu, B., & Kidane, D. (2014). The Implication of Integrated Watershed Management for Rehabilitation of Degraded Lands: Case Study of Ethiopian Highlands. *Journal of Agriculture and Biodiversity Research*, 78–90.

Bamutaze, Y., Thiemann, S., & Förch, G. (2014). *Integrated Watershed Management-A Tool for Urban Water Security*. Workshop Results from Mbale, Uganda. Mbale: Makerere University, IWM Expert GmbH and Freie Universität Berlin.

Biswas, A. K. (2004). Integrated water resources management: A reassessment. *Water International*, *29*(2), 248–256. doi:10.1080/02508060408691775

Cassman, K. G. (1999). Ecological intensification of cereal production systems: Yield potential, soil quality, and precision agriculture. *Proceedings of the National Academy of Sciences of the United States of America*, *96*(11), 5952–5959. doi:10.1073/pnas.96.11.5952 PMID:10339523

Cyber-nook.com. (2011). *Quotes from Poor Richard's Almanac, 1746*. Available on: http://www. cyber-nook.com

Dore, T., Makowski, D., Malezieux, E., Munier-Jolain, N., Tchamitchian, M., & Tittonell, P. (2011). Facing up to the paradigm of ecological intensification in agronomy: Revisiting methods, concepts and knowledge. *European Journal of Agronomy*, *34*(4), 197–210. doi:10.1016/j.eja.2011.02.006

FAO. (2001). *Sustainable land management*. Available on: http://www.fao.org/nr/land /sustainable-land-management/en

Förch, G., Winnegge, R., & Thiemann, S. (Eds.). (2005). *DAAD Alumni Summer School 2005: Topics of Integrated Water Resources Management*. Weiterbilding in Siegen 18. FWU Water Resources Publications, Vol. 03/2005. Retrieved from http://fwu.fb10.unisiegen.de/bkd/summerschool.htm

Förch, G., Winnegge, R., & Thiemann, S. (Eds.). (2006). *DAAD Alumni Summer School 2006: Participatory Water Resources Management Plan.* Weiterbilding in Siegen 20, FWU Water Resources Publications, Vol. 05/2006. Retrieved from http://fwu.fb10.unisiegen. de/bkd/summerschool.htm

Förch, G., Winnegge, R., & Thiemann, S. (Eds.). (2007). DAAD Alumni Summer School: Participatory Monitoring of Sub-Catchment Management Planning – Bwathonaro Watershed, Kenya, Final Report, August 2007. Meru, Universitat Siegen and Kenyatta University.

Förch, G., Winnegge, R., & Thiemann, S. (Eds.). (2008). DAAD Alumni Summer School: Water Demand in Participatory Watershed Management – Ngaciuma-Kinyaritha Watershed, Kenya, Final Report, November 2008. Meru: Universität Siegen and Kenyatta University.

Förch, N., & Luwesi, C. N. (2009). *Major Challenges and Outcomes of the DAAD Alumni Summer Schools of Kenya (2006-2008).* A Conference poster presented during the CICD/ Universität Siegen's Conference 2009 on "Knowledge transfer in Development Co-Operation – Study, Research and Consulting", Siegen.

Förch, N., & Luwesi, C. N. (2010). *Challenges and Opportunities of Participatory Evaluation in the Context of North-South Cooperation: The Case of Capacity Building in Integrated Watershed Management in Kenya.* A Conference poster presented during the UK Evaluation Society (UKES) Annual Conference 2010 "Evaluation in a Turbulent World: Challenges, Opportunities and Innovation Practice", Birmingham.

Förch, N., & Luwesi, C. N. (2011). *Kenya - Capacity Building for Participatory Management of Water Resources.* Retrieved from http://afrik4r.org/profiles/blogs/kenya-capacity-building-for?xg_source=activity

Förch, N., & Ngonzo, C. (2009). *Capacity Building in Integrated Watershed Management in Kenya: An Independent Evaluation.* Paper presented during the 2009 Global Assembly Conference of the International Development Evaluation Association (IDEAS) on "Capacity Building Evaluation", Johannesburg, South Africa. Available on: www.ideas-int.org

GoK. (2002). Water Act 2002. *Kenya Gazette.*

GoK. (2007). *Kenya Vision 2030.* Kenya Government Printer.

GoK. (2010). The Constitution of Kenya 2010. *Kenya Gazette Supplement, 55.*

GoK. (2012). *Masterplan for the Conservation and Sustainable Management of Water Catchment Areas in Kenya.* Ministry of Environment and Mineral Resources.

GoK. (2013). *The National Water Master Plan 2030, Ministry f Environment, Water and Natural Resources, Water Resources management Authority*. JICA.

GWP. (2015). *Global Water Partnership, Integrated Water Resources management in Eastern Africa: Coping with 'Complex' hydrology*. Technical Focus Paper. Available from: www.gwp.org

Hoff, H. (2011). *Understanding the Nexus*. Background Paper for the Bonn2011 Conference: The Water, Energy and Food Security Nexus. Stockholm Environment Institute.

ICWE. (1992). *The Dublin statement and report of the conference*. International Conference on Water and Environment: Development Issues for the 21st century, Dublin, Ireland.

Jaglin, S. (2002). The Right to water versus cost recovery: Participation, urban water supply and the poor in sub-Saharan Africa. *Environment and Urbanization*, *4*(1), 231–245. doi:10.1177/095624780201400119

Jensen, K. M. (2013). Viewpoint–swimming against the Current: Questioning development policy and practice. *Water Alternatives*, *6*, 276–283.

Lal, R. (2015). World Water Resources and Achieving Water Security. *Agronomy Journal*, *107*(4), 1526–1532. doi:10.2134/agronj15.0045

Lenton, R., & Muller, M. (2009). *Intergrated water resources management in practice. Better water management for development*. London: Earthscan.

Luwesi, C. N., Förch, N., Mutiso, M. N., Akombo, R. A., Wambua, P. P., & Badr, E. (2014). A Performance Assessment and Evaluation of IWM Capacity Building Activities on Participatory Water Resource Management in Kenya. *Journal of Agri-Food and Applied Sciences*, 2(4), 93-108.

Luwesi, C. N., Shisanya, C. N., & Obando, J. A. (2012). Warming and Greening: The Dilemma facing Green Water Economy under Changing Hydro-Climatic Conditions in Muooni Catchment (Machakos, Kenya). Saarbrüken: Lambert Academic Publishing.

Meire, P., Coenen, M., Lombardo, C., & Robba, M. (2008). *Integrated Water Management - Practical Experiences and Case Studies*. Dordrecht, The Netherlands: Springer.

Merrey, D. J., Drechsel, P., Penning de Vries, F. W. T., & Sally, H. (2005). Integrating "livelihoods" into integrated water resources management: Taking the integration paradigm to its logical next step for developing countries. *Regional Environmental Change*, *5*(4), 197–204. doi:10.100710113-004-0088-5

Molle, F. (2008). Nirvana concepts, narratives and policy models: Insight from the water sector. *Water Alternatives, 1*(1), 131–156.

Ngige, A. & Macharia. (2006). *Kenya Water Policy Overview Paper.* IT Power East Africa.

NSF. (2016). *USAID INFEWS Workshops 2016, "Food, energy and water nexus".* The Next Einstein Forum, Dakar.

Obando, J. A., Luwesi, C. N., Mathenge, J. M., Kinuthia, W., Wambua, P. P., Mutiso, M. N., & Bader, E. O. (2015). Performance Assessment and Evaluation of Community Participation in Water Sector Governance: The case of Ngaciuma-Kinyaritha catchment, Mount Kenya Region. In A. Beyene (Ed.), *Agricultural Water Institutions in East Africa, Current African Issues, No. 63* (pp. 23–41). The Nordic Africa Institute.

Ondieki, C. M. (2013). Hydrology and Integrated Water Resource Management for Sustainable Watershed Management in Kenya. In H. Muga & K. Thomas (Eds.), *Cases on the Diffusion and Adoption of Sustainable Development Practices* (pp. 352–375). Hershey, PA: Information Science Reference. doi:10.4018/978-1-4666-2842-7.ch013

Rockström, J., Falkenmark, M., Karlberg, L., Hoff, H., Rost, S., & Gerten, D. (2009). Future water availability for global food production: The potential of green water for increasing resilience to global change. *Water Resources Research, 45*(7), W00A12. doi:10.1029/2007WR006767

Savenije, H. H. G., Hoekstra, A. Y., & van der Zaag, P. (2014). Evolving water science in the Anthropocene. *Hydrology and Earth System Sciences, 18*, 319–332.

Shah, T., & van Koppen, B. (2006). Is India ripe for integrated water resources management IWRM: Fitting water policy to national development context. *Economic and Political Weekly, 41*(31), 3413–3421.

Shisanya, C., Obando, J., & Onywere, S. (2010). Towards Sustainable Water Resource Management for Food Security in Bwathonaro Catchment, Kenya. Report to Commission for Higher Education (CHE) for Project: Enhancing Sustainable Water Resource Management for Food Security: A Case Study of Bwathonaro Catchment, Kenya.

UN. (2016). *Sustainable Development Goals.* Retrieved from Http://www.un.org/sustainabledevelopment /development-agenda/

van der Bliek, J., McCormick, P., & Clarke, J. (2014). *On target for people and planet: Setting and achieving water related sustainable development goals. Colombo: International Water Management Institute*. Sri Lanka: IWMI.

WRMA. (2010). *Enforcement of Water Use Charges and Water Quality Thresholds in Kenya*. WRMA Evaluation Workshop, Meru.

ENDNOTE

1 Also check on available posts on http://www.water-energy-food.org/en/news/view__1612/the-nexus-approach-vs-iwrm-gaining-conceptual-clarity.html

Chapter 13
Advancing New Pathways for Sustainable Deployment of Agricultural Ground Water With Gender-Sensitive Data Sharing

Cush Ngonzo Luwesi
University of Kwango, DRC & United Nations Economic Commission for Africa (UNECA), Ethiopia

Paa Kofi Osei-Owusu
Conservation Alliance International, Ghana

Amos Yesutanbul Nkpeebo
University of the Free State, South Africa

Raphael Muamba Tshimanga
University of Kinshasa, DRC

ABSTRACT

At the basin level, watershed resources (water, land, and ecosystem services) are often managed in "silos," whereby gender-sensitive data and analytics are employed by different stakeholders without regard to existence of similar systems in other parts of the same sub-catchment. These cases create missed opportunities and insights for harnessing better innovative water interventions from baseline scenarios by examining the relevant frameworks, protocols, and tools for data sharing as well as the means of disaggregating water and climate change data into gender-sensitive formats transferable to similar sub-catchments. Innovative water solutions would enable optimal water services provision and financing without farmers depending wholly on external organizations to provide solutions to their financial challenges through taking part in technological development of viable water infrastructure.

DOI: 10.4018/978-1-7998-0163-4.ch013

BACKGROUND

An "ideal" system for agricultural water provision would need key technical, organizational and financing issues to be addressed prior to establishing an operating groundwater provision technology. An integrated Holiyas, Grundfos and MPesa (HGM) technology is suggested to be among such solutions filling the gap in agricultural water supply through a groundwater recharge sub-system (Holiyas), groundwater pumping sub-system (GrundfosLifelink) and ICT based water management sub-system for both withdrawals and payments (MPesa mobile money). Enhanced coherence, safety and productive capacity are required to cover all water rights at anytime and anywhere in the area of interest. Public-private partnerships (PPPs) are eagerly recommended to ensure sustainable financing of the new system through cost sharing to attract further financial support for future development of groundwater resources.

INTRODUCTION

In the next 20 years, increasing the productivity and incomes from smallholder crop, livestock, fishery and forestry production systems will be very crucial to achieving global food security (FAO, 2013; FAO, 2014; CCAFS, UNFAO, 2014). Largely, the world's rural communities are directly or indirectly dependent on agriculture which is provisioned, supported or complemented by water ecosystem services. Indeed, balancing the requirements of the aquatic environment and other uses has become of critical importance in Sub-Saharan Africa (SSA). However, the ways in which many smallholder farmers are already intensifying their production is extremely ecologically compromising making it necessary for very urgent for a fundamental shift away from unsustainable agricultural intensification using appropriate technologies that can be applied in the local context to achieve Sustainable Agricultural Intensification (SAI) or more specifically, Climate Smart Agriculture.

Most parts of SSA is beginning to experience increasingly frequent, deeper and more persistent droughts coupled with simultaneous intense rainfall and flooding events. Studies indicate that climate change affects not only the mean yield of the crops, but also induces variability in crop yields which means a significant increase in household vulnerabilities to climate change. There is no blueprint for an "ideal" system for agricultural water provision but key technical, organizational and financing issues need first to be addressed prior to establishing an operating groundwater provision technology (Luwesi et al., 2012). An integrated Holiyas, Grundfos and MPesa (HGM) technology is suggested to be among such solutions filling the gap in agricultural water supply through a groundwater recharge sub-system (Holiyas), groundwater pumping sub-system (GrundfosLifelink) and ICT based water management sub-system for both withdrawals and payments (MPesa mobile money). Enhanced coherence, safety and productive capacity are required to cover all water rights at anytime and anywhere in the area of interest. Public-private partnerships (PPPs) are eagerly recommended to ensure sustainable financing of the new system through cost sharing to attract further financial support for future development of groundwater resources.

Climate change also leaves strong implication for livestock feed quality and quantity and thus, their productivity. According to recent climate adaptation models, this variability is likely to become greater in the future. It is therefore very noble that the post-2015 United Nations Sustainable Development Goals (SDGs) allocates extensive focus to water management under its Goal 6 (ensure availability and sustainable management of water and sanitation for all, UN, 2014) and placing a premium on water cooperation. At the national levels, water resource management can open up opportunities for mainstreaming climate change adaptation, aimed at achieving development objectives such as increased water security and agro-ecological resilience (Luwensi et al, 2017).

The African continent faces the triple facet of water crisis, namely: Availability, Accessibility and Affordability (the "3As"). These issues are being escalated by the changing patterns of the global macroclimate, demographic growth and urbanization, as well as some of the human interventions on the environment. That is the reasons why the international community has shown solidarity to the people living in developing countries by designing various agronomic and engineering methods to mitigate some of the climate impacts on our warming macroclimate. The expected results would be the prevention of escalating natural disasters, water and energy crises, food insecurity and poverty outreach. Other interventions include regulatory approaches that enable people's behaviour change for their adaptation to the changing natural environment. In recognition of the huge irrigation potential and the pressing need to address food security demands in Africa, there are regional and national ambitions to expand and develop irrigation on the continent. It is in that context that many African countries are investing in multipurpose water dams to fill the huge gap between the available water resources and the actual water demands for domestic use and production. However, the drive to expand irrigation is not without challenges.

First, irrigation is the largest water consumer in most developing countries, with water demands often rising above 80% of the total consumption (Luwesi et al.,2017). Yet, farming water is vulnerable to climate change and its unsustainable use result in enormous inefficiencies and massive losses in the agricultural production. there is a water crisis looming ahead of acute climatic impacts and thus threatening the future and relevance of irrigation in the water sector. This water scarcity is being essentially deepened by what is described as a 'crisis of governance' (Mathenge, et al, 2014). Beyond multi-purpose water dams, flexible irrigation canals and other infrastructure, irrigation development needs a participatory governance system, which is essential for the mitigation of the socio-cultural and ecological impacts of its intensive cultivation on local communities as well as other actors and institutions involved in irrigation. For this reason, the water discourse has evolved focusing more on a diversity of issues in water governance that include new legal frameworks and organizational structures, financing instruments and best practices; and a new thinking regarding the achievement of coherence in water policies (Rosegrant, Cai and Cline, 2002). However, there is currently a shift toward the analytical context of this new crisis, which puts an emphasis on financing challenges facing water development (Luwesi et al., 2017).

Farmer need water but they can neither have access to it at the "right" time, in the "right" quantities, at the "right" places, nor afford adequate funding to get the appropriate technological solution for their water problem. It is therefore time to think of the most readily available path to meet future water demands in agriculture by achieving water savings (Gleick, 2001). Economic instruments have been recommended as the best alternative for achieving the two major objectives identified for smallholder irrigators: improved cost recovery and economic efficiency (Obando et al., 2017 Obando et al., 2017). Farmers need a fair price to cover the cost of water services supply and foster their economic efficiency, especially technical efficiency for water allocation. This definitely improves their financial resources, thus enabling them to achieve higher values for water services and their uses in agriculture through technological innovation. It also encourages ecological conservation by providing signals to induce behavioral change, especially with regard to the misconception of agricultural water right: the "manna from heaven" (Walsh, 2012; Kassam et al., 2007).

Applying a price to water is not only done because of cost recovery but is equally important as a tool to change behaviour and make sure that water is distributed more fairly (Walsh, 2012; Gleick, 2000}.

However, policy makers and water users alike are ambivalent about how far water should be treated as a basic right, whether it should be provided free or with a subsidy, or whether it is a commercial service to be charged for. The result is often that water services are priced below economic levels and the sector is chronically underfinanced, which in turn deepens the agricultural water crisis (Luwesi et al., 2017). There is thus a global consensus that water resources should no longer sustain if treated as simple *res communes*. Rather, water users shall accept to pay for their use and misuse, if they appreciate the full cost of renewing these "God–given provisions" or necessities of life (Goetz, 2013). Happily, studies have shown that consumers are more likely eager to pay off the services they value than those without a value added. Moreover, they would always appreciate the net benefits from projects that promote new technologies than those just using natural resources without adding new inputs (Rosegrant and Cline, 2003; Pretty, 1999).

There is growing commitment for innovative water projects in the Arid and Semi-Arid Lands (ASALs) to improve the conditions of our hard-working small-scale farmers. Benefits accrue from such projects have actually resulted in a massive support for the user-pays and polluter-pays principles in the ASALs (Haas and Nagarajan, 2011). Several institutions have funded projects that promote new technologies for groundwater provision in agriculture. This chapter illustrates three of such projects implemented in East and West Africa, namely *GrundfosLifelink*'s groundwater pumping technology (from Denmark) coupled with *MPesa* mobile money transfer technology applied in Kenya and the whole of East and Central Africa; and *Holiya* groundwater storage technology (also known as "*Bhungroo*") developed along the edges of Gujarat State desert of India and which is currently being tested in Ghana, West Africa (Luwesi et al., 2017). This study projects an integration of these technologies into a Climate Smart Agriculture (CSA) water design so as to provide a basis for groundwater development and services financing in Africa. The authors would appreciate your feedback for the improvement of this proposed design of the groundwater provision technology, to bring about the desired change in the agricultural water sub-sector of the African ASALs. Tiffen, M. and Mortimore, M. (2002). Agroecosystems regaining the high ground: Reviving the Hillsides of Machakos. World Resources 2000-2001: 149-158.

UNDERSTANDING THE WATER INNOVATION SETTING IN THE TROPICS

Tropical ASALs face enormous water challenges due to recurrent droughts. For example, Tiffen and Mortimore (2002) reveal that Machakos County, in southeastern Kenya, has long been a theatre of dry spells, which kill hundreds of thousands of people in a single year. With an area of 5,953 km^2 and a population of 192,369 capita, the county has been prone to natural disasters, El Niño and La Niña adding to the burden, such as the ones experienced in the years 1983-84 and 2009-10. UNEP (2009) relates how the Government of Kenya constructed demonstration sites and built several dams that can supply water to towns and other earthdams that serve for agricultural purposes. However, due to increasing water demands and decreasing water availability, many women and their children have to walk on average 3 kilometres to fetch water during normal periods. During major droughts, when many demonstration sites are dry and/or carry salty water, farmers receive water in forms of humanitarian aid. They also have to look for alternative drinking water sources and use indigenous solutions to ensure soil moisture to sustain their crops. However, in most areas watercourses do no longer exist, except a few seasonal flooded areas and swamps. Thus, the environment as well as peoples' health and livelihoods suffer from water scarcity due to erratic rainfall received in the county. This is perceived to be the primary cause of loss of potential yields, food insecurity and socio-economic underdevelopment in Machakos County. This constitutes a threat to the life of over 75,000 people living in Machakos Town and its environs, owing to the inefficiency of existing irrigation systems and the lack of water engineering infrastructure (Luwesi, 2010).

Shah et al. (2016)· and Singh and Lal (2003) report similar situations in the tropical Indian peninsula. Despite the fact that visitors are welcomed by a surprising high diversity of dishes served across the country, the "Jewel of Western India", Gujarat State, is generally featured in media by a theatre of dry spells (Figure 1). Gujarat State has an area of about 196,024 km^2 with a coastline of 1,600 Km and a population of over 60 million people. In the state of Gujarat, one can enjoy Hindu vegetarian specialty, while in the North of India one can be served with *moghol* (*halal*) meat from Muslim origin. Lunch highly varies from Hindus, Muslims, Sikhs, Jains, Parses and Christians, and you cannot eat the same thing twice. Yet, the north-west part of Gujarat State is facing the most acute water crises of all times.

In such dry environment water solutions abound and are welcomed by the people. The types of gifts sold in the market are usually water efficient. Recently, some businessmen have taken advantage of a prolonged drought to attract clients by offering "free" water during the remaining of the summer. According to the Times of India, some would offer up to 500 litres of water, for four days a week and during two months of summer, in exchange of the purchase of big housing electronic devices such as electric ovens, refrigerators, washing machines and TV sets. At Rajkot, several businessmen have confirmed that water offer served as their permit for multiplying their sales threefold. But where do they get that water?

Subramanian (2001), indicates that "*Water problems in developed countries have been solved only by reservoir storage at different stages of the river flow*". So, if water shortage problems are solved more quickly in the ASALs of India rather than humid regions, it is simply because of the adaptation of traditional water storage systems to drought, mainly using deep ponds, storage wells and underground tanks. It is in such context that Paul Biplab, an Indian scientist, adapts and disseminates a traditional technological solution known as *Bhungroo* by some and *Holiya* by others but simply meaning a "straw". in Gujarati, It is a flood water storage technology that frees excess water from farmlands during the rainy season, using a "small stick". Then, this excess water is naturally stored underground in unsaturated soil layers, so as to be pumped for farming during the dry season (Osei-Owusu, 2014; Biplab, 2012).

Figure 1. Dryspells in Gujarat State and across the Indian Peninsula
(Times of India, 2015)

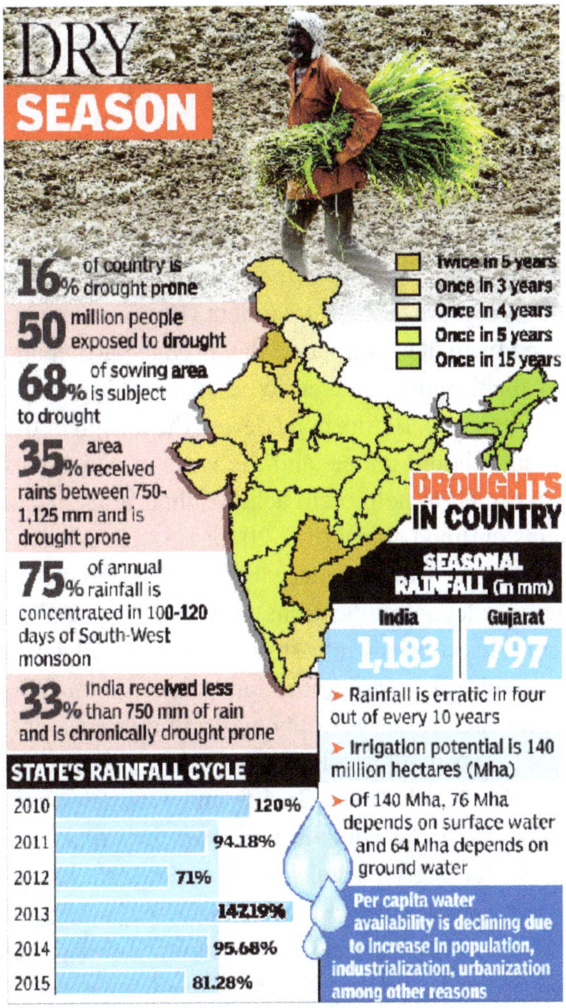

Holiya technology has proven to be effective by providing water to more than 3,000 families of farmers in some of the most affected areas of the Gujarat State (Shah et al., 2016 ; Biplab, 2012). In Northern Ghana, the Holiyas is being tested by Conservation Alliance International (CAI) at Jagsi, Kpesenkpe and Weisi communities of Walewale area under the funding of the CGIAR research program on Water, Land and Ecosystem (WLE). These communities face acute water constraints and challenges for domestic and agricultural purposes during the dry season while their lands are flooded during the rainy season. It is our hope that this new technology will free many farms from waterlogging issues in the ASALs and that farming activities will thrive during the dry season as the most viable irrigation solutions will be available to farmers who do not have access to agricultural water during the dry season (WLE, 2015).

Irrigation Financing Issues and Its Shifting Paradigm

Supplying water of adequate quality and in sufficient quantities is one of the major challenges facing our modern society. The water sector is challenged by population growth, rapid urbanization, ageing infrastructure and climate change. In most countries, public authorities or state-owned organizations are responsible for the drinking water supply and wastewater treatment (Rosegrant and Cline, 2003). This chapter does proceed from the assumption that private participation can sometimes help extend access to public goods and services at reasonable prices, if the arrangement governing it is well designed. Before considering the design of these arrangements, however, governments should understand the ways in which private participation can or cannot be expected to help them achieve their aims, including tackling the underlying problems outlined above (EC, 2016; USAID, 2014).

Water services are on the boundary between economic infrastructure (i.e. transport, electricity, telecommunications) and purely social infrastructure (i.e. health and education). In economic infrastructure there is either a high degree of user charging (i.e. power, public transport, ports, and telecommunications) or substantial public budgetary provision (roads) (Koppenjan and Bert Enserink, 2009). In social infrastructure there is normally exclusive or heavy reliance on public finance, generally known as the "3Ts": Taxes, Tariffs and Transfer (Wambuaet al., 2017). The ubiquitous nature of investments in water services make it less attractive for the private finance sector. Agricultural water development has particularly not yet offered incentives for private investors to be attracted, even if there is a growing significant interest in projects that require large capital disbursement (Prasad, 2006).

It shall be noted that irrigation, which is the largest consumer of water in most developing countries, with water demands often exceeding 80% of the total consumption, has the most readily available path for attracting new investments in the water sector to achieve water savings and meet future demands for food production (Russell et al., 2007 ; Rosegrant, Cai and Cline, 2002; Ringler, 2001; Rosegrant et al., 2000). This is justified by the fact that irrigated agriculture is increasingly put under pressure to demonstrate and improve both its efficiency and effectiveness (Wichelns, 2002). Without new investments in efficient irrigation technologies, there is a likelihood that water will have to be diverted more and more to meet the needs of urban and industrial areas (Russell et al., 2007 ; Wichelns, 2002 ; Rosegrant et al., 2000). Without compromising the needs of the rising population while fostering agricultural growth and food production, innovative irrigation systems are required to improve water use efficiency.

Agricultural water efficiency is measured in terms of cost recovery, improved productivity and equitable allocation of the resources at the technical, managerial and institutional levels (Kassam et al., 2007; Wallace, 2000). The scope for improving both water cost, allocational and technical efficiencies in agriculture is still considerable but requires substantial funding to finance innovative investments. Such relevance is illustrated here through integration of the Holiyas, GrundfosLifelink's pumps, and MPesa (HGM) Technologies in Irrigation.

According to United Nations Secretary-General's High-Level Panel on Global Sustainability (Bigas, 2012), population will grow by 30% between 2000 and 2025 and up to 50% between 2000 and 2050 leading to an increase of the global population from 7 to 9.5 billion by the year 2050. Food production will need to have doubled to meet the needs of such a large population. Worldwide, agriculture is the biggest consumer of freshwater resources and accounts for 70-90% of freshwater withdrawal. It has been observed that climate change in most African urban and peri-urban landscapes results in erratic rainfall patterns with a cycle of excess water conditions (flooding) and limited water conditions (drought) in most semi-arid and dry lands (in the short term), leading to crop failure (in the medium term) and water, energy and food crises (in the long term) (Cooper et al., 2008). Many farms in ASALs are flooded and face waterlogging issues, which limit farming activities during the rainy season. With the advent of water management technologies, farmers now have the opportunity to choose from a variety of water management solutions that best fit their farming conditions (Carriger & Vallée, 2007).

It is a well-known fact that the pressure on existing water supplies will continue to increase due to growing water demand for agricultural purposes, while the need for water for natural ecosystem processes has often not adequately addressed (GWP, 2005). The adequate allocation of water for the environment is critical in supporting biodiversity and the production of ecosystem services. A failure of adequate water supply for environmental services might result in a decline in the environmental capacity to provide water and nutrients that support agricultural practices (Van Hofwegen, 2006). Population growth, changes in consumer behaviour and the negative impacts of climate change on our water resources will thus require more efficient water use by the agricultural sector so that it can increase food production with less water (GWP, 2005).

In developing countries, Kenya for instance, most smallholder farms do not have access to modern irrigation schemes due to constraints related to finance and technical know-how for the maintenance of the schemes (Monteiro et al, 2010). These elements are at the same time the determinants of private participation in water finance. Private participation changes the water sector by introducing an operator that is independent of the government and has a strong incentive to be profitable. A private provider cannot be directed in the same way as a public provider and its profit incentive (EC, 2016; Hechler-Fayd'herbe, 2009). If given access to microfinance, smallholder farms will thus have an opportunity to use irrigation facilities and therefore extend their cropping activities during the dry season. Such independence and the profit incentive may also help the government achieve its financial targets in the water sector. In fact, private participation has been considered to have effects in the utility investment's decisions as well as in public policy and its enforcement. This enables financial institutions to play an important role in shaping a sustainable water sector (McGranahan and Satterthwaite, 2006.

It is important to reiterate that the profit incentives have a high propensity of leading the private provider to make better investment decisions (Augsburg and Fouillet, 2010). For emphasis, a private provider may build fewer "white elephants" given the enabling environment. S/he may miss fewer profitable opportunities to expand the business, such as extending access to unconnected households or farmers that want service and can pay for it (Ng and Loosemore, 2007). If the private provider has some responsibility for determining and financing investment, if prices cover costs and can be expected to do so in the future, and if the operator keeps some of the profits that result from increasing access, the private provider can be expected to invest in increasing water access for all (McGranahan and Satterthwaite, 2006; Ng & Loosemore, 2007; McGranahan & Satterthwaite, 2006). With a few exceptions though, private participation mostly offers the prospect of changing policy in a way that alleviates the fundamental problem set out earlier: The politics of water pricing leads to prices being set below costs, frustrating the extension of access (Cooper et al., 2008; Ng & Loosemore, 2007). Private investment in water services place premium on the rules for setting prices because those rules determine the return on investment. In the case of agricultural water supply investment decisions may call for contingency valuation. However, if stable rules about pricing and subsidies are achieved, that should encourage investment and thus help the government to improve access and efficiency in water management (Kramer, 2011; Rogers et al., 2002). On the other hand, if extending access to poor households will cost the private provider more than it gains in revenues and subsidies, or avoids in contractual penalties, it can't be expected to increase access. Amidst such a conundrum, private participation in agricultural water development has not exactly been an overly attractive sector for individual investors. However, the wind is steadily turning.

A CALL FOR INNOVATIVE AGRICULTURAL WATER DEVELOPMENT IN DRYLANDS

Private participation in agricultural water provision is beginning to show significant application when it comes to addressing critical water needs, especially, in projects that will require large capital disbursement (Prasad, 2006). Such relevance could be tested through an Integrated Holiyas-Grundfos-MPesa (HGM) groundwater technology. Figure 2 illustrates the design of an integrated HGM technology, which is a sort of climate smart agriculture design based on three components: (i) Holiya groundwater storage technology, also known as "Bhungroo Irrigation technology" along the edges of Gujarat State desert, in India, (ii) GrundfosLifelink's groundwater pumping technology from Denmark, and MPesa mobile money transfer technology from Vodafone consortium but mostly applied in Kenya And the rest of East and Central Africa.

Holiya Technology for Groundwater Storage (HGS)

India's farmers are experts in dealing with monsoonal climate, which leads to extreme rainfall during the main growing season (Biermann, 2001). When an entire year is dry, water comes in just a few frantic weeks of rainfall, thus devastating harvest

Figure 2. An integrated Holiya-Grundfos irrigation technology with MPESA money transfer
(Haas & Nagarajan, 2011)

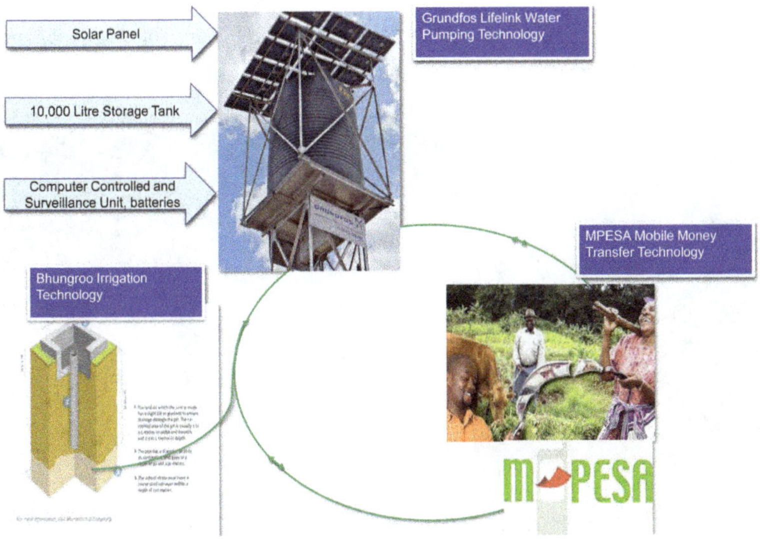

overnight. In such conditions, Peasant farmers across the peninsula have mastered the art of storing excess water during the monsoon to access it during the dry season using in-depth technological knowledge of the "stick", Holiya (Biermann, 2001).

Also locally known as *Pompi*, in northern Ghana, HGS technology is a socio-technical solution developed by Indian farmers to suck excess water from their water logged farm plots in monsoon periods and inject it underground with the aim of using it for sowing paddy rice during the dry season (CA, 2016;Goyal, 2014). *Holiyas* are relatively simple groundwater recharge and subsurface storage systems (Figure 3).

HGS technologies were mainly promoted in Gujarat State in the mid-1990s by a number of NGOs and international development agencies to support a few farmers who invested in the approach. The systems were diverse and context based, because the design and installation of the technologies were based on the prevailing soil, vegetation, rock and socio-cultural contexts of the area, as well as the prevailing market conditions for the materials needed to construct the systems. But there were four major factors (CA and IWMI, 2016) that informed farmers' decision to select a specific HGS technology for installation, namely: (i) the surface flooding features; (ii) sub-surface or geological formations that support the storage of flood water; (iii) agricultural land use practices; and, (iv) the willingness of the target community to engage in dry season farming. In essence, *holiyas* are just vertical pipes dug into the ground to an average depth of 9m where the water table is shallow, and 30 to 100 m in dry areas. The pipe is usually made of a 10cm-diameter PVC which is roughly perforated either at the bottom third part of it or aside such that water drains down to the pipe into the unsaturated soil layers up to the aquifer, in the form of lenses, when it rains (see Figure 3), This enables reducing the amount of water logged on

Figure 3. Holiya systems designed for Ghana (Right) from the Gujarat one (Left) *(CA, 2016; Goyal, 2014)*

the surface, and securing groundwater replenishment for irrigation during the dry season (CA, 2016; Goyal, 2014).

The HGS technology is as a farmer friendly technology but its deployment requires intensive agronomic and hydrogeological surveys for irrigation to take place in the target drylands. It has proven its technological capability in some areas of India that were the most affected by annual floods and droughts, including the Gujarat State (Bunsen & Rathod, 2016; Shah et al., 2006). If this technology could free more than 3,000 farmlands from water logging and drought effects, much more is expected in Africa. It is precisely the reason why it is being tested in Northern Ghana, where farmlands have been named "overseas" in West Mamprusi District. These farms are being submerged by excess water during the rainy season and then when drought comes, the soils are bare (CA, 2016; Goyal, 2014). In 2015, Conservation Alliance International (NGO) then took the initiative to install three (3) Bhungroo based HGS systems in three communities of northern Ghana, notably Kpasenkpe and Jhasiin West Mamprusi District, and Wiesi in Bulsa South District. During the construction process fivefold stages were strictly followed. They encompassed Bhungroo siting; drilling; Bhungroo design and erection; filtration chamber construction; and community sensitization and work handover (CA, 2015).

- **Siting:** This stage entailed several technical assessments of potential sites to ascertain the feasibility for a Bhungroo installation. The siting also dealt with a large set of data collection for initial assessments by soil scientists and the hydrogeologist. The selection of the potential sites for further assessments and the final installation of the Bhungroo had to be approved by the beneficiary community leadership.
- **Drilling:** This stage involved creating a vertical hole underground using a borehole drilling machine. The depth of the drilling varied in the three communities due to the differences in altitudes, the water table depths and the characteristics of unsaturated soil layers. The HGS depth for Kpasenkpe was 45m, while Wiesi and Jhasi recorded 54m and 74m, respectively.
- **Bhungroo Design and Erection:** This stage dealt with arranging the Bhungroo pipes according to the soil strata identified during the drilling process. In the design process, layers of soils that had no potential capacity to hold water were assigned unperforated pipes, popularly referred to as plain pipes, while perforated pipes were assigned to unsaturated water storage zones, with fix strainers or screens. The main reason was to enhance the flow of injected water from the unsaturated layer to the aquifer. When the pipes have been fixed underground, wire meshes were coiled around the Bhungroo pipe from the top soil to 3m above.

- **Filtration Chamber Construction:** This A chamber of 1.5 to 1.55m in length, breadth and depth was created around the Bhungroo for the cleaning of flood waters. This chamber was to provide a housing for the filtration materials of the Bhungroo. The filtration materials included stones of different size, sand and wire mesh. Once the chamber was completely dug out, a wall made of cement blocks was set in all the four sides of the chamber. The various filtration materials were then laid in the chamber, layer by layer.
- **Community Sensitization and Work Handover:** Prior to the siting up to completion of the Bhungroo the beneficiary communities were involved in the process. The research and engineering works were done in close conjunction with local communities and other relevant stakeholders living in Kpasenkpe and Jhasi in West Mamprusi district, and Wiesi in the Bulsa South district. The installation process of the Bhungroo was made possible after a series of engagements with the target communities. Then, upon completion, communities decided on the allocation of water and land around the Bhungroo based on a participatory approach that was gender balanced. Finally, the work was ended over to them for exploitation.

To date, HGS and other similar technological systems have been tried not only in northern Ghana but also in a few villages in the south by Conservation Alliance under a FTF-ATT USAID based grant. There are in total 9 Bhungroos tested successfully around the Walewale area of Northern Ghana, which generate roughly 1,000 m^3 during the dry season to serve about 800 families of farmers. Others have not been successful in some locations, but they succeeded, they could have been useful for preventing surface water logging and thus enable cultivation of monsoon crops in other locations (CA & IWMI, 2016; CA, 2015). An Environmental Impact Assessment (EIA) done in Banaskantha District, Gujarat in 2015 under the IWMI-Tata Water Policy Program confirm that fact. It indicates that Holiyas don't damage the environment, don't require any deep technological knowledge and are relatively cheap to build, even though they are not widely used (Bunsen and Rathod, 2016; Osei-Owusu, 2014).

The research team interviewed 41 individual and village communities of farmers about the impact of Holiya construction (Bunsen and Rathod, 2016; Shah et al., 2006). Respondents said that *Holiyas* also have a lower environmental impact that either boreholes or canals, yet only a quarter of farmers used them. They found the *technology* useful and reported improvements in flood mitigation, drought resilience and cropping. The structures enabled farmers to grow crops during the winter and, in some instances, grow green fodder for livestock during the summer. But farmers also said that despite canals shortcoming of shifting the burden of excess water to different locations, they preferred canals to HGS fix system as the cost of building canals and maintaining them is less than for holiyasBut HGS remains avery effective flood control infrastructure, which has to complement canals to capture and store flood water.

GRUNDFOSLIFELINK'S PUMPING AND MPESA MONEY TRANSFER IN KENYA

The secret conduit in integrating HGS technology within a HGM design is the combined GrundfosLifelink's Solar pump and MPesa mobile transfer technologies. MPesa ("M" for "mobile", and "Pesa" for "money" is Swahili language) is a mobile phone apps that enables the transfer of money from user to user and various microfinance services including money deposit, withdrawal and transfer as well as payment of goods and services e.g. "Lipa naMPesa", (Haas and Nagarajan, 2011; Karago, 2009). Figure 4 illustrates the potential application of MPesa for mobilizing finance for water services and water development projects such as the integrated HGM technology.

Launched in 2007 by Vodafone for Safaricom, the largest mobile network operator in Kenya, MPesa has proved that GrundfosLifelink's technology does not

Figure 4. MPESA money transfer process for water payments

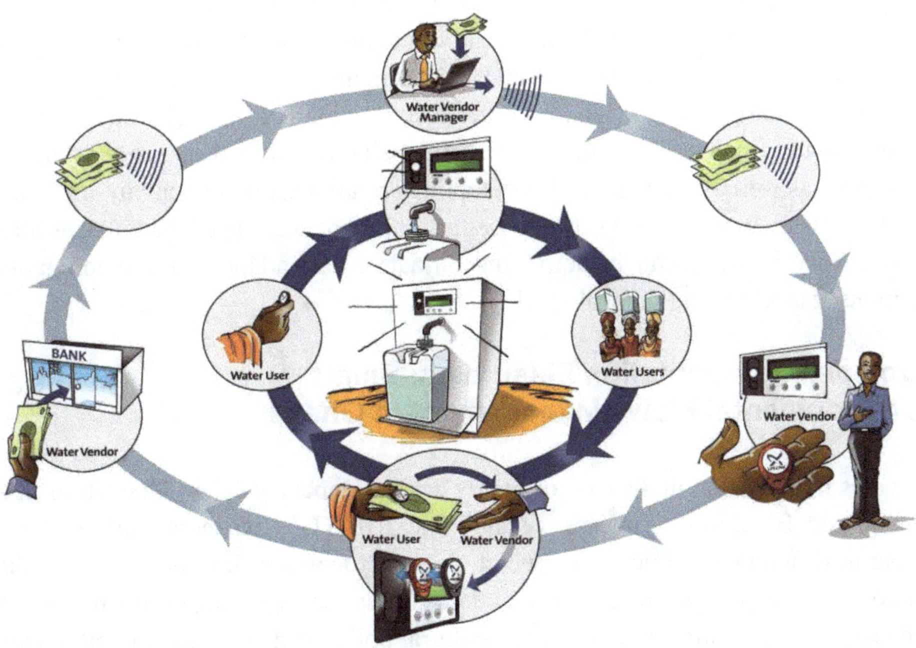

have to be the neglected orphan of the financial world. In 2009, these ICT and green energy based technologies have been applied at Musingini in Machakos County to improve drinking water supply. The technology uses a solar pump activated by a mobile phone to withdraw groundwater (Luwesiet al.,2015). The latter enables the charging of every drop that flows from the tap at the consumption site. Such an exemplary application of modern technologies for water provision and efficient payment has yet to be made possible in irrigation though the partnership between the Danish pump making company (GrundfosLifelink) and the Kenyan based mobile operator (Safaricom) has improved drinking water supply at Musingini. However, there exist some low grade but effective locally developed irrigation technologies, which are farmer friendly and that can be adapted or integrated in the HGM model. These included among others drip irrigation kits such as Amiran farmers kits, 'KrishakBandhu' (KB) buckets and quarter acre kits, etc.(Haasand Nagarajan, 2011; Karago, 2009). IDE-Ghana is currently marketing this irrigation equipment, and smallholders farm using KB micro irrigation equipment earn annual average extra income of US$ 400 on a quarter-acre plot (Luwesiet al.,2015).

HGS systems under development in Northern Ghana may be integrated within the Grundfos-MPesa framework with support of Amiran or KB micro irrigation equipment to make a full Holiya-Grundfos-MPesa (HGM)system. This shall be carefully designed to enable smallholders afford full water supply on small scale farming plots of 20 square meters throughout the dry season. Once integrated, HGM technology could be scaled out to water stressed areas and make a significant impact in farmers' livelihoods in the ASALs of Ghana, Kenya and other drylands of Africa (Luwesi et al., 2017). But, the impediments to the wide deployment is not just dependent on the technological investment and its viability. The success of this technology also depend on other factors including natural water resource endowments, finance availability, human and social capital, governance and political economy as well as the legacy of development relations, population buy-in, values and behavior (Lagger, 2011). This poses a significant challenge which makes local governance instrumental in delivering climate smart adaptation and mitigation options, especially at the local level.

Compelling Adaptation Fissures in Smart Environmental Flows Management in Ghana

Studies indicate that in the case of highly context dependent CSA innovations like Integrated Bhungroo-Grundfos Technologies (IBGTs), expanding the evidence base in different localities is a crucial element for measuring the effects of this innovation on ecosystem resilience and ecosystem services advancement. One of the key pieces of information that agricultural policy makers need, and often lack, are the current and near-future projected effects of climate change in their country and their implications for the agricultural priorities and programs of the country. Factors such as increasing rainfall variability, delayed onset of the rainy season, and increasing seasonal maximum temperatures, are all examples of climate change impacts that are already being realized and different responses are needed for effective adaptation for each of these (Biplab, 2013; Osei-Owusu, 2014). Identifying effective adaptation options to maintain and enhance the capacity of agriculture to generate productivity and income increases needed for food security is thus one fundamental objective of the CSA evidence base. The role of local communities in gathering data to support the role out of climate smart options thus becomes more essential. Given the requires capacity boost, local communities will be able to collaborate with academia, research centers, policy makers, the media and government departments in generating quality data on various climate phenomena (FoE-Ghana 2014. Lake Bosomtwe Project Brief, Ghana). This collaboration is useful in developing recommendations on the type of practice or change in agricultural systems needed for climate smart options like IBGTS.

Evidently, the increased mobile penetration in Ghana has had a significant effect on the capacity of local communities to support the gathering of data on various ecosystem variables and scenarios offering collaborative solutions based on networks (of people, of ideas, of sensors), enabling new forms of social and environmental innovation. Ground observations indicate that this capacity could be enhanced by developing communication decision support system built with instant messaging mobile applications to capture and transmit data in real time. Creating such a virtual innovation platform could facilitate simple icon-based and image-rich data capturing and real-time or near real-time information sharing on the ecosystem services from climate smart innovations like IBGTs technologies. It will be supported with coordinated field verification trips to validate data. This data will be used to generate quarterly GIS maps on the scales patterns of ecosystem services from IBGTS projects. It will contribute to improving decision making speed and facilitate fact-based decision making for private sector organisations, government departments, communities and other external actors.

IBGTS Technology Domain Mapping for Adoption in Northern Ghana

The IPCC identifies economic wealth, technology, information and skills, infrastructure, institutions and equity as the principal determinants of adaptive capacity (IPCC, 2001). In recent assessments, it has been postulated that social factors, in particular power relations – e.g. ‘social capital’, governance structures and the role and functions of institutions – have been underplayed in earlier studies

(IPCC, 2007). Based on the findings of ACCRA's consultative process, local adaptive capacity comprises five intertwining features that are conducive to adaptive capacity. These are: the asset base, institutions and entitlements, knowledge and information, innovation, and flexible forward-looking decision-making. In the process design of IBGTS application, it is important conduct detailed domain analysis. This is necessary because it is extremely unlikely for any type of adaptive action to be taken in light of climate change alone (Huq and Reid, 2004; Handmer et al., 1999). This suggests that a domain analysis of SEF could be essential. It can uncover underlying opportunities or challenges which are usually glossed over by broader feasibility studies. For example, *Magee* et al. (2013), on the Porto Alegre's Cities Programme project, it noted that themes such as cultural and political relations go far beyond generalized development problems and places emphasis on the subdomain of engagement and identity.

Water Productivity and Associated Capacity and Capability Challenges

The key principles for improving water productivity at field, farm and basin level, which apply regardless of whether the crop is grown under rainfed or irrigated conditions, are: (i) increase the marketable yield of the crop for each unit of water transpired by it; (ii) reduce all outflows (e.g. drainage, seepage and percolation), including evaporative outflows other than the crop stomatal transpiration; and (iii) increase the effective use of rainfall, stored water, and water of marginal quality. The effectiveness of IBGTS will depend largely on how these principles apply. However, very little is still known about water productivity in the context of IBGTS in Northern Ghana. There are now numerous examples of climate change risks being incorporated into existing policies, programs or decision-making processes related to resource management, community development, livelihood enhancements, coastal zone management, sustainable development and risk management. Practical climate change adaptation initiatives are invariably integrated with other programs, and often aim to enhance adaptive capacity. In many drylands in developing countries, water productivity has been improved significantly a set of irrigation kits either locally developed or adopted from other innovations. Drip irrigation has been found to have a higher productivity function. Drip irrigation, which is also known as *trickle irrigation* or *microirrigation*, is an irrigation method that allows a grower to control the application of water and fertilizer by allowing water to drip slowly near the plant roots through a network of valves, pipes, tubing, and emitters. For many crops, switching from a conventional flood/furrow or sprinkler system to drip irrigation can reduce water use by 50 percent or more. However, there is also the question of farmer-capacity to procure, install and use drip irrigation systems, especially in the case of Northern Ghana where most of the farmers are smallholder poor households. It is not uncommon for smallholder farmers to be poor. Tailor made solutions can therefore be derived from the experiences of other similar cases, especially from India. In India, the International Development Enterprises (IDE) has developed simple, affordable low-cost drip irrigation systems for smallholder vegetable growers that could be adapted to the situation in Northern Ghana. These systems include: bucket kit, family nutrition kit, drum kit, customized system combo kit.

A popular innovation in low-cost drip irrigation by IDE is the KrishakBandu (KB) or "Farmer's Friend" which uses lay-flat lateral drip lines with a wall thickness of only 0.125-0.25 mm, which expand to 16-mm in diameter when filled with water. Microtubes are used as emitters to provide uniform water application. The cost in India is around US $600 per ha for closely spaced crops. The inlet pressure head for the KB Drip system can range from 0.5 to 3 meters. KB Drip is popular due to its lower cost, small package sizes, ability to operate at very low pressure, ease of installation and use, and uniformity of water distribution. This system may be deemed more suitable to for implementation alongside the IBGTS in Northern Ghana. However, it is important to do contextual assessment of its applicability in a different geographic setting. Figure 5 illustrates the critical domains of influence and the key issues under each domain as well as the relationships between and within the domains. The critical quadrats are the policy domain and the enterprise domain. These two domains have the largest influence in relation to the implementation of IBGTS.

Figure 5. Domain mapping of low pressure irrigation incorporated into IBGTS in Northern Ghana
Source: Authors, 2019

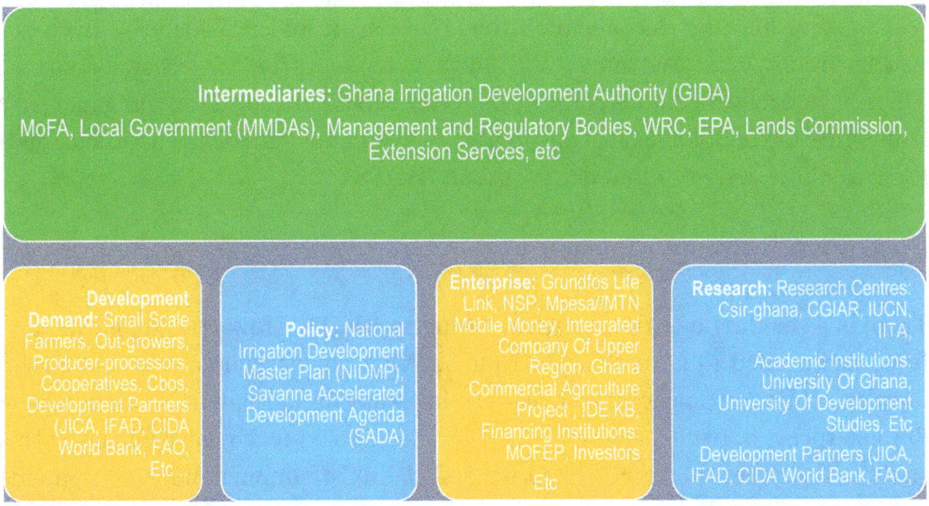

Policy Domain Analysis

There are small scale irrigation activities practiced along the banks of the tributaries of the White Volta Basin. These involve the use of the traditional system whereby water is accessed from surface flows by means of buckets tied to strings, and applied to irrigated vegetables. There are no notable large scale irrigation activities in the dry season. There is, however, tremendous potential for the development of an irrigation industry involving the conjunctive use of surface and groundwater resources if the appropriate infrastructure is put in place to support farmers and the youth. The Water Resources Commission recently granted abstraction license to the Integrated Water Management and Agriculture Development Limited, IWAD, to abstract surface water resources from the Kulpawn for large scale irrigation activities close to the study area. The activities of IWAD are supported by both the World Bank and the Savannah Accelerated Development Authority, SADA, to provide opportunities for the youth in agriculture and to assist in enhancing food security in the area. Per capita water allocation for domestic activities, based on responses to structured questionnaires, is currently lower than 30 liters per capita per day. It is much more precarious in the dry season when the populations compete with their animals for ponds and streams of the White Volta Basin in the process of drying out.

Traditionally, women cannot own land in northern Ghana because they are not *qualified* to inherit properties. Women are usually allocated lands to farm by their husbands or household heads. They can seek for permission from their husbands to farm on those lands. Studies indicate that most often than not, women are given degraded lands for crops production. Given that the IBGTS seeks to strongly empower women in sustainable water management, this circumstance may pose a threat to the IBGTS undertaking. It will have a big influence determining scales in the demand domain for IBGTS and by extension the enterprise domain. A win-win outcome may therefore require strong multi-stakeholder dialogues with chiefs, women groups, local government and others in the enterprise and research domains. This is where intermediaries come in. their role should include developing inclusive facilitating methodologies that ensure that women have entrenched access to lands for agriculture activities in order to sustain IBGTS.

Enterprise Domain Analyses

There is a standing question on water lifting in IBGTS. How can groundwater be lifted and availed to farmers in their plots? What other technology would enable farmers achieve efficiency in the use of this water in agriculture? This is where the enterprise domain provides a solution to a socio-technical challenge. This study will support the combined GrundfosLifelink's pumping and MPesa mobile charging technologies. In effect, these ICT and green energy based technologies have been applied at Musingini in Machakos County, Kenya, to improve on drinking groundwater withdrawal using a solar pump activated by a mobile phone. The latter enables charging every drop that flows from the tap to the consumption site using mobile money technology. Such an exemplary application of modern technologies for water provision and efficient use has not yet been made possible in irrigation though the partnership between the Danish pump making company (GrundfosLifelink) and the Kenyan based mobile operator (Safaricom) has improved drinking water supply in Musingini. to make this solution even more effective, a third locally developed farmer friendly technology is also suggested; the KB irrigation kit. The KB can be recommended for IBGTS projects in northern Ghana owing to its low cost element. IDEI is currently mass marketing this irrigation equipment which has been carefully designed to be affordable and adapted to the small scale of farm plots without compromising on quality. The KB systems can be customized for plots as small as 20 square meters. On average, the smallholder farmer using KB micro irrigation equipment earns an extra $400 a year on a quarter-acre plot. This is making significant impacts in farmers lives in India and could be out scaled to water stressed areas like northern Ghana on pilot basis.

A New Theory of Change

At the basin level, watershed resources (water, land and ecosystem services) are often managed in "silos." In some cases, gender sensitive data and analytics are employed by different stakeholders without the knowledge that similar systems might already exist in other parts of the same sub-catchment. These cases create missed opportunities as water managers and users can gain insights and harness better intelligence through data and analytics compiled from various technological applications in the basin. As such, a first step should be to investigate the level of gender-sensitivity of existing data sharing frameworks, protocols, tools and techniques adopted for recent, present and proposed adaptation interventions.

As a primary input, innovative water interventions will need to develop baseline scenarios identifying, quantifying and examining the relevant frameworks, protocols and tools for data capture and sharing as well as the means of disaggregating climate change data into gender sensitive formats workable for farming communities, government decision makers and local data providers in White Volta Basin. Such water interventions will also need to identify and quantify the current state of, and impacts on, water for farming, people and nature and how these are changing with time within the project locations. To this end, application of the DPSIR model will to allow a comprehensive assessment of the issues through examination of the relevant Driving-forces and Pressures on the environment, the consequent State of the environment and its Impacts, and the Responses undertaken as well as the interlinkages between each of these elements. This will have to be supported by local experience sharing sessions to synthesize local expert insights and new knowledge. International Experience sharing Seminars will also be organised under innovative water interventions to draw from the various experiences of individual projects and research activities currently on going on artificial water recharge technology and environmental implications of maximizing production. Figure 6 depicts such an intervention logic.

In the case of Ghana, such innovative water interventions will have to work with the Women in Agriculture Directorate (WIAD) of the Ministry of Food and Agriculture to provide support for the implementation of Ghana's Gender in Agriculture Development Strategy II (GADS II). This will advance technological

Figure 6. A new theory of change

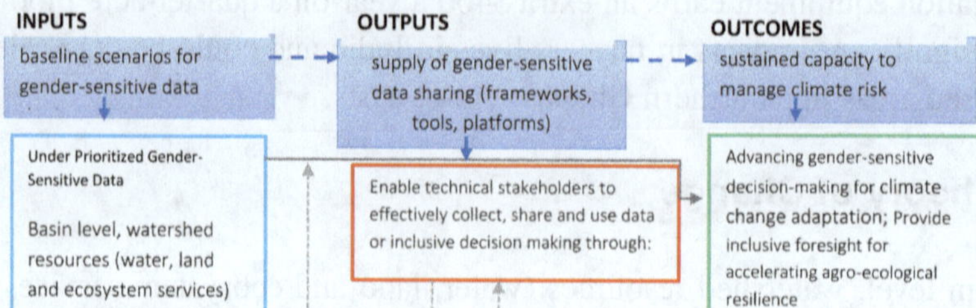

GADS II= Gender in Agriculture Development Strategy II; CoP= Communities of Practice, BIT=BhungrooIrrigation Technology; IBG=Integrated Bhungroo-GrundfosTechnology; LACA = Local Adaptive Capacity, VCA= Vulnerability Capacity Assessment

transfer as a means of accelerating agro-ecological resilience, advancing gender-sensitive data sharing, supporting women-led environmental flow restoration, water security. It will also enhance women's access to water, land and information by harnessing the potential of vulnerable farmers for climate data sharing along the agricultural value chain in ultra-critical sub-catchments.

Given the significant role of women in agriculture in Sub-Saharan Africa, it is no longer a notion but a fact that building a sustainable future entails connecting and sharing the knowledge, skills and leadership of women in climate action. Innovative water interventions need to place women as lead agents of climate smart agriculture intensification through gender sensitive data sharing. Thus, outputs from these interventions should include Cases for Gender-Sensitive Data Sharing in critical sub-catchments as Innovative cooperation platforms for adaptation action, E-Symposiums and Best Practices. These will result in major improvements for sustainable agricultural intensification, environmental flow restoration and; support effective management of infrastructure particularly the Pwalugu Multi-Purpose Dam (PMD); (See Pawlugu Multipurpose Dam Project; http://www.vra.com/about_us/pawlugu_multipurpose_dam.php). It will promote water and ecosystem-based solutions across White Volta Basin by advancing basin-level decision-making processes, and IWM processes and gender-sensitive datasets through Integrated Agro-ecological Technologies.

CONCLUSION AND POLICY RECOMMENDATIONS

Conclusion

Water scarcity has been described as 'management crisis', following decades of climate change, increased deforestation and other land use activities as well as the disturbance of life supporting ecosystems and their services. Hence, the Water Vision 2000 (Cosgrove and Rijsberman, 2000) could rightly conclude that "Water crises are not about too little water.... but about managing water badly such that billions of people and the environment suffer badly." The recent hydro-technological developments and ICT have been propounded to be a relief for water developers and managers of all the sectors of life, including the services and management sub-sectors. However, to sustain such technological solutions in the above described 'management crisis' context, there is a great need for reshaping local stakeholders' mindset toward the "user-pays" and "polluter-pays" principles.

There is a great need to invest in the financing water technologies to improve water services and enable water users have full access to the service at the "right" time, in the "right" quantity, and at the "right" place. An integration of Holiyas, GrundfosLifelink's pump and MPesa (mobile money) is suggested to be among the solutions needed to fill the gap in agricultural water supply. This system can be designed to pump groundwater for irrigation using a solar powered pump and ICT based management of both water withdrawals and payments by customers while recharging groundwater through Holiyas' wells.

The successful application of Holiyas in the Gujarat State of India and in Ghana coupled with the success of the GrundfosLifelink's pump and MPesa (mobile money) in drinking water supply at Musingini area of Machakos County in Kenya are a prerequisites for an integrated HGM system to work elsewhere in the African semi-arid and dry lands. Such integrated hydro-tech and ICT systems may vary in breadth depth and intensity but are eagerly needed to provide water solutions during the dry season and charge every drop flowing from the tap. These are exemplary technologies and may revolutionize agricultural development by improving water access and its use efficiency in many ASALs. These integrated hydro-tech innovative systems constitute rising cases of a new thinking on the use of technological implements and best organizational practices to deliver new water values and manage the available water resources while ensuring cost recovery.

Recommendations of the Study

Establish Innovative Partnerships for Water Solutions

Since, there is no blueprint for an "ideal" system for agricultural water provision, technical, organizational and financing issues need to be addressed prior to establishing an operating groundwater provision technology and this requires innovative partnerships. The public sector can form partnerships with private firms and other organizations from civil society to raise funds for specific projects. Such partnerships have been widely used for financing public services in some countries. However, they need to demonstrate comparative advantage by the respective partners, efficient risk allocation between them and clear accountability for the results. For that reason, they will require bringing local communities onboard to buy-in the innovation and establish partnerships with the innovators.

Develop Communication-Based Decision Support Systems for Climate Information

One of the key pieces of information that agricultural policy makers need and often lack, are the current and near-future projected effects of climate change in their country, and the implications for the agricultural priorities and programmes of the country. The increased mobile penetration in Africa has had a significant effect on the capacity of local communities to support the gathering of data on various environmental variables and set scenarios for collaborative solutions based on networks (of people, of ideas, of sensors), enabling new forms of social innovation. It has been observed that this capacity could be enhanced by developing communication based decision support systems built with instant messaging mobile applications like WhatsApp, Skype, Google-Hangout, Facebook Messenger, opentokbox, etc. It is advisable to use applications that have massive popularity and usage among rural mobile phone users to create such a virtual innovation platform. If a simple icon-based and image-rich data capturing could facilitate sharing real-time or near real-time information on environmental trends, then it is a climate smart innovation that can be integrated with other technologies to enhance the HGM system. It would best be supported by coordinated field verification trips to validate data. These data may be used to generate quarterly GIS maps on the scales and patterns of ecosystem services and other environmental trends on which depend HGM systems. This would thus contribute to improving decision making speed and facilitate fact-based decision making for private sector organisations, government departments, communities and other external actors.

Interventions Need to Promote Gender-Sensitive Data Sharing

Most rural women in drought-prone areas remain more vulnerable to climate change as they have poor access to land, poor access to education, capital and other economic resources. At the same time, women often offer valuable insights and solutions into a better management of climate change and related risks. At the rural household level, women have been observed to be important decision-makers. Women are also early adopters of many new agricultural techniques and best practices and often first responders in crises.

Given these facts, women must be recognized as vital agents of change so that their needs and their contributions will be a part of the solution. Yet existing climate vulnerability and adaptation frameworks have consistently under-prioritized the practical needs of women and girls in terms of climate actions due to the general paucity and inadequacy of gender-sensitive data. As a result, the contributions of women and girls is often overlooked in climate actions and disaster management in most parts of Africa. New vulnerability and adaptation frameworks are therefore required to be tested and established in order to prioritize the role of women and girls in in climate actions. As such, a first step should be to investigate the level of gender-sensitivity of existing data sharing frameworks, protocols, tools and techniques adopted for recent, present and proposed adaptation interventions. As a primary input, innovative water interventions will need to develop baseline scenarios identifying, quantifying and examining the relevant frameworks, protocols and tools for data capture and sharing as well as the means of disaggregating climate change data into gender sensitive formats workable for farming communities, government decision makers and local data providers in ultra-critical basins. Such water interventions will also need to identify and quantify the current state of, and impacts on, water for farming, people and nature and how these are changing with time within the project locations.

REFERENCES

Abu-Zeid, M. (2001). Water pricing in Irrigated agriculture. *International Journal of Water Resources Development*, *17*(4), 527–538. doi:10.1080/07900620120094109

AGRA. (2009). *AGRA: Early Accomplishments, Foundation for Growth*. Available at: http://www.agra-alliance.org

Alfonso, G. S. (2015). *How to promote territorial development through the CSO-LA thematic programme*. Workshop on Decentralisation Reforms - Local Governance - Local and Territorial Development, Brussels, Belgium.

Altieri, M. A. (2002). Agroecology: The science of natural resource management for poor farmers in marginal environments. *Agriculture, Ecosystems & Environment*, *93*(1), 1–24. doi:10.1016/S0167-8809(02)00085-3

Arnold, C., & Tony, A. (2005). Privatization of public water services: The states' role in ensuring public accountability. *Pepperdine Law Review*, *32*(3), 561–604.

Augsburg, B., & Fouillet, C. (2010). Profit Empowerment: The Micro-Credit Institution's Mission Drift. *Perspectives on Global Development and Technology*, *9*(3-4), 327–355. doi:10.1163/156914910X499732

Backeberg, G. R., & Sanewe, A. J. (2006). The research and development strategy for water utilisation in agriculture–responding to diverse needs of farmers in South Africa. *Irrigation and Drainage*, *55*(3), 281–290. doi:10.1002/ird.232

Bazzani, G. M. (2005). An integrated decision support system for irrigation and water policy design: DSIRR. *Environmental Modelling & Software*, *20*(2), 153–163. doi:10.1016/j.envsoft.2003.12.017

Biermann, F. (2001). Big science, small impacts—In the South?The influence of global environmental assessments on expert communities in India. *Global Environmental Change*, *11*(4), 297–309. doi:10.1016/S0959-3780(01)00008-5

Bigas, H. E. (2012). *The global water crisis: Addressing an urgent security issue.* Hamilton, Canada: UNU-INWEH.

Biplab, K. P. (2012). *Towards green Gujarat: End of water woes.* Gujarat, India: Government of Gujarat, Gujarat Ecology Commission.

Bunsen, J., & Rathod, R. (2016). Pipe assisted underground taming of surface floods The experience with Holiyas in north Gujarat. Colombo, Sri Lanka: IWMI-Tata Water Policy Program, IWMI-Water Policy Research Highlights, Issue No 2.

CA & IWMI. (2016). *Pre-selection of suitable sites for Bhungroo and PAVE Irrigation technology in Northern Ghana.* Accra, Ghana: IFDC. (unpublished)

CA. (2015). *Enhancing ecosystem services for smallholder women farmers through an integrated Bhungroo based research.* First progress report to the WLE. Accra, Ghana: IWMI. (unpublished)

CA. (2016). *Bhungroo Project: Water management solutions to support diversified cropping systems for men and women in Northern Ghana.* Second progress report to the WLE. Accra, Ghana: IWMI. (unpublished)

Cai, X., Ringler, C., & You, J.-Y. (2008). Substitution between water and other agricultural inputs: Implications for water conservation in a River Basin context. *Ecological Economics*, *66*(1), 38–50. doi:10.1016/j.ecolecon.2008.02.010

Câmara, M. R. O., Kalungu, J. W., & Coelho, R. D. (2010). Irrigation technology in South Africa and Kenya. *Ciência Rural*, *40*(10), 2218–2225. doi:10.1590/S0103-84782010005000175

Carriger, S., & Vallée, D. (2007). More crop per drop. *Rice Today*, *6*(2), 10–13.

CCAFS & FAO. (2014). Climate-Smart Agriculture: What is it? Why is it needed? Rome: FAO Secretariat.

Clarke, R., & King, J. (2004). *The Atlas of Water.* London, UK: Earthscan Publications Ltd.

Cooper, P. J. M., Dimes, J., Rao, K. P. C., Shapiro, B., Shiferawa, B., & Twomlow, S. (2008). Coping better with current climatic variability in the rain-fed farming systems of sub-Saharan Africa: An essential first step in adapting to future climate change? *Agriculture, Ecosystems & Environment*, *126*(1), 24–35. doi:10.1016/j.agee.2008.01.007

Cosgrove, W. J., & Rijsberman, F. R. (2000). *World Water Vision, Making Water Everybody's Business. World Water Council.* London, UK: Earthscan Publications Ltd.

DeFries, R., & Rosenzweig, C. (2010). Toward a whole-landscape approach for sustainable land use in the tropics. *Proceedings of the National Academy of Sciences of the United States of America*, *107*(46), 19627–19632. doi:10.1073/pnas.1011163107 PMID:21081701

DWAF. (2008). *WMA charges 2008*. Pretoria, South Africa: Department of Water Affairs and Forestry, Republic of South Africa.

EC. (2016). Supporting decentralisation, local governance and local development through a territorial approach Directorate-General for International Cooperation and Development. Brussels: EC.

FAO. (1997). Irrigation potential in Africa - a basin approach. Rome: FAO, FAO Land and Water Bulletin No. 4.

FAO. (2003). The irrigation challenge - increasing irrigation contribution to food security through higher water productivity canal irrigation systems. Rome: FAO, IPTRID Secretariat, IPTRID/FAO Issue paper No. 4.

FAO. (2010). *Climate-Smart Agriculture Policies, Practices and Financing for Food Security, Adaptation and Mitigation*. Rome: FAO Secretariat.

FAO. (2014). *The State of Food and Agriculture*. Rome: FAO Secretariat.

Foster, T. (2012). *Private sector provision of rural water services*. A Desk Study for Water For People.

Frantz, C., & Mayer, S. (2009). The emergency of climate change: Why are we failing to take action? *Analyses of Social Issues and Public Policy (ASAP)*, *9*(1), 205–222. doi:10.1111/j.1530-2415.2009.01180.x

Garnett, T., Appleby, M. C., Balmford, A., Bateman, I. J., Benton, T. G., Bloomer, P., ... Godfray, H. C. J. (2013). Sustainable Intensification in Agriculture: Premises and Policies. *Science*, *341*(6141), 33–34. doi:10.1126cience.1234485 PMID:23828927

Gleick, P. (2001). Making every drop count. *Scientific American, 284*(2), 40–45. doi:10.1038cientificamerican0201-40

Gleick, P. H. (2000). The changing water paradigm – A look at twenty-first century water resources development. *Water International, 25*(1), 127–138. doi:10.1080/02508060008686804

Goetz, M. (2013). The paradox of value: Water rates and the law of diminishing marginal utility. *Journal AWWA, 105*(9), 1-6.

Goyal, S. (2014). *World Water Day 2014: The Water-Energy-Food Nexus in India*. Academic Press.

GWP. (2005). *ToolBox for Integrated Water Resource Management*. Stockholm: GWP Secretariat. Available at: www.gwp.org

Haas, S., & Nagarajan, G. (2011). *Water Delivery through Payment Platform – MPESA Pushes the Rural Frontier*. Financial Services Assessment.

Hechler-Fayd'herbe, N. (2009, June). Micro Finance: Three Perspectives on the Achievements, Challenges and Future of Microfinance. *Credit Suisse Research Quarterly*.

Huq, S., Lim, B., Spanger-Siegfried, E., Burton, I., & Malone, E. (2004). *Adaptation Policy Frameworks for Climate Change: Developing Strategies, Policies and Measures*. Cambridge, UK: Cambridge University Press.

International, I. D. E. (2010). *Business Plan. IDE Microirrigation Services Central America*. Edinburgh, UK: IDE International.

IPCC. (2014). Summary for policymakers. In *Climate Change 2014: Impacts, Adaptation, and Vulnerability. Part A: Global and Sectoral Aspects. Contribution of Working Group II to the Fifth Assessment Report of the Intergovernmental Panel on Climate Change* (pp. 1–32). Cambridge, UK: Cambridge University Press.

Karago, C. (2009, December). A lifeline of water. *The Option*, 12-15.

Kassam, A. H., Molden, D., Fereres, E., & Doorenbos, J. (2007). Water productivity: Science and practice-introduction. *Irrigation Science, 25*(3), 185–188. doi:10.100700271-007-0068-x

Kenya, A. (2017). *Amiranfarmers kits*. Available at: www.amirankenya.com

Koppenjan, J. F. M., & Enserink, B. (2009). Public–private partnerships in urban infrastructures: Reconciling private sector participation and sustainability. *Public Administration Review*, *69*(2), 284–296. doi:10.1111/j.1540-6210.2008.01974.x

Kramer, M.R. (2011). Creating shared value. *Harvard Business Review, 89*(2), 62-77.

Lagger, A. (2011). *Access to affordable irrigation technology for small farmers in northern Ghana-An example of marketing to the base of the pyramid (Bop Version)* (Masters' Thesis). Münster, Germany: University of St. Gallen (HSG).

Luwesi, C. N. (2010). *Hydro-economic Inventory in Changing Environment– An assessment of the efficiency of farming water demand under fluctuating rainfall regimes in semi-arid lands of South-East Kenya*. Saarbrüken, Germany: Lambert Academic Publishing.

Luwesi, C. N., Akudugu, M. A., Beyene, A., Komba, B. B., & Nour, S. S. (2017). An introduction to financial sustainability in the water sector: Understanding the basic economic principles subtending financial flows. In Innovative Water Finance in Africa: A guide for Water Managers – Vol. 2: Economics and Principles of Water Finance Innovations. Academic Press.

Luwesi, C. N., Kinuthia, W., Mutiso, M. N., Akombo, R. A., Doke, D. A., & Ruhakana, A. (n.d.). Climate Change, Pro-Poor Schemes and Water Inequality—Strengths and Weaknesses of Kauti Irrigation Water Users' Association, Kenya. In A. Beyene (Ed.), *Agricultural Water Institutions in East Africa*. Uppsala, Sweden: NordiskaAfrikainstitutet.

Luwesi, C. N., Nkpeebo, A. Y., Osei-Owusu, Y., & Osei-Owusu, P. K. (2017). Towards integrated innovative technologies for sustainable provision and financing of agricultural groundwater in African Drylands. In Hydrology and Best Practices for Managing Water Resources in Arid and Semi-Arid Lands. IGI Global.

Luwesi, C.N., Obando, J.A., & Shisanya, C. (2012). *Hydro-economic inventory for sustainable livelihood in Kenyan ASALs: The case of MuooniDam*. CICD Series 9.

Luwesi, C. N., Obando, J. A., & Shisanya, C. A. (2017). The impact of a warming micro-climate on Muooni farmers of Kenya. *Agriculture*, *7*(3), 1–20.

Mathenge, M. M., Luwesi, C. N., Shisanya, C. A., Mahiri, I., Akombo, R. A., & Mutiso, M. N. (2014). Community participation in water sector governance in Kenya: A performance based appraisal of community water management systems in Ngaciuma-Kinyaritha catchment, Tana basin, Mount Kenya region. *International Journal of Innovative Research and Development*, *3*(5), 783–792.

McGranahan, G., & Satterthwaite, D. (2006). *Governance and getting the private sector to provide better water and sanitation services to the urban poor*. London, UK: IIED.

NEPAD. (2002). *Comprehensive Africa Agriculture Development Programme*. Rome: FAO Corporate Document Repository. Available at: http://www.fao.org/docrep/005/y6831e/y6831e00.htm#TopOfPage

Ng, A., & Loosemore, M. (2007). Risk allocation in the private provision of public infrastructure. *International Journal of Project Management*, *25*(1), 66–76. doi:10.1016/j.ijproman.2006.06.005

Obando, J. A., Luwesi, C. N., Beyene, A., Tshimanga, R. M., & Ruhakana, A. (2017). The core business of integrated water management: Achieving water governance performance. In Innovative Water Finance in Africa: A guide for Water Managers – Vol. 1: Water Finance Innovations' Context. Academic Press.

Osei-Owusu, P. K. (2014). *Overview of Bhungroo*. Accra, Ghana: CA. Available at: http://conservealliance.org/overview-of-bhungroo/#

Pérard, E. (2009). Water supply: Public or private? An approach based on cost of funds, transaction costs, efficiency and political costs. *Policy and Society*, *27*(3), 193–219. doi:10.1016/j.polsoc.2008.10.004

Perret, S., & Geyser, M. (2007). The full financial costs of irrigation services: A discussion on existing guidelines and implications for smallholder irrigation in South Africa. *Water S.A.*, *33*(1), 67–78.

Prasad, N. (2006). Privatisation results: Private sector participation in water services after 15 years. *Development Policy Review*, *24*(6), 669–692. doi:10.1111/j.1467-7679.2006.00353.x

Pretty, J. (1999). Can sustainable agriculture feed Africa? New evidence on progress, processes and impacts. *Environment, Development and Sustainability*, *1*(3-4), 253–274. doi:10.1023/A:1010039224868

Ringler, C. (2001). *Optimal Water Allocation in the Mekong River Basin. ZEF Discussion papers on Development Policy 38*. Bonn: University of Bonn, ZEF.

Rogers, P., De Silva, R., & Bhatia, R. (2002). Water is an economic good: How to use prices to promote equity, efficiency, and sustainability. *Water Policy*, *4*(1), 1–17. doi:10.1016/S1366-7017(02)00004-1

Romeo, L. G. (2005). A Territorial Approach to Local Development; The missing decentralization-development link? In *Workshop on Decentralization Reforms, Local Governance, Local and Territorial Development*. Brussels: New York University, International LLC.

Rosegrant, M. W., Cai, X., & Cline, S. A. (2002). *World water and food to 2025: dealing with scarcity. IFPRI-2020 Vision - International Water Management book*. Washington, DC: IFPRI.

Rosegrant, M. W., & Cline, S. A. (2003). Global food security: Challenges and policies. *Science, 302*(5652), 1917–1919. doi:10.1126cience.1092958 PMID:14671289

Rosegrant, M. W., Ringler, C., McKinney, D. C., Cai, X., Keller, A., & Donosso, G. (2000). Integrated economic-hydrologic water modelling at the basin scale: The Maipo river basin. *Agricultural Economics, 24*, 33–46.

Russell, C. S., Clark, C. D., & Schuck, E. C. (2007). Economic Instruments for Water management in the Middle East and North Africa. *International Journal of Water Resources Development, 23*(4), 659–677. doi:10.1080/07900620701488596

Shah, T., Singh, O. P., & Mukherji, A. (2006). Some aspects of South Asia's groundwater irrigation economy: Analyses from a survey in India, Pakistan, Nepal Terai and Bangladesh. *Hydrogeology Journal, 14*(3), 286–309. doi:10.100710040-005-0004-1

Shah, T., Verma, S., Durga, N., Rajan, A., Goswami, A., & Palrecha, A. (2016). *HarKhetKoPani(WatertoEveryFarm)RethinkingPradhanMantriKrishiSinchaiYojana (PMKSY)*. Colombo: IWMI, IWMI-Tata Water Policy Program, IWMI-TATA Policy Paper.

Singh, K. S., & Lal, R. B. (2003). *People of India - Gujarat (Part 1): Anthropological Survey of India*. Delhi: Popular Prakashan.

Smith, P., Ashmore, M. R., Black, H. I. J., Burgess, P. J., Evans, C. D., Quine, T. A., ... Orr, H. G. (2013). The role of ecosystems and their management in regulating climate, and soil, water and air quality. *Journal of Applied Ecology, 50*(4), 812–829. doi:10.1111/1365-2664.12016

Subramanian, V. (2001). Water Quality and Quantity. In V. Subramanian & A. Ramanathan (Eds.), Ecohydrology. New Delhi: Capital Publishing Company.

Tardieu, H., & Préfol, B. (2002). Full cost or "sustainable cost" pricing in irrigated agriculture. Charging for water can be effective but is it sufficient? *Irrigation and Drainage, 51*(2), 97–107. doi:10.1002/ird.44

The Times of India. (n.d.).Retrieved from http://timesofindia.indiatimes.com/

Tiffen, M., & Mortimore, M. (2002). Agroecosystems regaining the high ground: Reviving the Hillsides of Machakos. *World Resources, 2000-2001*, 149–158.

Trabacchi, C., & Mazza, F. (2015). *Emerging solutions to drive private investment in climate resilience*. Venice: CPI. Available at: http://climatepolicyinitiative.org/publication/emerging-solutions-to-drive-privateinvestment-in-climate-resilience/

UN. (2016). *Multi-stakeholder forum on science, technology and innovation for the Sustainable Development Goals: Summary by the Co-Chairs Note by the President of the Economic and Social Council*. High-level political forum on sustainable development Convened under the auspices of the Economic and Social Council. New York, NY: UN Economic and Social Council. Retrieved from http://www.un.org/ga/search/view_doc.asp?symbol=E/HLPF/2016/6&Lang=E

UN. (2017). Role of technology in implementing and monitoring the SDGs. Concept Note. In *6th Economic and Social Council Youth Forum*. New York, NY: UN Headquarters, Trusteeship Council Chamber. Available at: https://www.un.org/ecosoc/sites/www.un.org.ecosoc/files/files/en/2017doc/Role-of-technology-in-implementing-the-SDGs.pdf

UNEP. (2009). *Atlas of Our Changing Environment*. DEWA.

USAID. (2014). *Local governance and decentralization assessment: implications of proposed reforms in Ukraine*. Washington, DC: USAID. Available at: http://pdf.usaid.gov/pdf_docs/pa00k59f.pdf

Van Hofwegen, P. (2006). *Enhancing Access to Finance for Local Governments and Financing Water for Agriculture*. Report no. 1 of the (Gurria) Task Force on Financing Water for All (WWC/ GWP).

Wallace, J. S. (2000). Increasing agricultural water use efficiency to meet future food production. *Agriculture, Ecosystems & Environment*, *82*(1-3), 105–119. doi:10.1016/S0167-8809(00)00220-6

Walsh, A. (2012). The Commodification of the Public Service of Water: A Normative Perspective. *Public Reason*, *3*(2), 90–106.

Wambua, P. P., Luwesi, C. N., Njuguna, J. W., & Ngoma, J.-B. N. (2017). Financing water management and developmemt: An Introduction to Financial Instruments and Operations. In Innovative Water Finance in Africa: A guide for Water Managers – Vol. 2: Economics and Principles of Water Finance Innovations. Academic Press.

Wichelns, D. (2002). An economic perspective on the potential gains from improvements in irrigation water management. *Agricultural Water Management*, *52*(3), 233–248. doi:10.1016/S0378-3774(01)00134-2

WLE. (2015). *Water, land and ecosystem in Africa (WLE in Africa) - 2015 project review and stakeholder engagement*. WLE workshop report. Available at: https:// waterlandandecosystems.wikispaces.com/

WUR. (2017). *'KrishakBandhu' (KB) buckets and quarter acre kits*. Available at: www.wur.nl

Compilation of References

Abida & Harikrishna. (2008). Study on the Quality of Water in Some Streams of Cauvery River. *Journal of Chemistry*, *5*(2), 377–384.

Abila, R., Muthangya, M., Mutuku, E., Mutati, K., Munguti, M., & Musyoka, C. M. (2012). Physico - chemical and bacteriological quality assessment of shallow wells in Kitui town, Kenya. *Jnl. Env. Sci. Wat. Res*, *1*(2), 27–33.

Abu-Zeid, M. (2001). Water pricing in Irrigated agriculture. *International Journal of Water Resources Development*, *17*(4), 527–538. doi:10.1080/07900620120094109

Adhikari, U., Nejadhashemi, A. P., & Herman, M. R. (2016). *A review of climate change impacts on water resources in East Africa.* American Society of Agricultural and Biological Engineers (ASABE). doi:10.13031/trans.58.10907

Aditya, S., Lal, M., Silva, N. S., & McCartney, M. (2017). Understanding the hydrological impacts of climate change in the Tana River Basin, Kenya. *IWMI Working Paper*, (178).

AGRA. (2009). *AGRA: Early Accomplishments, Foundation for Growth.* Available at: http://www.agra-alliance.org

Akkermans, J., Brenninkmeijer, V., Huibers, M., & Blonk, R. W. B. (2013). Competencies for the contemporary career: Development and preliminary validation of the Career Competencies Questionnaire. *Journal of Career Development*, *40*(3), 245–267. doi:10.1177/0894845312467501

Akotsi, E., & Gachanja, M. (2004). *Changes in Forest Cover in Kenya"s five "water towers" 2000-2003.* Kenya Forests Working Group.

Albert, M. J. (2004). *Hydraulic Analysis and Double mass curves of the middle Rio Grande from Cochiti to San Marcial, New Mexico* (Masters Thesis). Colorado State University. Fort Collins, CO.

Albert, M. G. K., Francis, W. Z., & Xuebin, Z. (2009). *Guidelines on analysis of extremes in a changing climate in support of informed decisions for adaptation.* World Meteorological Organization.

Alemu, B., & Kidane, D. (2014). The Implication of Integrated Watershed Management for Rehabilitation of Degraded Lands: Case Study of Ethiopian Highlands. *Journal of Agriculture and Biodiversity Research*, 78–90.

Alexander, L.V., Zhang, X., Peterson, T.C., Caesar, J., Gleason, B., Klein Tanka, M.G., … Vazquez-Aguirre, J.L. (2006). Global observed changes in daily climate extremes of temperature and precipitation. *J Geophys Res, 111*(D5), 109. doi:10.1029/2005JD006290

Alfonso, G. S. (2015). *How to promote territorial development through the CSO-LA thematic programme*. Workshop on Decentralisation Reforms - Local Governance - Local and Territorial Development, Brussels, Belgium.

Altieri, M. A. (2002). Agroecology: The science of natural resource management for poor farmers in marginal environments. *Agriculture, Ecosystems & Environment, 93*(1), 1–24. doi:10.1016/S0167-8809(02)00085-3

Ames, D. P., Horsburgh, J., Goodall, J., Whiteaker, T., Tarboton, D., & Maidment, D. (2009). Introducing the open source CUAHSI Hydrologic Information System desktop application (HIS Desktop). In *18th World IMACS Congress and MODSIM09 International Congress on Modelling and Simulation, Modelling and Simulation Society of Australia and New Zealand and International Association for Mathematics and Computers in Simulation* (pp. 4353-4359). Academic Press.

Amougu, O., Descroix, L., Yero, K. S., Le Brreton, E., Mamadou, I., Ali, A., ... Belleudy, P. (2010). Increasing Riverflow in the Sahel. *Water (Basel), 2*(2), 170–199. doi:10.3390/w2020170

Andrews, J. D. (1994) Community participation in rural water village water supply in Botswana. In Proceedings: Integrated water resources management workshop. Gaborone: University of Botswana.

Anyamba, A., & Tucker, C. J. (2005). Analysis of Sahelian Vegetation Dynamics using NOAA-AVHRR NDVI data from 1981 -2003. *Journal of Arid Environments, 63*(3), 596–614. doi:10.1016/j.jaridenv.2005.03.007

APHA (American Public Health Association). (1992). *Standard methods for the examination of water and wastewater* (18th ed.). Washington, DC: American Public Health Association.

Arnold, C., & Tony, A. (2005). Privatization of public water services: The states' role in ensuring public accountability. *Pepperdine Law Review, 32*(3), 561–604.

Arntzen, R., Chanda, R., Nkambwe, M., Ringrose, S., Sefe, F.T.K., & Vanderpost, C. (1994). *Desertification and possible solutions in the mid-Boteti River area: A Botswana case study for the Intergovernmental Convention to Combat Desertification (INCD). Findal Report Volume 1 – The Desertification process and its impact*. Ministry of Agriculture, Government of Botswana.

Artan, G., Gadain, H. M., Muthusi, F. M., & Muchiri, P. W. (2007). *Improving Flood Forecasting and Early Warning in Somalia, Feasibility Study, Technical Report No W-10*. Nairobi, Kenya: FAO-SWALIM.

ASCE. (1993). Criteria for Evaluation of Watershed Models. Journal of Irrigation and Drainage Engineering. *ASCE, 119*(3), 429 - 442.

ASCE. (1993). Criterion for Evaluation of Watershed Models. *Journal of Irrigation and Drainage Engineering, 119*(3), 429 - 442.

Augsburg, B., & Fouillet, C. (2010). Profit Empowerment: The Micro-Credit Institution's Mission Drift. *Perspectives on Global Development and Technology, 9*(3-4), 327–355. doi:10.1163/156914910X499732

Awuor, C. B., Orindi, V. A., & Adwera, A. O. (2008). Climate change and coastal cities: The case of Mombasa, Kenya. *Environment and Urbanization, 20*(1), 231–242. doi:10.1177/0956247808089158

Ayuyo, I. O., & Sweta, L. (2014). Land cover and land use mapping and change detection of Mau complex in Kenya using geospatial technology. *International Journal of Science and Research, 3*(3), 767–178.

Backeberg, G. R., & Sanewe, A. J. (2006). The research and development strategy for water utilisation in agriculture–responding to diverse needs of farmers in South Africa. *Irrigation and Drainage, 55*(3), 281–290. doi:10.1002/ird.232

Bakr, N., Weindorf, D. C., Bahnassy, M. H., Marei, S. M., & El-Badawi, M. M. (2010). Monitoring land cover changes in a newly reclaimed area of Egypt using multi-temporal Landsat data. *Journal of AppliedGeography, 30*, 592–605.

Baldyga, T. J., Miller, N. S., Driesse, L. K., & Gichaba, N. C. (2007). Assessing land coverchange in Kenya's Mau Forest region using remotelysensed data. *African Journal of Ecology*. doi:10.1111/j.1365-2028.2007.00806.x

Bamba, I. (2010). *Anthropisation et dynamique spatio-temporelle de paysages forestiers en République Démocratique du Cong (PhD thesis)*. Brussels: Free University of Brussels.

Bamutaze, Y., Thiemann, S., & Förch, G. (2014). *Integrated Watershed Management- A Tool for Urban Water Security*. Workshop Results from Mbale, Uganda. Mbale: Makerere University, IWM Expert GmbH and Freie Universität Berlin.

Barima, Y. S. S. (2009). *Dynamique, fragmentation et diversité végétale des paysages forestiers en milieux de transition forêt-savane dans le Département de Tanda (Côte d'Ivoire)* (PhD thesis). Brussels: Free University of Brussels.

Bates, B. C., Kundzewicz, Z. W., Wu, S., & Palutikof, J. P. (2008). *Climate change and water*. IPCC Technical Paper VI. Geneva, Switzerland: IPCC. Retrieved from http://ipcc.ch/pdf/technicalpapers/climate-change-water-en.pdf

Batram, J. B., & Balance, R. (1996): Water quality monitoring: a practical guide to the design and implementation of freshwater quality studies and monitoring programmes. Published on behalf of UNESCO, WHO and UNEP by E$FN SPON.

Bazzani, G. M. (2005). An integrated decision support system for irrigation and water policy design: DSIRR. *Environmental Modelling & Software*, *20*(2), 153–163. doi:10.1016/j.envsoft.2003.12.017

Beimers, P. B., van Eick, A. J., Lam, K. S., & Roos, B (2001b). Building sand-storage dams, SASOL Foundation Kitui District, Kenya. Practical work report, Delft University of Technology.

Beimers, P. B., van Eick, A. J., Lam, K. S., & Roos, B. (2001a). Improved design sand-storage dams, Kitui District, Kenya. Project report, Delft University of Technology.

Béland, M., Goïta, K., Bonn, F., & Pham, T. T. H. (2006). Assessment of land-cover changes related to shrimp aquaculture using remote sensing data: A case study in the GiaoThuy District, Vietnam. *International Journal of Remote Sensing*, *27*(8), 1491–1510. doi:10.1080/01431160500406888

Bendsen, H., & Meyer, T. (2002). *The dynamics of land use systems in Ngamiland; changing livelihood options and strategies*. Harry Oppenheimer Okavango Research Centre, University of Botswana, Maun.

Beven, K. (2001). *Rainfall-runoff modelling: The Primer*. John Wiley & Sons Ltd.

Beven, K., & Germann, P. (1982). Macropores and water flow in soils. *Water Resources Research*, *18*(5), 1311–1325. doi:10.1029/WR018i005p01311

Bhatt, L. R., Lacoul, P., Lekhak, H. D., & Jha, P. K. (1999). Physicochemical characteristics and phytoplankton of Taudha Lake Kathmandu. *Pollution Research*, *18*(14), 353–358.

Biermann, F. (2001). Big science, small impacts—In the South?The influence of global environmental assessments on expert communities in India. *Global Environmental Change*, *11*(4), 297–309. doi:10.1016/S0959-3780(01)00008-5

Bigas, H. E. (2012). *The global water crisis: Addressing an urgent security issue*. Hamilton, Canada: UNU-INWEH.

Biplab, K. P. (2012). *Towards green Gujarat: End of water woes*. Gujarat, India: Government of Gujarat, Gujarat Ecology Commission.

Biswas, A. K. (2004). Integrated water resources management: A reassessment. *Water International*, *29*(2), 248–256. doi:10.1080/02508060408691775

Bogaert, J., & Hong, S. K. (2009). Landscape ecology: monitoring landscape dynamics using spatial pattern metrics. In S. K. Hong, J. A. Lee, B. S. Ihm, A. Farina, Y. Son, E. S. Kim, & J. C. Choe (Eds.), *Ecological Issues in a Changing World* (pp. 109–131). Dordrecht: Kluwer Academic Publishers.

Bogaert, J., & Mahamane, A. (2005). Ecologie du paysage: Cibler la configuration et l"échelle spatiale. *Annales des Sciences Agronomiques du Bénin*, *7*, 39–68.

Bolstad, T. (1999). Coordinating methodologies for scaling landcover classifications from site-specific to global: Steps toward validating global map products. *Remote Sensing of Environment*, *70*(1), 16–28. doi:10.1016/S0034-4257(99)00055-3

Bonn, F. and Rochon, G. (1992). Précis de télédétection. *Principes et méthodes, 1.*

Booker, J. F., Howitt, R. E., Michelsen, A. M., & Young, R. A. (2012). Economics and the modeling of water resources and policies. *Natural Resource Modeling*, *25*(1), 168–218. doi:10.1111/j.1939-7445.2011.00105.x

Borst, L., & de Haas, S. A. (2006). *Hydrology of Sand Storage Dams - A case study in the Kiindu catchment, Kitui District, Kenya.* Free University of Amsterdam.

Bossenbroek, J., & Timmermans, T. (2003). *Setting up a measuring program at Kisayani, to measure the affected area by sand storage dams, Traineeship Report.* Delft University of Technology.

Botes, A., Henderson, J., Nakale, T., Nantanga, K., Schachtschneider, K., & And Seely, M. (2003). Ephemeral rivers and their development: Testing an approach to basin management committees on the Kuiseb River, Namibia. *Physics and Chemistry of the Earth Journal*, *28*(20-27), 853–859. doi:10.1016/j.pce.2003.08.028

Botts, M., Percivall, G., Reed, C., & Davidson, J. (2006, October). OGC® sensor web enablement: Overview and high level architecture. In *International conference on GeoSensor Networks* (pp. 175-190). Springer Berlin Heidelberg.

Brauman, K. A., Daily, G. C., Duarte, T. K., & Mooney, H. A. (2007). The nature and value ecosystem services: An overview highlighting hydrologic services. *Annual Review of Environment and Resources*, *32*(1), 67–98. doi:10.1146/annurev.energy.32.031306.102758

Brink, A. B., & Eva, H. D. (2009). Monitoring 25 Years of Land Cover Change Dynamics in Africa: A Sample based Remote Sensing Approach. *Applied Geography (Sevenoaks, England)*, *29*(4), 501–512. doi:10.1016/j.apgeog.2008.10.004

Bunsen, J., & Rathod, R. (2016). Pipe assisted underground taming of surface floods The experience with Holiyas in north Gujarat. Colombo, Sri Lanka: IWMI-Tata Water Policy Program, IWMI-Water Policy Research Highlights, Issue No 2.

Burger, A. S., Malda, W., & Winsemius, H. C. (2003). *Research to Sand-storage dams in Kitui district.* Delft University of Technology.

Busulwa, H. S., & Bailey, R. G. (2004). Aspects of the physico-chemical environment of the Rwenzori Rivers, Uganda. *African Journal of Ecology*, *42*(1), 87–92. doi:10.1111/j.1365-2028.2004.00467.x

CA & IWMI. (2016). *Pre-selection of suitable sites for Bhungroo and PAVE Irrigation technology in Northern Ghana.* Accra, Ghana: IFDC. (unpublished)

CA. (2015). *Enhancing ecosystem services for smallholder women farmers through an integrated Bhungroo based research*. First progress report to the WLE. Accra, Ghana: IWMI. (unpublished)

CA. (2016). *Bhungroo Project: Water management solutions to support diversified cropping systems for men and women in Northern Ghana*. Second progress report to the WLE. Accra, Ghana: IWMI. (unpublished)

Cai, X., Ringler, C., & You, J.-Y. (2008). Substitution between water and other agricultural inputs: Implications for water conservation in a River Basin context. *Ecological Economics, 66*(1), 38–50. doi:10.1016/j.ecolecon.2008.02.010

Câmara, M. R. O., Kalungu, J. W., & Coelho, R. D. (2010). Irrigation technology in South Africa and Kenya. *Ciência Rural, 40*(10), 2218–2225. doi:10.1590/S0103-84782010005000175

Campbell, J. B. (2002). *Introduction to remote sensing* (3rd ed.). New York, NY: The Guilford Press.

Carriger, S., & Vallée, D. (2007). More crop per drop. *Rice Today, 6*(2), 10–13.

Cashdan, E. (1987). Trade and Its Origins on the Botletli River, Botswana. *Journal of Anthropological Research, 43*(2), 121–138. doi:10.1086/jar.43.2.3630221

Cassman, K. G. (1999). Ecological intensification of cereal production systems: Yield potential, soil quality, and precision agriculture. *Proceedings of the National Academy of Sciences of the United States of America, 96*(11), 5952–5959. doi:10.1073/pnas.96.11.5952 PMID:10339523

CCAFS & FAO. (2014). Climate-Smart Agriculture: What is it? Why is it needed? Rome: FAO Secretariat.

Centeno, J. A., Cook, A., & Weinstein, P. (2006). Environmental Toxicology and exposure to natural dust: The role of trace elements. *Chinese Journal of Geochemistry, 25*(1), 222–223. doi:10.1007/BF02840174

Central Statistics Office. (2003). *National statistical tables, August 2001. Central statistics Office, Botswana*. Gaborone: Department of Printing and Publishing Services.

Chahine, M. T. (2007). The hydrological cycle and its influence on climate. *Nature, 359*(6394), 373–380. doi:10.1038/359373a0

Chapman, D. (1996). Water quality assessments; a guide to the use of biota, sediments and water in environmental monitoring (2nd ed.). Published on behalf of UNESCO, WHO and UNEP by E$FN SPON.

Charkhabi, A. H., & Sakizadeh, M. (2006). Assessment of spatial variation of water quality parameters in the most polluted branch of the Anzali Wetland, Northern Iran. *Polish Journal of Environmental Studies, 15*(3), 395–403.

Chave, Andalo, Brown, Cairns, Chambers, Eamus, … Folster. (2005). *Comparison of various remote sensing data sources in the retrieval of forest stand*. Academic Press.

Chiwa, R. (2012). *Effects of land Use and Land Cover Changes on the Hydrology of Weruweru-Kiladeda Sub-catchment in Pangani River Basin Tanzania* (Unpublished Master's Thesis). Kenyatta University, Nairobi, Kenya.

Chow, Maidement, & Mays. (1988). *Applied Hydrology*. McGraw Hill.

Chowdhury, J. U., Stedinger, J. R., & Lu, L.-H. (1991). Goodness of fit tests for regional generalized extreme value flood distributions. *Water Resources Research*, *27*(7), 1765–1776. doi:10.1029/91WR00077

Ciza, K. (2015). Identification des moteurs de déforestation dans la région d'Isangi. République démocratique du Congo.

Clarke, R., & King, J. (2004). *The Atlas of Water*. London, UK: Earthscan Publications Ltd.

Collier, P., Conway, G., & Venables, T. (2008). Climate change and Africa. *Oxford Review of Economic Policy*, *24*(2), 337–353. doi:10.1093/oxrep/grn019

Congalton, R. G., & Green, K. (1999). *Assessing the accuracy of remotely sensed data: Principles and practices*. Boca Raton, FL: Lewis Publishers.

Conner, L. G., Ames, D. P., & Gill, R. A. (2013). HydroServer Lite as an open source solution for archiving and sharing environmental data for independent university labs. *Ecological Informatics*, *18*, 171–177. doi:10.1016/j.ecoinf.2013.08.006

Cooper, P. J. M., Dimes, J., Rao, K. P. C., Shapiro, B., Shiferawa, B., & Twomlow, S. (2008). Coping better with current climatic variability in the rain-fed farming systems of sub-Saharan Africa: An essential first step in adapting to future climate change? *Agriculture, Ecosystems & Environment*, *126*(1), 24–35. doi:10.1016/j.agee.2008.01.007

Cosgrove, W. J., & Rijsberman, F. R. (2000). *World Water Vision, Making Water Everybody's Business. World Water Council*. London, UK: Earthscan Publications Ltd.

County Government of Bomet. (2014). *Medium Term County Fiscal Strategy Paper*. County Government of Bomet.

Cox, S. (2007). *OGC Implementation Specification 07-022r1: Observations and Measurements-Part 1-Observation schema*. Open Geospatial Consortium. Tech. Rep.

Cyber-nook.com. (2011). *Quotes from Poor Richard's Almanac, 1746*. Available on: http://www.cyber-nook.com

Daolan, Z., David, O.W., & Zhanqing, H. (1997). *Rates and patterns of landscape change between 1972 and 1988 in the Schangai Mountain area of China and North Korea*. Academic Press.

DeFries, R., & Rosenzweig, C. (2010). Toward a whole-landscape approach for sustainable land use in the tropics. *Proceedings of the National Academy of Sciences of the United States of America*, *107*(46), 19627–19632. doi:10.1073/pnas.1011163107 PMID:21081701

Desclée, B., Bogaert, P., & Defourny, P. (2006). Forest change detection by statistical object-based method. *Remote Sensing of Environment*, *102*(1-2), 1–11. doi:10.1016/j.rse.2006.01.013

Desclée, B., deWasseige, C., Bogaert, P., & Defourny, P. (2006). Tropical forest monitoring by object-based detection: Towards an automated method in an operational perspective. *Proceedings of the 1st international conference on object-based image analysis (OBIA 06).*

Descroix, L., Esteves, M., Soulex, Y. K., Rajot, J. L., Malain, M. A., Boubkrois, S., . . . Mamadou, I. (2011). Runoff Evolution accoeding to Land use Change in a Small Sahelian Catchment. *Hydrology and Earth System Science Discussion Paper, 8*, 1569-1607.

Dessu, S. B., & Melesse, A. M. (2012). Modelling the rainfall–runoff process of the Mara River basin using the Soil and Water Assessment Tool. *Hydrological Processes*, *26*(26), 4038–4049. doi:10.1002/hyp.9205

Dessu, S. B., & Melesse, A. M. (2013). Impact and uncertainties of climate change on the hydrology of the Mara River basin, Kenya/Tanzania. *Hydrological Processes*. doi:10.1002/hyp.9434

Dessu, S. B., Melesse, A. M., Bhat, M. G., & McClain, M. E. (2014). Assessment of water resources availability and demand in the Mara River Basin. *Catena*, *115*, 104–114. doi:10.1016/j.catena.2013.11.017

DFID. (2003). *Handbook for the Assessment of Catchment Water Demand and Use*. HR Wellingford U.K.

Dhembere, A. J., Pandhe, G. M., & Singh, C. R. (1998). Groundwater characteristics and their significance with special reference to public health in Pravara area. *Maharastra. Poll. Res.*, *17*(1), 87–90.

DHI Water, . (2008). *Environment, Inc. and Lane Council of Governments June*. McKenzie River Mike Basin Model.

DHI. (2003). *Mike Basin: Rainfall–Runoff Modelling Reference Manual*. Copenhagen, Denmark: Danish Hydraulic Institute.

DHI. (2006). *Mike Basin Users Manual*. Copenhagen, Denmark: DHI Water and Environment.

Di Gregorio & Jansen. (1998). New Concepts for a Land Cover Classification System. *Land (Basel)*, *1*, 55–65.

Doerr, S. H., Shakesby, R. A., & Walsh, R. P. D. (2000). Soil water repellency: Its causes, characteristics and hydro-geomorphological significance. *Earth-Science Reviews*, *51*(1-4), 33–65. doi:10.1016/S0012-8252(00)00011-8

Dore, T., Makowski, D., Malezieux, E., Munier-Jolain, N., Tchamitchian, M., & Tittonell, P. (2011). Facing up to the paradigm of ecological intensification in agronomy: Revisiting methods, concepts and knowledge. *European Journal of Agronomy*, *34*(4), 197–210. doi:10.1016/j.eja.2011.02.006

Dosio, A. (2017). Projections of temperature and heat waves for Africa with an ensemble of CORDEX Regional Climate Models. *Climate Dynamics*, *49*(1-2), 493–519. doi:10.100700382-016-3355-5

Dosio, A., & Panitz, H. J. (2015). Climate change projections for CORDEX-Africa with COSMOCLM regional climate model and differences with the driving global climate models. *Climate Dynamics*. doi:10.100700382-015-2664-4

Dracup, J. A., Lee, K. S., & Paulson, E. G. Jr. (1980). On the definition of Droughts. *Water Resources Research*, *16*(2), 297–302. doi:10.1029/WR016i002p00297

Droogers, P., Immerzeel, W. W., Terink, W., Hoogeveen, J., Bierkens, M. F. P., van Beek, L. P. H., & Debele, B. (2012). Water resources trends in Middle East and North Africa towards 2050. *Hydrology and Earth System Sciences*, *16*(9), 3101–3114. doi:10.5194/hess-16-3101-2012

DTRP. (1996). National Settlement Policy. Department of Town and Regional Planning. Government Printer.

DWAF. (2008). *WMA charges 2008*. Pretoria, South Africa: Department of Water Affairs and Forestry, Republic of South Africa.

EC. (2016). Supporting decentralisation, local governance and local development through a territorial approach Directorate-General for International Cooperation and Development. Brussels: EC.

Edwards, W. M., Shipitalo, M. J., & Owens, L. B. (1993). Gas, water and solute transport in soils containing macropores: A review of methodology. *Geoderma*, *57*(1-2), 31–44. doi:10.1016/0016-7061(93)90146-C

Elmoustafa, A.M. (2013). Sustainable water management for seasonal rivers deltas, case study: Coporolo river, Angola. *IJRRAS, 15*(3).

Falkenmark, M., & Widstrand, C. (1992). Population and Water Resources: A delicate balance. Population Bulletin. Population Reference Bureau.

Falkenmark, M. (1989). The massive water scarcity now threatening Africa: Why isn't it being addressed? *Ambio*, 112–118.

Fallkenmark, M. (2003). Freshwater as shared between society and ecosystems: From divided approaches to integrated challenges. *Philosophical Transactions of the Royal Society of London. Series B, Biological Sciences*, *358*(1440), 2037–2049. doi:10.1098/rstb.2003.1386 PMID:14728797

Famien, A. M., Janicot, S., Ochou, A. D., Vrac, M., Defrance, D., Sultan, B., & Noël, T. (2017). *A bias-corrected CMIP5 dataset for Africa using CDF-t method. A contribution to agricultural impact studies*. Earth Syst. Dynam. Discuss. doi:10.5194/esd-2017-111

FAO. (1981). *Arid zone Hydrology for Agricultural Development*. Irrigational Drainage Paper.

FAO. (1997). Irrigation potential in Africa - a basin approach. Rome: FAO, FAO Land and Water Bulletin No. 4.

FAO. (2001). *Sustainable land management*. Available on: http://www.fao.org/nr/land/sustainable-land-management/en

FAO. (2003). The irrigation challenge - increasing irrigation contribution to food security through higher water productivity canal irrigation systems. Rome: FAO, IPTRID Secretariat, IPTRID/FAO Issue paper No. 4.

FAO. (2010). *Climate-Smart Agriculture Policies, Practices and Financing for Food Security, Adaptation and Mitigation*. Rome: FAO Secretariat.

FAO. (2011). *AQUASTAT online database*. Rome: Food and Agriculture Organization. Retrieved from http:// www.fao.org/nr/water/aquastat/data

FAO. (2014). *The State of Food and Agriculture*. Rome: FAO Secretariat.

Finegan, B., & Nasi, R. (2004). The biodiversity and conservation potential of shifting cultivation landscapes. In G. Schroth, G. A. B. da Fonseca, C. A. Harvey, C. Gascon, H. L. Vasconcelos, & A. M. N. Izac (Eds.), Agroforestry and biodiversity conservation in tropical landscapes. Island Press.

Flint. (2001). Fundamental concepts of recharge in the desert South West: A regional modeling perspective. In *Ground water hydrology recharge in desert environment; South Western United States*. Academic Press.

Foley, J. A., DeFries, R., Asner, G. P., Barford, C., Bo-nan, G., Carpenter, R. S., ... Coe, M. T. (2005). Global Con- sequences of Land Use. *Science*, *309*(5734), 570–574. doi:10.1126cience.1111772 PMID:16040698

Food and Agriculture Organization of the United Nations (FAO). (2011). *FAO's Preparedness and Response to the Horn of Africa Drought*. Retrieved from URL: http://www.fao.org/emergencies/emergency-types/drought/en/

Förch, G., Winnegge, R., & Thiemann, S. (Eds.). (2005). *DAAD Alumni Summer School 2005: Topics of Integrated Water Resources Management*. Weiterbilding in Siegen 18. FWU Water Resources Publications, Vol. 03/2005. Retrieved from http://fwu.fb10.unisiegen.de/bkd/summerschool.htm

Förch, G., Winnegge, R., & Thiemann, S. (Eds.). (2006). *DAAD Alumni Summer School 2006: Participatory Water Resources Management Plan*. Weiterbilding in Siegen 20, FWU Water Resources Publications, Vol. 05/2006. Retrieved from http://fwu.fb10.unisiegen. de/bkd/summerschool.htm

Förch, G., Winnegge, R., & Thiemann, S. (Eds.). (2007). DAAD Alumni Summer School: Participatory Monitoring of Sub-Catchment Management Planning – Bwathonaro Watershed, Kenya, Final Report, August 2007. Meru, Universitat Siegen and Kenyatta University.

Förch, G., Winnegge, R., & Thiemann, S. (Eds.). (2008). DAAD Alumni Summer School: Water Demand in Participatory Watershed Management – Ngaciuma-Kinyaritha Watershed, Kenya, Final Report, November 2008. Meru: Universität Siegen and Kenyatta University.

Förch, N., & Luwesi, C. N. (2009). *Major Challenges and Outcomes of the DAAD Alumni Summer Schools of Kenya (2006-2008).* A Conference poster presented during the CICD/ Universität Siegen's Conference 2009 on "Knowledge transfer in Development Co-Operation – Study, Research and Consulting", Siegen.

Förch, N., & Luwesi, C. N. (2010). *Challenges and Opportunities of Participatory Evaluation in the Context of North-South Cooperation: The Case of Capacity Building in Integrated Watershed Management in Kenya*. A Conference poster presented during the UK Evaluation Society (UKES) Annual Conference 2010 "Evaluation in a Turbulent World: Challenges, Opportunities and Innovation Practice", Birmingham.

Förch, N., & Luwesi, C. N. (2011). *Kenya - Capacity Building for Participatory Management of Water Resources*. Retrieved from http://afrik4r.org/profiles/blogs/kenya-capacity-building-for?xg_source=activity

Förch, N., & Ngonzo, C. (2009). *Capacity Building in Integrated Watershed Management in Kenya: An Independent Evaluation*. Paper presented during the 2009 Global Assembly Conference of the International Development Evaluation Association (IDEAS) on "Capacity Building Evaluation", Johannesburg, South Africa. Available on: www.ideas-int.org

Foster, T. (2012). *Private sector provision of rural water services*. A Desk Study for Water For People.

Frantz, C., & Mayer, S. (2009). The emergency of climate change: Why are we failing to take action? *Analyses of Social Issues and Public Policy (ASAP)*, *9*(1), 205–222. doi:10.1111/j.1530-2415.2009.01180.x

Fuller, W. H. (1974). *Desert Soils in Desert Biology*. Academic Press.

Fürst, J., Olang, L., Herrnegger, M., Wimmer, D., Omonge, P., & Kipkoech, E. (2015). *MaMa-Hydro: Exploring water resources planning and management options in the Nyangores headwater catchment of the vulnerable Maasai Mara River Basin in Kenya*. Final Technical Report No. 1. Vienna: Kommission für Entwicklungsforschung (KEF).

Fürst, J. (2004). *GIS in Hydrologie und Wasserwirtschaft*. Heidelberg, Germany: Herbert Wichmann.

Gao, J., & Skillcorn, D. (1998). Capability of SPOT XS data in producing detailed landcover maps at the urban – rural periphery. *International Journal of Remote Sensing*, *19*(15), 2877–2891. doi:10.1080/014311698214325

Garnett, T., Appleby, M. C., Balmford, A., Bateman, I. J., Benton, T. G., Bloomer, P., ... Godfray, H. C. J. (2013). Sustainable Intensification in Agriculture: Premises and Policies. *Science*, *341*(6141), 33–34. doi:10.1126cience.1234485 PMID:23828927

Ghazanfar, B., & Moat. (2003). Flora of tropical East Africa: quantitative analyses of the flora and its composition. *Proceedings of the XVIII session of AETFAT*.

Gichuki, F., Liniger, H., MacMillan, L. C., Schwilch, G., & Gikonyo, J. K. (1998). Scarce Water: Exploring Resource Availability, Use and Improved Management. *Eastern and Southern Africa. The Geographical Journal*, *8*, 15–27.

Githui, F., Mutua, F., & Bauwens, W. (2009). Estimating the impacts of land-cover change on runoff using the soil and water assessment tool (SWAT): Case study of Nzoia catchment, Kenya. *Hydrological Sciences Journal*, *54*(5), 899–908. doi:10.1623/hysj.54.5.899

Glantz, M. (Ed.). (2001). *Once Burned, Twice Shy: Lessons Learned from the 1997-98 El Niño*. Tokyo: The United Nations University.

Gleick, P. (2001). Making every drop count. *Scientific American*, *284*(2), 40–45. doi:10.1038c ientificamerican0201-40

Gleick, P. H. (1998). *The World's Water: The Biennial Report on Freshwater Resources 1998–1999*. Island Press.

Gleick, P. H. (1998). Water in crisis: Paths to sustainable water use. *Ecological Applications*, *8*(3), 571–579. doi:10.1890/1051-0761(1998)008[0571:WICPTS]2.0.CO;2

Gleick, P. H. (2000). The changing water paradigm – A look at twenty-first century water resources development. *Water International*, *25*(1), 127–138. doi:10.1080/02508060008686804

Godard, V. (2005). Typologie des paysages forestiers du sud du massif de Fontainebleau après la tempête de Décembre 1999. Évaluation des dégâts forestiers à l"aided"un semis de points et d"imageriesatellitale optique. *Revue Internationale de Géomatique*, *15*(3), 281–302. doi:10.3166/ rig.15.281-302

Goetz, M. (2013). The paradox of value: Water rates and the law of diminishing marginal utility. *Journal AWWA*, *105*(9), 1-6.

GoK (Government of Kenya). (2007). Water Resources Management Rules, 2007. LN 171, Government Press.

GoK. (2002). Water Act 2002. *Kenya Gazette*.

GoK. (2007). *Kenya Vision 2030*. Kenya Government Printer.

GoK. (2010). The Constitution of Kenya 2010. *Kenya Gazette Supplement, 55*.

GoK. (2012). *Masterplan for the Conservation and Sustainable Management of Water Catchment Areas in Kenya*. Ministry of Environment and Mineral Resources.

GOK. (2012). *National Policy for Sustainable development of Northern Kenya and other arid lands*. Nairobi, Kenya: GOK.

GoK. (2013). *The National Water Master Plan 2030, Ministry f Environment, Water and Natural Resources, Water Resources management Authority*. JICA.

GOK. (2013). *The National Water Master Plan 2030, Ministry of Environment, Water and Natural Resources, Water Resources management Authority, JICA*. GOK.

Gordon, C., Cooper, C., Senior, C., Banks, H., Gregory, J. M., Hagemann, S., ... Wiltshire, A. J. (2013). Climate change impact on available water resources obtained using multiple global climate and hydrology models. *Earth Syst Dynam*, *4*(1), 129–144. doi:10.5194/esd-4-129-2013

Government of Kenya (GoK). (2008). *Sector Plan for Environment, Water and Sanitation, 2008-2012*. Author.

Government of Kenya (GoK). (2015). *Second National Communication to the UNFCCC*. Author.

Goyal, S. (2014). *World Water Day 2014: The Water-Energy-Food Nexus in India*. Academic Press.

GWP, Global Water Partnership. (2015). *Integrated Water Resources Management in Eastern Africa: Coping with "Complex Hydrology". Technical Focus Papers (TFP) No 7*. Stockholm: Elanders.

GWP. (2005). *ToolBox for Integrated Water Resource Management*. Stockholm: GWP Secretariat. Available at: www.gwp.org

GWP. (2015). *Global Water Partnership, Integrated Water Resources management in Eastern Africa: Coping with 'Complex' hydrology*. Technical Focus Paper. Available from: www.gwp.org

Haas, S., & Nagarajan, G. (2011). *Water Delivery through Payment Platform – MPESA Pushes the Rural Frontier*. Financial Services Assessment.

Hake, G., Grünreich, D., & Meng, L. (2002). *Kartographie: Visualisierung raum-zeitlicher Informationen*. Berlin: Walter de Gruyter.

Hastenrath, S. (1984). *The Glaciers of Equatorial East Africa*. Dordrecht: Reidel. doi:10.1007/978-94-009-6251-4

Hastenrath, S. (1991). *Climate Dynamics of the Tropics*. Dordrecht: Kluwer. doi:10.1007/978-94-011-3156-8

Hastenrath, S. (1995). Glacier recession on Mount Kenya in the context of the global tropics. *Boletín del Instituto Francés de Estudios Andinos*, *24*, 633–638.

Hastenrath, S. (2001). Variations of East African climate during the past two centuries. *Climatic Change*, *50*(1/2), 209–217. doi:10.1023/A:1010678111442

Hastenrath, S., & Greischar, L. (1997). Glacier recession on Kilimanjaro, East Africa, 1912–89. *Journal of Glaciology*, *43*(145), 455–459. doi:10.1017/S0022143000035048

Hastenrath, S., Rostom, R., & Caukwell, R. A. (1989). Variations of Mount Kenya's glaciers 1963–87. *Erdkunde*, *43*, 202–210.

Hay, J. E., & Mimura, N. (2005). Sea-level rise: Implications for water resources management. *Mitigation and Adaptation Strategies for Global Change*, *10*(4), 717–737. doi:10.100711027-005-7305-5

Hechler-Fayd'herbe, N. (2009, June). Micro Finance: Three Perspectives on the Achievements, Challenges and Future of Microfinance. *Credit Suisse Research Quarterly*.

Hellwig, D. H. R. (1973). Evaporation of water from sand.1: Experimental set-up and climatic influences. *Journal of Hydrology (Amsterdam)*, *18*(2), 93–108. doi:10.1016/0022-1694(73)90096-6

Hepworth, N., Agol, D., Von-Lehr, S., & O'Grady, K. (2011). *Exploring the value of water stewardship in Africa; Kenya case study*. Alliance for Water stewardship/Marks Spencer/GIZ.

Hill, S. (1998). The Value of El Niño Forecast Methods for the U.S. Winter Wheat Producer: Do They Differ? *Proceedings of the American Agricultural Economic Association (AAEA)*.

Hoff, H. (2011). *Understanding the Nexus*. Background Paper for the Bonn2011 Conference: The Water, Energy and Food Security Nexus. Stockholm Environment Institute.

Horsburgh, J. S., Tarboton, D. G., Maidment, D. R., & Zaslavsky, I. (2008). A relational model for environmental and water resources data. *Water Resources Research*, *44*(5). doi:10.1029/2007WR006392

Hosking, J. R. M., & Wallis, J. R. (1997). *Regional frequency analysis: an approach based on L-moments*. Cambridge, UK: Cambridge University Press. doi:10.1017/CBO9780511529443

Hulme, M., Doherty, R., Ngara, T., New, M., & Lister, D. (2001). African climate change: 1900-2100. *Climate Research*, *17*(2), 145–168. doi:10.3354/cr017145

Hung, N. Q., Babel, M. S., Weesakul, S., & Tripathi, N. K. (2009). An artificial neural network model for rainfall forecasting in Bangkok, Thailand. *Hydrology and Earth System Sciences*, *13*(8), 1413–1425. doi:10.5194/hess-13-1413-2009

Huq, S., Lim, B., Spanger-Siegfried, E., Burton, I., & Malone, E. (2004). *Adaptation Policy Frameworks for Climate Change: Developing Strategies, Policies and Measures*. Cambridge, UK: Cambridge University Press.

Hurni, H. (1988). *Ecological issues in the creation of famines in Ethiopia*. Paper presented at the National Conference on Disaster Prevention and Preparedness Strategy for Ethiopia.

Hussain, I., Hussain, Z., & Sial, M. H. (2011). Water balance, supply and demand and irrigation efficiency of Indus basin. *Pakistan Economic and Social Review*, *49*(1), 13–38.

Ichoku, C., Ellison, L. T., Willmot, K. E., Matsui, T., Dezfuli, A. K., Gatebe, C. K., ... Okonkwo, C. (2016). Biomass burning, land-cover change, and the hydrological cycle in Northern sub-Saharan Africa. *Environmental Research Letters*, *11*(9), 095005. doi:10.1088/1748-9326/11/9/095005

Ickowitz, Vinceti, Termote, Powell, Kehlenbeck, & Hunter. (2006). *The Contribution of Forests and Trees to Sustainable Diets*. Academic Press.

ICWE. (1992). *The Dublin statement and report of the conference*. International Conference on Water and Environment: Development Issues for the 21st century, Dublin, Ireland.

IFAD. (2012). *Challenges and opportunities for agricultural water management in West and Central Africa: lessons from IFAD experience*. International Fund for Agricultural Development.

Iloweka, E.M. (2004). *The deforestation of rural areasin the low Congo province*. Monit Assess.

International, I. D. E. (2010). *Business Plan. IDE Microirrigation Services Central America*. Edinburgh, UK: IDE International.

IPCC. (1996). *Climate change 1995: the science of climate change. Summary for Policy-makers. Contribution of working group 1 to the Second Assessment Report of the Intergovernmental Panel on Climate Change*. Cambridge, UK: CambridgeUniversity Press.

IPCC. (2000). *IPCC special report: Emission scenarios. Summary for policymakers*. Geneva, Switzerland: IPCC.

IPCC. (2001). IPCC Third Assessment Report of Working Group 1. In *Impacts, Adaptation and Vulnerability: Summary for Policy Makers*. IPCC.

IPCC. (2007). *Climate change 2007: the physical science basis: contribution of Working Group I to the Fourth Assessment Report of the Intergovernmental Panel on Climate Change*. Cambridge, UK: Cambridge University Press.

IPCC. (2007). *The physical Science Basis-summary for policy makers*. Geneva Switzerland: IPCC.

IPCC. (2013). Annex I: Atlas of Global and Regional Climate Projections. In *Climate Change 2013: The Physical Science Basis. Contribution of Working Group I to the Fifth Assessment Report of the Intergovernmental Panel on Climate Change*. Cambridge, UK: Cambridge University Press.

IPCC. (2014). Summary for policymakers. In *Climate Change 2014: Impacts, Adaptation, and Vulnerability. Part A: Global and Sectoral Aspects. Contribution of Working Group II to the Fifth Assessment Report of the Intergovernmental Panel on Climate Change* (pp. 1–32). Cambridge, UK: Cambridge University Press.

IPCC. (2014a). Climate change 2014: impacts, adaptation, and vulnerability. In Part B: regional aspects. Contribution of Working Group II to the Fifth Assessment Report of the Intergovernmental Panel on Climate Change. Cambridge University Press.

IPCC. (2014b). Climate change 2014: Synthesis report. In R. K. Pachauri & L. A. Meyer (Eds.), *Contribution of WorkingGroups I, II, and III to the Fifth Assessment Report of the IPCC*. Geneva, Switzerland: IPCC.

Ireson, A., Makropoulos, C., & Maksimovic, C. (2004). *Water Resources Modelling under Data Scarcity: Coupling MIKE BASIN and ASM Groundwater Model*. Academic Press.

IUCN Eastern and Southern Africa Programme. (2009). *The Pangani River Basin: A Situation Analysis* (2nd ed.). Nairobi, Kenya: International Union for Conservation of Nature.

IWMI, International Water Management Institute. (2007). *Water for Food, Water for Life - A Comprehensive Assessment of Water Management in Agriculture (Summary Edition by David Molden)*. London: Earthscan.

Jaglin, S. (2002). The Right to water versus cost recovery: Participation, urban water supply and the poor in sub-Saharan Africa. *Environment and Urbanization*, *4*(1), 231–245. doi:10.1177/095624780201400119

Jain, C. K. (2004). Groundwater quality of District Dehradun, Uttranchal. *Ind. J. Env. & Ecoplan.*, *8*(2), 475–484.

Jansen, R. (2002). *The Okavango Delta management plan project: application of an ecosystem-based planning approach*. Paper presented at the seventeenth Global Biodiversity Forum, Valencia, Spain.

Jensen, J. R. (2004). *Introductory digital image processing – A remote sensing perspective* (3rd ed.). Upper Saddle River, NJ: Prentice Hall.

Jensen, K. M. (2013). Viewpoint–swimming against the Current: Questioning development policy and practice. *Water Alternatives*, *6*, 276–283.

JICA. (2013). *Project on the Development of National Water Mater Plan 2030 for Kenya. Nippon Koei Co*. Ltd.

Jones, A., Breuning-Madsen, H., Brossard, M., Dampha, A., Deckers, J., Dewitte, O., ... Robert, Z. (2013). *Soil Atlas of Africa*. Luxembourg: Publications Office of the European Union.

Joseph, P. M. (2016). *Climate Change Impacts on Water Resources over The Upper Tana Catchment of Kenya* (Doctoral dissertation). University of Nairobi.

Juston, J., Jansson, P.-E., & Gustafsson, D. (2014). Rating curve uncertainty and change detection in discharge time series: Case study with 44-year historic data from the Nyangores River, Kenya. *Hydrological Processes*, *28*(4), 2509–2523. doi:10.1002/hyp.9786

Kadlec, J. (2014). *Hydrodata. Info: A Web Service For Hydrological Time Series Visualization*. Academic Press.

Kadlec, J., & Ames, D. P. (2011, June). Design and development of web services for accessing free hydrological data from the Czech Republic. In *International Symposium on Environmental Software Systems* (pp. 581-588). Springer Berlin Heidelberg. 10.1007/978-3-642-22285-6_63

Kagoda, P. A., & Ndiritu, J. G. (2009). *Forecasting of Daily Streamflow in the Luvuvhu River Catchment Using Artificial Neural Networks*. 10th WARFSA/WaterNet Symposium, Entebbe, Uganda.

Karago, C. (2009, December). A lifeline of water. *The Option*, 12-15.

Karanja, A., Kotut, K., & Gitonga, N. M. (2015). Assessment of Physico-chemical Properties of Kasarani Section of River Thome, Nairobi. *Journal of Geography. Environment and Earth Science International*, *3*(4), 1–5.

Kaser, G., & Noggler, B. (1991). Observations on Speke Glacier, Rwenzori Range, Uganda. *Journal of Glaciology*, *37*(127), 313–318. doi:10.1017/S0022143000005736

Kaser, G., & Noggler, B. (1996). Glacier fluctuations in the Rwenzori Range (East Africa) during the 20th century — a preliminary report. *Zeitschrift für Gletscherkunde und Glazialgeologie*, *32*, 109–117.

Kaser, G., & Osmaston, H. (2002). *Tropical Glaciers*. Cambridge, UK: Cambridge University Press.

Kassam, A. H., Molden, D., Fereres, E., & Doorenbos, J. (2007). Water productivity: Science and practice-introduction. *Irrigation Science*, *25*(3), 185–188. doi:10.100700271-007-0068-x

Kassian, L. M., Tenywa, M., Liwenga, E. T., Dyer, K. W., & Bamutaze, Y. (2017). Implication of climate change and variability on stream flow in Iringa region, Tanzania. *Journal of Water and Climate Change*, *8*(2), 336–347. doi:10.2166/wcc.2016.238

Kaufman, Y.J., Koren, I., Remer, L.A., Tanre, D., Ginoux, P., & Fan, S. (2005). Dust transport and deposition observed from the Terra-Modis spacecraft over the Atlantic Ocean. *Jour. Geographical Research, 110*(D10S12), 1-16.

Kawabata, Y., Kawa, M., Yamada, M., Nwona-Agyeman, S.O, Aparin,V., Ollibekov, B.J., … Kataya, M. A. (2012). Seasonal Changes in Water Quality of Rivers and Ground Water in Karakalpakstan, Uzbekistan. *Journal of Arid Land Studies*, 171-174.

Kenya, A. (2017). *Amiranfarmers kits*. Available at: www.amirankenya.com

Keyantash, J. A., & Dracup, J. A. (2002). The quantification of drought: An evaluation of drought indices. *Bulletin of the American Meteorological Society*, *83*(8), 1167–1180. doi:10.1175/1520-0477-83.8.1167

Kgathi, D. L., Ngwenya, B. N., & Wilk, J. (2007). *Shocks and rural livelihoods in the Okavango Delta* (Vol. 289). Botswana: Development Southern Africa.

KibsonConsult. (2010*). Appendix 8-3 Hydrological Assessment Report on Mukurumudzi River*. Author.

Kilonzo, F., Masese, F. O., Van Griensven, A., Bauwens, W., Obando, J., & Lens, P. N. L. (2014). Spatial-temporal variability in water quality and macro-invertebrate assemblages in the Upper Mara River basin, Kenya. *Physics and Chemistry of the Earth*, *67-69*, 93–104. doi:10.1016/j.pce.2013.10.006

Kiragu, G. M. (2009). *Assessment of suspended sediment loadings and their impact on the environmental flows of upper transboundary Mara River, Kenya* (MSc. Thesis). Jomo Kenyatta University of Agriculture and Technology (JKUAT), Nairobi, Kenya.

Kisaka, M. O., Mucheru-Muna, M., Ngetich, F. K., Mugwe, J. N., Mugendi, D., & Mairura, F. (2015). Rainfall variability, drought characterization, and efficacy of rainfall data reconstruction: Case of Eastern Kenya. *Advances in Meteorology.*

Kitheka, J. U. (2014). Assessment of modification of the Tana River runoff due to developments in the Upper Tana Basin. *Proceedings of the 2nd Hydrological Society of Kenya (HSK) Workshop: "Hydrology in Water Cooperation and Security for Sustainable Economic Development". Sub-Theme: Hydrology and Sustainable Development*, 178-191.

Kitheka, J. U. (2013). The hydrologic alteration of the Tana river and the impacts in the Lower Tana Basin. *Proceedings of the National Conference on the Tana River organized by the University of Nairobi and National Environment Management Authority (NEMA).*

Kithia, S. M. (1997). Landuse changes and their effects on sediment transport and soil erosion within the Athi drainage basin, Kenya. In *Human Impact on Erosion and Sedimentation Proceedings of Rabat Symposium S6*. IAHS Publ.no. 245.

Komakech, H. C., & van der Zaag, P. (2011). Understanding the emergence and functioning of river committees in a catchment of the Pangani basin, Tanzania. *Water Alternatives*, *4*(2), 197–222.

Komakech, H. C., van der Zaag, P., & van Koppen, B. (2012). The Last Will Be First: Water Transfers from Agriculture to Cities in the Pangani River Basin, Tanzania. *Water Alternatives*, *5*(3), 700–720.

Koning, N., & Ross, J. C. (1999). The continued influence of organic pollution on the water quality of the turbid Modder River. *Water S.A.*, *25*(3), 285–292.

Kooke, S. O., Sumba, L. A., & Kurgat, P. (2012). Rain water harvesting for drought mitigation and food security. In Water Dependencies: Systems under Stress and Societal Responses. Hydrological Society of Kenya proceedings. Kenyatta University.

Koppenjan, J. F. M., & Enserink, B. (2009). Public–private partnerships in urban infrastructures: Reconciling private sector participation and sustainability. *Public Administration Review*, *69*(2), 284–296. doi:10.1111/j.1540-6210.2008.01974.x

Kothari, C.R. (2004). *Research Methodology Methods and Techniques*. New Age International (P) Limited, Publishers.

Kramer, M.R. (2011). Creating shared value. *Harvard Business Review, 89*(2), 62-77.

Krause, P., Boyle, D. P., & Base, F. (2005). *Comparisons of Different Efficiency Criteria for Hydrological Model Assessment*. Advances in Geosciences. doi:10.5194/adgeo-5-89-2005

Krummel, J. R., O'Neill, R. V., Gardner, R. H., Sugihara, G., De Angelis, D. L., Milne, B. T., ... Graham, R. L. (1988). Indices of landscape pattern. *Landscape Ecology*, (3), 153–162.

Kumarswami, N. (1991). An approach towards assessment of dug well water quality by physico-chemical characteristics-a case study. *Pollution Research*, *10*(1), 13–20.

Kundu, P. M., China, S. S., & Chemelil, M. C. (2008). Automated extraction of morphologic and hydrologic properties for River Njoro catchment in Eastern Mau, Kenya. *Journal of Science, Technology. Education and Management, 1*, 14–27.

Lagger, A. (2011). *Access to affordable irrigation technology for small farmers in northern Ghana-An example of marketing to the base of the pyramid (Bop Version)* (Masters' Thesis). Münster, Germany: University of St. Gallen (HSG).

Lal, R. (2015). World Water Resources and Achieving Water Security. *Agronomy Journal, 107*(4), 1526–1532. doi:10.2134/agronj15.0045

Lane, J. (1977). *Transmission losses. National Engineering Handbook, Hydrology.* USDA, SCS.

Lange, J. (2005). Dynamics of transmission losses in a large arid stream channel. *Journal of Hydrology (Amsterdam), 306*(1-4), 112–126. doi:10.1016/j.jhydrol.2004.09.016

Lasage, R., Aerts, J., Mutiso, G. C. M., & de Vries, A. (2008). Potential for community based adaptations to droughts: Sand dams in Kitui, Kenya. *Physics and Chemistry of the Earth, 33*(1-2), 67–73. doi:10.1016/j.pce.2007.04.009

Lengstorf, J., & Hansen, T. B. (2009). *PHP for absolute beginners* (Vol. 1). Apress.

Lenton, R., & Muller, M. (2009). *Intergrated water resources management in practice. Better water management for development.* London: Earthscan.

Lillesand, T. M., & Kiefer, R. W. (1994). *Remote Sensing and Image Interpretation* (4th ed.). New York: John Wiley & Sons.

Lind, E. W., & Morrison, M. E. S. (1974). *East African Vegetation.* London: Longman.

Linsley, K. R., Kohler, A. M., & Paullus, H. L. (1982). *Hydrology for Engineers (3rd ed.).* New York: McGraw Hill Company.

Linsley, R. K., & Franzini, J. (1989). *Water Resources Engineering, 3rd.* New York: McGraw-Hill Inc.

Linsley, R. K., Kohler, M. A., & Franzini, J. B. (1975). *Hydrology for Engineers* (2nd ed.). New York: McGraw Hill Book Co.

Linsley, R. K., Kohler, M. A., & Paulhus, J. L. H. (1988). *Hydrology for engineers.* McGraw-Hill Book Company.

Liya, M. (2010). *Mango.* Modeling the Effect of Land Use and Climate Change Scenarios on the Water Flux of the Upper Mara River Flow.

Lomba, B. L. (2007). *Contribution à l"étude de la phytodiversité de la réserve forestière de la Yoko.* (Unpublished Master thesis). Kisangani: University of Kisangani, Faculty of Sciences.

Lu, D., Mausel, P., Brondizio, E., & Moran, E. (2004). Change detection technique. *International Journal of Remote Sensing, 25*(12), 2365–2407. doi:10.1080/0143116031000139863

Luwesi, C. N., Akudugu, M. A., Beyene, A., Komba, B. B., & Nour, S. S. (2017). An introduction to financial sustainability in the water sector: Understanding the basic economic principles subtending financial flows. In *Innovative Water Finance in Africa: A guide for Water Managers – Vol. 2: Economics and Principles of Water Finance Innovations.* Academic Press.

Luwesi, C. N., Förch, N., Mutiso, M. N., Akombo, R. A., Wambua, P. P., & Badr, E. (2014). A Performance Assessment and Evaluation of IWM Capacity Building Activities on Participatory Water Resource Management in Kenya. *Journal of Agri-Food and Applied Sciences, 2*(4), 93-108.

Luwesi, C. N., Nkpeebo, A. Y., Osei-Owusu, Y., & Osei-Owusu, P. K. (2017). Towards integrated innovative technologies for sustainable provision and financing of agricultural groundwater in African Drylands. In *Hydrology and Best Practices for Managing Water Resources in Arid and Semi-Arid Lands.* Hershey, PA: IGI Global.

Luwesi, C. N., Shisanya, C. N., & Obando, J. A. (2012). Warming and Greening: The Dilemma facing Green Water Economy under Changing Hydro-Climatic Conditions in Muooni Catchment (Machakos, Kenya). Saarbrüken: Lambert Academic Publishing.

Luwesi, C.N., Obando, J.A., & Shisanya, C. (2012). *Hydro-economic inventory for sustainable livelihood in Kenyan ASALs: The case of MuooniDam.* CICD Series 9.

Luwesi, C. N. (2010). *Hydro-economic Inventory in Changing Environment– An assessment of the efficiency of farming water demand under fluctuating rainfall regimes in semi-arid lands of South-East Kenya.* Saarbrüken, Germany: Lambert Academic Publishing.

Luwesi, C. N., Kinuthia, W., Mutiso, M. N., Akombo, R. A., Doke, D. A., & Ruhakana, A. (n.d.). Climate Change, Pro-Poor Schemes and Water Inequality—Strengths and Weaknesses of Kauti Irrigation Water Users' Association, Kenya. In A. Beyene (Ed.), *Agricultural Water Institutions in East Africa.* Uppsala, Sweden: NordiskaAfrikainstitutet.

Luwesi, C. N., Obando, J. A., & Shisanya, C. A. (2017). The impact of a warming micro-climate on Muooni farmers of Kenya. *Agriculture, 7*(3), 1–20.

Luwesi, C. N., Shisanya, C. A., & Obando, J. A. (2011). Toward a hydro-economic approach for risk assessment and mitigation planning for farming water disasters in semi-arid Kenya. In M. Savino (Ed.), *Risk Management in Environment, Production and Economy* (pp. 27–46). Rijeka: InTech.

M¨olg, T., Georges, C., & Kaser, G. (2003a). The contribution of increased incoming shortwave radiation to the retreat of the Rwenzori Glaciers, East Africa, during the 20th century. *International Journal of Climatology, 23*(3), 291–303. doi:10.1002/joc.877

Magole, L., & Thapelo, K. (2005). The impact of extreme flooding of the Okavango River on the livelihood of the molapo farming community of Tubu Village, Ngamiland Sub-district, Botswana. *Botswana Notes and Records, 37*, 125–137.

Mahamane, M., Hochschild, V., Schutz, A., & Nkokuma, J. (2016). *Fragmentation rate and landscape structure of the Tillabery landscape (Sahel) Region with reference to desertification.* Academic Press.

Mahananda, M. R. (2010). Physico-Chemical analysis of surface water and ground water of Bargarh District, Orissa, India. *International Journal of Research and Review in Applied Sciences*, *2*(3), 284–295.

Maidment, R. D. (Ed.). (1993). Handbook of Hydrology. McGraw Hill.

Makana, J.-R., & Thomas, S. C. (2005). Effects of light-gaps and litter removal on seedling performance in six African timber tree species. *Biotropica*, *37*(2), 227–237. doi:10.1111/j.1744-7429.2005.00030.x

Mango, L. M., Melesse, A. M., McClain, M. E., Gann, D., & Setegn, S. G. (2011). Land use and climate change impacts on the hydrology of the upper Mara River Basin, Kenya: Results of a modelling study to support better resource management. *Hydrology and Earth System Sciences*, *15*(7), 2245–2258. doi:10.5194/hess-15-2245-2011

Manning, N., & Seely, M. (2005). Forum for Integrated Resource Management (FIRM) in Ephemeral Basins: Putting communities at the centre of the basin management process. *Physics and Chemistry of the Earth Journal*, *30*(11-16), 886–893. doi:10.1016/j.pce.2005.08.035

Manzungu, E. (2004). *Water for all: Improving water resources governance in southern Africa.* The Gatekeeper Series No. 113 on the Natural Resources Group of the International Institute for Environment and Development.

Manzungu, E., & Kujinga, K. (2002). The theory and governance of water resources in Zimbabwe. *Zambezia*, *29*(2), 195–216.

Mariotti, A., Pan, Y., Zeng, N., & Alessandri, A. (2015). Long-term climate change in the Mediterranean region in the midst of decadal variability. *Climate Dynamics*, *44*(5-6), 1437–1456. doi:10.100700382-015-2487-3

Massdam, R., & Smith, D. G. (1994). New Zealand's national river water quality network. 2. Relationships between physico -chemical data and environmental factors. *New Zealand Journal of Marine and Freshwater Research*, *28*(1), 37–54. doi:10.1080/00288330.1994.9516595

Mathenge, M. M., Luwesi, C. N., Shisanya, C. A., Mahiri, I., Akombo, R. A., & Mutiso, M. N. (2014). Community participation in water sector governance in Kenya: A performance based appraisal of community water management systems in Ngaciuma-Kinyaritha catchment, Tana basin, Mount Kenya region. *International Journal of Innovative Research and Development*, *3*(5), 783–792.

Mati, B. M., Mutie, S., Gadain, H., Home, P., & Mtalo, F. (2008). Impacts of land-use/cover changes on the hydrology of the trans-boundary Mara River, Kenya/Tanzania. *Lakes and Reservoirs: Research and Management*, *13*(2), 169–177. doi:10.1111/j.1440-1770.2008.00367.x

Mattias, T. (2010). *Studies of the waterscape of Kilimanjaro, Tanzania Water management in hill furrow irrigation* (Doctoral Thesis). Norwegian University of Science and Technology.

Mazvimavi, D., & Mmopelwa, G. (2006). Access to water in gazetted and ungazetted rural settlements in Ngamiland, Botswana. *Physics and Chemistry of the Earth Journal, 31*(15-16), 713–722. doi:10.1016/j.pce.2006.08.036

Mazvimavi, D., & Wolski, P. (2006). Long-term variations of annual flows of the Okavango and Zambezi Rivers. *Physics and Chemistry of the Earth Journal, 31*(15-16), 944–951. doi:10.1016/j.pce.2006.08.016

McCarthy, T. S., Cooper, G. R. J., Tyson, P. D., & Ellery, W. N. (2000). Seasonal flooding in the Okavango Delta, Botswana – recent history and future prospects. *South African Journal of Science, 96*, 25–33.

McGarigal, K., & Marks, B. J. (1994). *Fragstats: Spatial Pattern Analysis Program for Quantifying Structure.* Department of Agriculture, Pacific Northwest Research Station General Technical Report PNW-GTR-351.

McGranahan, G., & Satterthwaite, D. (2006). *Governance and getting the private sector to provide better water and sanitation services to the urban poor.* London, UK: IIED.

McMahon, T. A. (1979). *Hydrological characteristics of arid zones. Proceedings the symbosium on the hydrology of areas of low precipitation.* IAHS Publication.

Meire, P., Coenen, M., Lombardo, C., & Robba, M. (2008). *Integrated Water Management - Practical Experiences and Case Studies.* Dordrecht, The Netherlands: Springer.

Mendelsohn, J., & El Obeid, S. (2004). *Okavango River: The flow of a lifeline.* Struik.

Mendelsohn, J., & Obeid, S. (2004). *Okavango River: The flow of lifeline.* Cape Town: Struik Publishers.

Merrey, D. J., Drechsel, P., Penning de Vries, F. W. T., & Sally, H. (2005). Integrating “livelihoods” into integrated water resources management: Taking the integration paradigm to its logical next step for developing countries. *Regional Environmental Change, 5*(4), 197–204. doi:10.100710113-004-0088-5

Mertz, J., Pons, T., & Moreaux, L. (2002). Photo-induced flip-flop of amphiphilic molecules in lipid bilayer membranes. *Physical Review Letters, 89.*

Miguel, V. P. (1989). *Engineering Hydrology Principles and Practices.* Prentice Hall.

Mikwa, N. (2012). *Assessment of forest degradation in the Yoko Reserve forest using remote sensing* (Unpublished Master thesis). Gent: University of Gent.

Milano, M., Ruelland, D., Fernandez, S., Dezetter, A., Fabre, J., & Servat, E. (2012). Facing global changes in the Mediterranean basin: How could the current water stress evolve by the medium-term? *Comptes Rendus Geoscience, 344*(9), 432–440. doi:10.1016/j.crte.2012.07.006

Minaya, V. G., McClain, M. E., Moog, O., Omengo, F., & Singer, G. A. (2013). Scale-dependent effects of rural activities on benthic macroinvertebrates and physico-chemical characteristics in headwater streams of the Mara River, Kenya. *Ecological Indicators*, *32*, 116–122. doi:10.1016/j.ecolind.2013.03.011

Mirault, A., Rossi, A., Dupin, N., Bruneval, P., Bel, A., Azarine, A., ... Surg, A. T. (2013). Recurrent tamponade and aortic dissection in syphilis. *Stansal Nov*, *96*, 111-113.

Mitchard, Saatchi, Lewis, Feldpausch, Gerard, Woodhouse, & Meir. (2011). *Comment on 'A first map of tropical Africa's above-ground biomass derived from satellite imagery'*. Academic Press.

Mogaka, H., Gichere, S., Davis, R., & Hirji, R. (2006). *Climate variability and water resources degradation in Kenya: Improving water resources development and management*. World Bank paper No.69.

Molle, F. (2008). Nirvana concepts, narratives and policy models: Insight from the water sector. *Water Alternatives*, *1*(1), 131–156.

Moorhead, J. E., Gowda, P. H., Singh, V. P., Porter, D. O., Marek, T. H., Howell, T. A., & Stewart, B. A. (2015). Identifying and Evaluating a suitable index for agricultural drought monitoring in the Texas high plains. *Journal of the American Water Resources Association*, 1–14.

Mora, C. F., Frazier, A. G., Longman, R. J., Dacks, R. S., Walton, M. M., Tong, E. J., ... Giambelluca, T. W. (2013). The projected timing of climate departure from recent variability. *Nature*, *502*(7470), 183–187. doi:10.1038/nature12540 PMID:24108050

Moran, E. F., Lambin, E. F., Turner, B. L. II, Geist, H., Agbola, S., Angelsen, A., ... Xu, J. (2001). The Causes of Land-Use and -Cover Change: Moving beyond the Myths. *Global Environmental Change*, *11*(4), 261–269. doi:10.1016/S0959-3780(01)00007-3

Mounir, Z. M., Ma, C. M., & Amadou, I. (2011). Application of Water Evaluation and Planning (WEAP): A Model to Assess Future Water Demands in the Niger River (In Niger Republic). *Modern Applied Science*, *5*(1), 38–49. doi:10.5539/mas.v5n1p38

MOWI. (2010). *Water sector status report*. Dar es Salaam, Tanzania: Ministry of Water and Irrigation.

Mozafari, G. A., Khosravi, Y., Abbasi, E., & Tavakoli, F. (2011). Assessment of Geostatistical Methods for Spatial Analysis of SPI and EDI Drought Indices. *World Applied Sciences Journal*, *15*, 474–482.

Mromba, C. (2012). *Effects of water abstraction on the riparian vegetation of Kiladeda River in Pangani river basin, Tanzania* (Unpublished Master's Thesis). Kenyatta University, Nairobi, Kenya.

MRWUA. (2011). *Nyangores River Sub Catchment Management Plan (SCMP)*. Bomet, Kenya: Mara River Water Users Association.

Muchiri, P. W. (2007). *Climate of Somalia. Technical Report No W-01*. Nairobi, Kenya: FAO-SWALIM.

Mugatsia, E. A. (2010). *Simulation and Scenario Analysis of Water Resources Management in Perkerra Catchment Using WEAP Model (Master's Thesis)*. Department of Civil and Structural Engineering, School of Engineering, Moi University. Retrieved from http://www.weap21.org

Mundia, C. N., & Ariya, M. (2006). Dynamics of Land use/cover changes and degradation of Nairobi city, Kenya. Land degradation development. *Wiley Online Library*, *17*(1), 97–108.

Munyao, J. N., Munyoki, J. M., Kitema, M. I., Kithuku, D. N., Munguti, J. M., & Mutiso, S. (2004). *Kitui sand dams: Construction and operation*. Nairobi, Kenya: SASOL Foundation.

Mutiga, J. K., Mavengano, S. T., Zhongbo, S., Woldai, T., & Becht, R. (2010). Water Allocation as a Planning Tool to Minimise Water Use Conflicts in the Upper EwasoNg'iro North Basin, Kenya. *Water Resources Management*, *24*(14), 3939–3959. doi:10.100711269-010-9641-9

Mutua, B. M., & Klik, A. (2007). Predicting daily streamflow in ungauged rural catchments: The case of Masinga catchment, Kenya. *Hydrological Sciences Journal*, *52*(2), 292–304. doi:10.1623/hysj.52.2.292

Mwendwa, G. (2014). *Rainwater harvesting in Mikuyuni* (Unpublished MSc thesis). Kenyatta University.

Nash, J. E., & Sutcliffe, V. J. (1970). River flow forecasting through conceptual models: Part 1. A discussion of principles. *Journal of Hydrology, 10*(3), 282 - 290.

Ndiiri, J. A. (2011). Application of the Water Evaluation and Planning (WEAP) Model to assess and plan future water demands in the Mara River Basin, Kenya. University of Dar es Salaam.

Neessen, D. (2004). Regional water balance modelling of a semi-arid catchment in South Kitui District, Kenya. Katholieke Universiteit Leuven Faculteit Landbouwkundige en Toegepaste Biologische Wetenschappen.

NEPAD. (2002). *Comprehensive Africa Agriculture Development Programme*. Rome: FAO Corporate Document Repository. Available at: http://www.fao.org/docrep/005/y6831e/y6831e00.htm#TopOfPage

New, M., Hewitson, B., Stephenson, D.B., Tsiga, A., Kruger, A., Manhique, A., … Lajoie, R. (2006). Evidence of trends in daily climate extremes over southern and West Africa. *J Geophys Res, 111*(D14), 102. doi:10.1029/2005JD006289

Ng, A., & Loosemore, M. (2007). Risk allocation in the private provision of public infrastructure. *International Journal of Project Management*, *25*(1), 66–76. doi:10.1016/j.ijproman.2006.06.005

Ngeno, E. K. (2016). *Assessing the impact of land use changes on the hydrology of the Nyangores sub-catchment, Bomet County, Kenya* (MSc. thesis). Kenyatta University.

Ngige, A. & Macharia. (2006). *Kenya Water Policy Overview Paper*. IT Power East Africa.

Ngigi, S. N., Savenije, H. H. G., Thome, J. N., Rockström, J., & de Vries, F. W. T. P. (2005, April). Agro-hydrological evaluation of on-farm rainwater storage systems to supplement irrigation in Laikipia District, Kenya. *Agricultural Water Management, 73*(1), 21–41. doi:10.1016/j.agwat.2004.09.021

Niang, I., Ruppel, O. C., Abdrabo, M. A., Essel, A., Lennard, C., Padgham, J., & Urquhart, P. (2014). Africa. In *Climate Change 2014: Impacts, Adaptation, and Vulnerability. Part B: Regional Aspects. Contribution of Working Group II to the Fifth Assessment Report of the Intergovernmental Panel on Climate Change* (pp. 1199–1265). Cambridge, UK: Cambridge University Press.

Nicholls, R. J., & Cazenave, A. (2010). Sea-level rise and its impact on coastal zones. *Science, 328*(5985), 1517–1520. doi:10.1126cience.1185782 PMID:20558707

Nolan, K. M., & Shields, R. R. (2000). *Measurement of stream discharge by wading*. US Department of the Interior, US Geological Survey. Retrieved from http://www.esd101.net/cms/lib

Notter, B., MacMillan, L., Viviroli, D., Weingartner, R., & Liniger, H. (2007). Impacts of Environmental change on water resources in the Mt. Kenya region. *Journal of Hydrology, 343*, 266 - 278.

NSF. (2016). *USAID INFEWS Workshops 2016, "Food, energy and water nexus"*. The Next Einstein Forum, Dakar.

Ntale, H. K., & Gan, T. Y. (2003). Drought indices and their application to East Africa. *International Journal of Climatology, 23*(11), 1335–1357. doi:10.1002/joc.931

Ntale, J. F., & Litondo, K. O. (2013). An investigation into the entrepreneurial behaviour and human capital formation among small scale farmers in Kenya. *African Journal of Social Sciences, 3*, 122–134.

Nyamai, C. M., Mathu, E. M., Opiyo-Akech, N., & Wallbrecher, E. (2003). A re-appraisal of the geology, geochemistry, structures and tectonics of the Mozambique belt in Kenya, east of the Rift System. African Journal of Science and Technology (AJST). *Science and Engineering Series, 4*(2), 51–71.

Obando, J. A., Luwesi, C. N., Beyene, A., Tshimanga, R. M., & Ruhakana, A. (2017). The core business of integrated water management: Achieving water governance performance. In Innovative Water Finance in Africa: A guide for Water Managers – Vol. 1: Water Finance Innovations' Context. Academic Press.

Obando, J. A., Luwesi, C. N., Mathenge, J. M., Kinuthia, W., Wambua, P. P., Mutiso, M. N., & Bader, E. O. (2015). Performance Assessment and Evaluation of Community Participation in Water Sector Governance: The case of Ngaciuma-Kinyaritha catchment, Mount Kenya Region. In A. Beyene (Ed.), *Agricultural Water Institutions in East Africa, Current African Issues, No. 63* (pp. 23–41). The Nordic Africa Institute.

Ogutu, J. O., Owen-Smith, N., Piepho, H. P., Kuloba, B., & Edebe, J. (2012). Dynamics of ungulates in relation to climatic and land use changes in an insularized African savanna ecosystem. *Biodiversity and Conservation*, *21*(4), 1033–1053. doi:10.100710531-012-0239-9

Ohowa, B. O., Mwashote, B. M., & Shimbira, W. S. (1997). Dissolved Inorganic Nutrient Fluxes from Two Seasonal Rivers into Gazi Bay, Kenya. *Estuarine, Coastal and Shelf Science*, *45*(2), 189–195. doi:10.1006/ecss.1996.0187

OKACOM. (2011a). *Cubango-Okavango River Basin Transboundary Diagnostic Analysis*. The Permanent Okavango River Basin Water Commission.

OKACOM. (2011b). *Strategic action programme (SAP) for the sustainable development and management of the Cubango-Okavango basin*. The Permanent Okavango River Basin Water Commission.

OKACOM. (2015). *Overview of the Okavango River Basin*. The Okavango River Basin.

Oki, T., & Kanae, S. (2006). Global hydrological cycles and world water resources. *Science*, *313*(5790), 1068–1072.

Oki, T., Caterina, V., & Kate, H. (2006). *Hydrology 2020: An integrating Science to meet world water challenges*. IAHS Publication 300.

Okoye & Kwamboka. (2016). Impact of urbanization on the morphology of Motoine/Ngong River channel, Nairobi River Basin. *JGRP*, *9*(4), 36-46.

Okpara, J. N., & Tarhule, A. (2015). Evaluation of Drought Indices in the Niger Basin, West Africa. *Journal of Geography and Earth Sciences*, *3*(2), 1–32.

Olang, L. O. (2009). *Analyses of land cover change impact on flood events using Remote Sensing (RS), GIS and Hydrological Models. A Case Study of the Nyando River Basin in Kenya* (PhD thesis). University of Natural Resources and Applied Life Sciences (BOKU), Vienna, Austria.

Olang, L. O., & Fürst, J. (2010). Effects of land cover change on flood peak discharges and runoff volumes: Model estimates for the Nyando River Basin, Kenya. *Hydrological Processes*. doi:10.1002/hyp.7821

Olang, L. O., & Kundu, P. M. (2011). Deforestation and land degradation of the Mau Forest Complex in Eastern Africa: a review for conservation and restoration purposes. In E. O. Ekundayo (Ed.), *Environmental Monitoring* (pp. 245–262). Rijeka: InTech Publishers.

Olang, L. O., Kundu, P. M., Ouma, G., & Fürst, J. (2012). Impacts of land cover change scenarios on storm runoff generation: A basis for management of the Nyando Basin, Kenya. *Land Degradation & Development*. doi:10.1002/ldr.2140

Oliver, H. R., & Oliver, S. A. (Eds.). (1994). *The Role of Water and the Hydrological Cycle in Global Change. NATO ASI Series. Series I: Global Environmental Change* (Vol. 31). London: Springer.

Oloo, F. O., Olang, L., & Strobl, J. (2015). Spatial modelling of solar energy potential in Kenya. *International Journal of Sustainable Energy Planning*, *6*, 17–30.

Omonge, P. O. (2016). *Simulation of water resource use and allocation in Nyangores sub-catchment of the upper Mara Basin, Bomet County, Kenya* (MSc. thesis). Kenyatta University.

Omwoyo, M. A., Muthama, N. J., Opere, A., & Onwonga, R. (2017). Simulating Streamflow in Response to Climate Change in the Upper Ewaso Ngiro Catchment, Kenya. *Journal of Climate Change and Sustainability*, *1*(2), 1229.

Ondieki, C. M. (1994). *The Hydrology of Arid and Semi arid lands; Case of Laikipia District* (Unpublished PhD thesis). Nairobi University.

Ondieki, C. M. (2012). Water Dependencies: Systems under Stress and Societal Responses. In *Hydrological Society of Kenya (HSK) Conference Proceedings*. Kenyatta University Press.

Ondieki, C. M. (2013). Hydrology and Integrated Water Resource Management for Sustainable Watershed Management in Kenya. In H. Muga & K. Thomas (Eds.), *Cases on the Diffusion and Adoption of Sustainable Development Practices* (pp. 352–375). Hershey, PA: Information Science Reference. doi:10.4018/978-1-4666-2842-7.ch013

Ondieki, C. M. (2015). Aspects of Water Quality and Water Pollution Control in Vulnerable Environments. *Journal of Geography and Earth Sciences*, *3*(2), 83–93. doi:10.15640/jges.v3n2a5

Onyando, J. O., Olang, L. O., & Chemelil, M. C. (2005). Regional Analysis of Conceptual Rainfall-Runoff models for runoff simulation in ungauged catchments of Kenya. *Journal of Civil Engineering Research and Practice*, *2*, 23–37. doi:10.4314/jcerp.v2i1.29135

Orimoloye, R. I., Mazinyo, P. S., Nel, W., & Iortyom, T. E. (2018). Assessing changes in climate variability observation and simulation of temperature and relative humidity: A case of East London, South Africa. *Research Journal of Environmental Sciences*, *12*(1), 1–13. doi:10.3923/rjes.2018.1.13

Osei-Owusu, P. K. (2014). *Overview of Bhungroo*. Accra, Ghana: CA. Available at: http://conservealliance.org/overview-of-bhungroo/#

Oyeabande, L. (1990). Aspects of Urban Hydrology and the Challenges for African Urban Environments. Urban Climatology in Africa. *African Urban Quarterly*, 39-68.

Parissopouluos, G. A., & Wheater, H. S. (1992). Effects of hysteresis on ground water recharge from ephemeral flows. *Water Resources Research*, *28*(11), 3055–3061. doi:10.1029/92WR01902

Paulo, A. A., & Pereira, L. S. (2006). Drought concepts and characterization, comparing drought indices applied at local and regional scales. *Water International*, *31*(1), 37–49. doi:10.1080/02508060608691913

Pellikka, P. K. E., Lötjönen, M., Siljander, M., & Lens, L. (2009). Airborne remote sensing of spatiotemporal change (1955-2004) in indigenous and exotic forest cover in the Taita Hills, Kenya. *International Journal of Applied Earth Observation and Geoinformation*, *11*(4), 221–232. doi:10.1016/j.jag.2009.02.002

Pérard, E. (2009). Water supply: Public or private? An approach based on cost of funds, transaction costs, efficiency and political costs. *Policy and Society*, *27*(3), 193–219. doi:10.1016/j.polsoc.2008.10.004

Perret, S., & Geyser, M. (2007). The full financial costs of irrigation services: A discussion on existing guidelines and implications for smallholder irrigation in South Africa. *Water S.A.*, *33*(1), 67–78.

Peter, F., Malchus, B., Carlton, B., Madelyn, C., & Kern, L. (2000). Land stewardship in the 21st. Century: The contributions of Watershed Management. *USDA, Forest Service Proceedings*.

Peters, J. E. (2013). Promoting Rainwater harvesting (RWH) in small Island Developing States (SIDS): A case in the Granadines. In Cases on the Diffusion and Adoption of Sustainable Development Practices (pp. 403-438). Academic Press.

Pilgrim, D. (1982). Assessment of derived rural and urban runoff coefficients. Institute of Engineers Australia. *Civil Engineering Transactions*, *24*, 235–241.

Prasad, N. (2006). Privatisation results: Private sector participation in water services after 15 years. *Development Policy Review*, *24*(6), 669–692. doi:10.1111/j.1467-7679.2006.00353.x

Pretty, J. (1999). Can sustainable agriculture feed Africa? New evidence on progress, processes and impacts. *Environment, Development and Sustainability*, *1*(3-4), 253–274. doi:10.1023/A:1010039224868

Prudhomme, C., Giuntoli, I., Robinson, E. L., Clark, D. B., Arnell, N. W., Dankers, R., ... Wisser, D. (2014). Hydrological droughts in the 21st century, hotspots and uncertainties from a global multimodel ensemble experiment. *Proceedings of the National Academy of Sciences of the United States of America*, *111*(9), 3262–3267. doi:10.1073/pnas.1222473110 PMID:24344266

Pulwarthy, R. S., & Sivakumar, M. V. K. (2014). Information Systems in a Changing Climate: Early warning and drought risk management. *Weather and Climate Extremes*, *3*, 14–21. doi:10.1016/j.wace.2014.03.005

Puttemans, S. (2004). *Potential for small scale irrigation from groundwater dams in South Kitui*. Katholieke Universiteit Leuven Faculteit Landbouwkundige en Toegepaste Biologische Wetenschappen.

Pyrce, R. (2004). Hydrological Low Flow Indices and Their Uses. Watershed Science Centre, Trent University.

Qadir, M., Wichelns, D., Raschid-Sally, L., McCornick, P. G., Drechsel, P., Bahri, A., & Minhas, P. S. (2010). The challenges of wastewater irrigation in developing countries. *Agricultural Water Management*, *97*(4), 561–568. doi:10.1016/j.agwat.2008.11.004

Quiring, S. M. (2009). Developing objective operational definitions for monitoring drought. *Journal of Applied Meteorology and Climatology*, *48*(6), 1217–1229. doi:10.1175/2009JAMC2088.1

Radiro, M. O., Sijali, I. V., & Maingi, P. M. (2003). Integrated soil management for sustainable use of salt affected soils of Taveta Division. Salt affected soils. *Proceedings of a workshop. Taveta. 22-27 August 2003. IDRP Workshop Report, Nairobi, Kenya. KARI Irrigation and Drainage Research Programme*, 7-14.

Republic of Kenya. (1992). *The Study on the National Water Master Plan*. Nairobi Climate Variability and Water Resources Degradation in Kenya.

Republic of Kenya. (1997). *National Development Plan*. Author.

Republic of Kenya. (1998a). *The Aftercare Study on the National Water Master Plan*. Tokyo: Nippon Koei Co., Ltd.

Republic of Kenya. (2001a). Economic Survey 2001. Nairobi: Author.

Republic of Kenya. (2001b). *National Disaster Management Policy*. Nairobi: Draft.

Republic of Kenya. (2001c). *EwasoNyiro North River Catchment Conservation and Water Resources Management Study*. Tokyo: Nippon Koei Co. Ltd.

Republic of Kenya. (2002). *National Development Plan 2002–2008*. Author.

Richards, J. A. (1993). *Remote sensing digital image analysis. An introduction (2nd ed.)*. Berlin: Springer-Verlag. doi:10.1007/978-3-642-88087-2

Ringler, C. (2001). *Optimal Water Allocation in the Mekong River Basin. ZEF Discussion papers on Development Policy 38*. Bonn: University of Bonn, ZEF.

Rockström, J., Falkenmark, M., Karlberg, L., Hoff, H., Rost, S., & Gerten, D. (2009). Future water availability for global food production: The potential of green water for increasing resilience to global change. *Water Resources Research*, *45*(7), W00A12. doi:10.1029/2007WR006767

Rockström, J., Falkenmark, M., Karlberg, L., Hoff, H., Rost, S., & Gerten, D. (2009). Future water availability for global food production: The potential of green water for increasing resilience to global change. *Water Resources Research*, 45.

Rogers, P., De Silva, R., & Bhatia, R. (2002). Water is an economic good: How to use prices to promote equity, efficiency, and sustainability. *Water Policy*, *4*(1), 1–17. doi:10.1016/S1366-7017(02)00004-1

Romeo, L. G. (2005). A Territorial Approach to Local Development; The missing decentralization-development link? In *Workshop on Decentralization Reforms, Local Governance, Local and Territorial Development*. Brussels: New York University, International LLC.

Rosegrant, M. W., Cai, X., & Cline, S. A. (2002). *World water and food to 2025: dealing with scarcity. IFPRI-2020 Vision - International Water Management book*. Washington, DC: IFPRI.

Rosegrant, M. W., & Cline, S. A. (2003). Global food security: Challenges and policies. *Science*, *302*(5652), 1917–1919. doi:10.1126cience.1092958 PMID:14671289

Rosegrant, M. W., Ringler, C., McKinney, D. C., Cai, X., Keller, A., & Donosso, G. (2000). Integrated economic-hydrologic water modelling at the basin scale: The Maipo river basin. *Agricultural Economics*, *24*, 33–46.

Rosenfield, G., & Fitzpatrick-Lins, K. (1986). A coefficient of agreement as a measure of thematic classification accuracy. *Photogrammetric Engineering and Remote Sensing*, *52*(2), 223–227.

Rose, S., & Peters, N. E. (2001). Effects of urbanization on streamflow in Atlanta area. *USA Hydrological Proceesses*, *15*(8), 1441–1457. doi:10.1002/hyp.218

Rozenga, W. (1995). *Ecologically sustainable land use in Lower Boteti, Central District, Botswana*. Utrecht, The Netherlands: Department of Geography of Developing Countries. Utrecht University.

Russell, C. S., Clark, C. D., & Schuck, E. C. (2007). Economic Instruments for Water management in the Middle East and North Africa. *International Journal of Water Resources Development*, *23*(4), 659–677. doi:10.1080/07900620701488596

Save the Rhino. (2018). *All Rhinos translocated to Tsavo East National Park have died*. Retrieved from https://www.savetherhino.org/africa/kenya/all-rhinos-translocated-to-tsavo-east-national-park-have-died/

Savenije, H. H. G., Hoekstra, A. Y., & van der Zaag, P. (2014). Evolving water science in the Anthropocene. *Hydrology and Earth System Sciences, 18*, 319–332.

Schapera, I. (1970). *Tribal innovators: Tswana and social change 1795-1940*. London: Athlone Press.

Schilling, J., Freier, K. P., Hertig, E., & Scheffran, J. (2012). Climate change, vulnerability and adaptation in North Africa with focus on Morocco. *Agriculture, Ecosystems & Environment*, *156*, 12–26. doi:10.1016/j.agee.2012.04.021

SCS. (1986). *National engineering handbook*. USDA.

Seely, M., Henderson, J., Heyns, P., Jacobson, P., Nakale, T., Nantanga, K., & Schachtscschneider, K. (2003). Ephemeral and endoreic river systems: Relevance and management challenges. In Transboundary rivers, sovereignty and development: Hydropolitical drivers in the Okavango River basin. African Water Issues Research Institute and Green Cross International.

Sefe, F. T. K. (1996). Maun groundwater development project: Phase 1 – exploration and resource assessment. Hydrological analyses. Eastend Investment Report to the Department of Water Affairs, Government of Botswana.

SEI. (2011). *Water Evaluation And Planning System: User Guide*. Stockholm Environment Institute, U.S. Center. Retrieved from http://www.weap21.org

Seth, A., Rauscher, S. A., Rojas, M., Giannini, A., & Camargo, S. J. (2011). Enhanced spring convective barrier for monsoons in a warmer world? *Climatic Change*, *104*(2), 403–414. doi:10.100710584-010-9973-8

Shah, T., Verma, S., Durga, N., Rajan, A., Goswami, A., & Palrecha, A. (2016). *HarKhetKoPani (Water to Every Farm) Rethinking PradhanMantriKrishiSinchaiYojana (PMKSY).* Colombo: IWMI, IWMI-Tata Water Policy Program, IWMI-TATA Policy Paper.

Shah, T., Singh, O. P., & Mukherji, A. (2006). Some aspects of South Asia's groundwater irrigation economy: Analyses from a survey in India, Pakistan, Nepal Terai and Bangladesh. *Hydrogeology Journal*, *14*(3), 286–309. doi:10.100710040-005-0004-1

Shah, T., & van Koppen, B. (2006). Is India ripe for integrated water resources management IWRM: Fitting water policy to national development context. *Economic and Political Weekly*, *41*(31), 3413–3421.

Shah, T., Verma, S., & Pavelic, P. (2013). Understanding smallholder irrigation in Sub-Saharan Africa: Results of a sample survey from nine countries. *Water International*, *38*(6), 809–826. doi:10.1080/02508060.2013.843843

Shaw, E. M. (1986). *Hydrology in Practice*. Chapman Hall.

Shaw, P. (1985). The Desiccation of Lake Ngami: An Historical Perspective. *The Geographical Journal*, *151*(3), 318–326. doi:10.2307/633012

Shisanya, C., Obando, J., & Onywere, S. (2010). Towards Sustainable Water Resource Management for Food Security in Bwathonaro Catchment, Kenya. Report to Commission for Higher Education (CHE) for Project: Enhancing Sustainable Water Resource Management for Food Security: A Case Study of Bwathonaro Catchment, Kenya.

Shongwe, M. E., van Oldenborgh, G. J., van den Hurk, B., & van Aalst, M. (2011). Projected changes in mean and extreme precipitation in Africa under global warming. Part II: East Africa. *Journal of Climate*, *24*(14), 3718–3733. doi:10.1175/2010JCLI2883.1

Sigleo, A., & Frick, W. (2015). Seasonal Variations in River Flow and Nutrient Concentrations in a Northwestern USA Watershed. Environmental Protection Agency, Western Ecology Division, Newport, and U.S. Environmental Protection Agency, Ecosystems Research Division.

Sijali, I. V., Radiro, M. O., & Maingi, P. M. (2003). *Salt affected soils: Proceedings of a workshop*. IDRP Workshop Report. Nairobi, Kenya: KARI Irrigation and Drainage Research Programme.

Singh, K. S., & Lal, R. B. (2003). *People of India - Gujarat (Part 1): Anthropological Survey of India*. Delhi: Popular Prakashan.

SMEC. (1987). Southern Okavango Integrated Water Development Phase 1: Final Report Technical Study, Volume III, Water resource and development. Report by Snowy Mountains Engineering Corporation to the Department of Water Affairs, Ministry of Mineral Resources and Water Affairs, Government of Botswana, Gaborone.

SMEC. (1991). *Botswana national water master plan study: final report. Prepared by Snowy Mountains Engineering Corporation Ltd, WLPU Consultants and Swedish Geological International AB*. Gaborone: Ministry of Mineral Resources and Water Affairs.

Smith, P., Ashmore, M. R., Black, H. I. J., Burgess, P. J., Evans, C. D., Quine, T. A., ... Orr, H. G. (2013). The role of ecosystems and their management in regulating climate, and soil, water and air quality. *Journal of Applied Ecology*, *50*(4), 812–829. doi:10.1111/1365-2664.12016

Smith, R. C. (2007). Land-use allocation protects the Peruvian Amazon. *Science*, *317*, 1233–1236. PMID:17690260

Sombroek, W. G., Braun, H. M. H., & van der Pouw, B. J. A. (1982). *The explanatory soil map and agro-climatic zone map of Kenya*. Kenya Soil Survey Report No. E1.

Stedinger, J. R., Vogel, R. M., & Foufoula-Georgiou, E. (1992). Frequency analysis of extreme events. In D. R. Maidment (Ed.), *Handbook of Hydrology*. New York: McGraw-Hill.

Story, M., & Congalton, R. G. (1986). Accuracy assessment: A user's perspective. *Photogrammetric Engineering and Remote Sensing*, *52*(3), 397–399.

Streamtec Pty Ltd 2011, Environmental Water Requirements of the Mukurumudzi River, Kwale, Kenya (Streamtec Report 8/2011). (2011). Ukunda, Kenya: Prep. by Davies PM, Davies LM and Craig IR for Base Titanium Ltd.

Streamtec Report 10/2011. (2011). *Measuring Flows in the Mukurumudzi River, Kwale, Kenya*. Author.

Subramanian, V. (2001). Water Quality and Quantity. In V. Subramanian & A. Ramanathan (Eds.), Ecohydrology. New Delhi: Capital Publishing Company.

Sungu, R. O. (2019). *An assessment of the influence of water allocation on sustainable water resources Management: A case study of the Nyando River Basin* (Unpublished PhD thesis). University of West Cape, South Africa.

Swatuk, L. (2005). Political challenges to implementing IWRM in sounthern Africa. *Physics and Chemistry of the Earth*, *30*(11-16), 872–880. doi:10.1016/j.pce.2005.08.033

Swatuk, L., & Rahm, D. (2004). Integrating policy, disintegrating practice: Water resources management in Botswana. *Physics and Chemistry of the Earth*, *29*(15-18), 1357–1364. doi:10.1016/j.pce.2004.09.011

Tarboton, D. G., Maidment, D., Zaslavsky, I., Ames, D. P., Goodall, J., & Horsburgh, J. S. (2010). *Cuahsi hydrologic information system 2010 status report*. Consortium of Universities for the Advancement of Hydrologic Science, Inc.

Tardieu, H., & Préfol, B. (2002). Full cost or "sustainable cost" pricing in irrigated agriculture. Charging for water can be effective but is it sufficient? *Irrigation and Drainage*, *51*(2), 97–107. doi:10.1002/ird.44

Taylor, P., Cox, S., Walker, G., & Valentine, D. (2012, July). Adapting standard information models for water data. In *10th International Conference in Hydroinformatics, HIC2012*. Hamburg, Germany: IWA Publishing.

Terink, W., Immerzeel, W. W., & Droogers, P. (2013). Climate change projections of precipitation and reference evapotranspiration for the Middle East and Northern Africa until 2050. *International Journal of Climatology, 33*(14), 3055–3072. doi:10.1002/joc.3650

The Guardian. (2018). *Eight of 14 Rhinos die after move to Kenya National Park.* Retrieved from ttps://www.theguardian.com/world/2018/jul/13/kenya-rhinos-die-after-being-moved-between-national-parks

The Times of India. (n.d.).Retrieved from http://timesofindia.indiatimes.com/

Thompson, L. G., Mosley-Thompson, E., Davis, M. E., Henderson, K. A., Brecher, H. H., Zagorodnov, V. S., ... Beer, J. (2002). Kilimanjaro ice core records: Evidence of Holocene climate change in tropical Africa. *Science, 298*(5593), 589–593. doi:10.1126cience.1073198 PMID:12386332

Thrupp, L. A. (1997). *The diversity and dynamics of shifting cultivation: myths, realities and policy implications.* Washington, DC: World Resources Institute.

Tiffen, M., & Mortimore, M. (2002). Agroecosystems regaining the high ground: Reviving the Hillsides of Machakos. *World Resources, 2000-2001*, 149–158.

Tolson, A. B., & Shoemaker, A. C. (2007). Cannonsville Reservoir watershed SWAT2000 model development, Callibration and Validation. *Journal of Hydrology, 337*, 68 - 86.

Tou, J. T., & Gonzalez, R. C. (1974). Pattern recognition principles. Reading, MA: Addison Wesley.

Trabacchi, C., & Mazza, F. (2015). *Emerging solutions to drive private investment in climate resilience.* Venice: CPI. Available at: http://climatepolicyinitiative.org/publication/emerging-solutions-to-drive-privateinvestment-in-climate-resilience/

Tramblay, Y., Jarlan, L., Hanich, L., & Somot, S. (2017). Future Scenarios of Surface Water Resources Availabilityin North African Dams. *Water Resources Management*. doi:10.100711269-017-1870-8

Tsholofelo, M. (2005). *The contribution of the Boteti river to the livelihoods of the Chanoga Community. Winter School report.* Harry Oppenheimer Okavango Research Centre.

Tucker, C. J. (1979). Red and photographic infrared linear combinations for monitoring vegetation. *Remote Sensing of Environment, 8*(2), 127–150. doi:10.1016/0034-4257(79)90013-0

Turner, B. L. II, & Meyer, W. B. (1994). Global land use and land cover change: an overview. In W. B. Meyer & B. L. Turner II, (Eds.), *Changes in Land Use and Land Cover: a global perspective* (pp. 3–10). Cambridge, UK: University of Cambridge.

Turubanova, P.S.A., Hansen, M.C., Adusei, B., Broich, M., Altstatt, A., Mane, L., & Justice, C.O. (2012). *Quantifying forest cover loss in Democratic Republic of the Congo, 2000– 2010, with Landsat ETM+ data*. Remote Sensing of Environment.

U.S. Department of the Interior. (2011). *Quality of water Colorado River Basin Progress Report No. 23*. Bureau of Reclamation.

UN. (2016). *Multi-stakeholder forum on science, technology and innovation for the Sustainable Development Goals: Summary by the Co-Chairs Note by the President of the Economic and Social Council*. High-level political forum on sustainable development Convened under the auspices of the Economic and Social Council. New York, NY: UN Economic and Social Council. Retrieved from http://www.un.org/ga/search/view_doc.asp?symbol=E/HLPF/2016/6&Lang=E

UN. (2016). *Sustainable Development Goals*. Retrieved from Http://www.un.org/sustainabledevelopment /development-agenda/

UN. (2017). Role of technology in implementing and monitoring the SDGs. Concept Note. In *6th Economic and Social Council Youth Forum*. New York, NY: UN Headquarters, Trusteeship Council Chamber. Available at: https://www.un.org/ecosoc/sites/www.un.org.ecosoc/files/files/en/2017doc/Role-of-technology-in-implementing-the-SDGs.pdf

UNEP. (2009). *Atlas of Our Changing Environment*. DEWA.

UNEP/Government of Kenya. (2000). *Devastating Drought in Kenya: Environmental Impacts and Responses*. Nairobi: United Nations Environment Programme.

United Republic of Tanzania (URT). (2012). 2012 Population and Housing Census. Dar es Salaam, Tanzania: National Bureau of Statistics.

UN-Water. (2013). *United Republic of Tanzania Water Country Brief (Water country report)*. United States Department of State.

Urama, K. C., & Ozor, N. (2010). Impacts of climate change on water resources in Africa: The role of adaptation. *African Technology Policy Studies Network*, *29*, 1–29.

URT. (2009). The national irrigation policy. Ministry of Water and Irrigation, Tanzania.

USAID. (2014). *Local governance and decentralization assessment: implications of proposed reforms in Ukraine*. Washington, DC: USAID. Available at: http://pdf.usaid.gov/pdf_docs/pa00k59f.pdf

Vaishali, P., & Punita, P. (2013). Assessment of seasonal variation in water quality of River Mini, at Sindhrot, Vadodara. *International Journal of Environmental Sciences, 3*(5).

Van Aalst, M. K., Fankhauser, S., Kane, S. M., & Sponberg, K. (2000). *Climate Informationand Forecasting for Development. Paper No. 79*. Washington, DC: Climate Change Series, The World Bank Environment Department.

van der Bliek, J., McCormick, P., & Clarke, J. (2014). *On target for people and planet: Setting and achieving water related sustainable development goals. Colombo: International Water Management Institute*. Sri Lanka: IWMI.

Van Hofwegen, P. (2006). *Enhancing Access to Finance for Local Governments and Financing Water for Agriculture*. Report no. 1 of the (Gurria) Task Force on Financing Water for All (WWC/ GWP).

Van Schaik, L. (2010). *The role of Macro pore flow from PLOT to catchment scale: A study in a semi arid area* (PhD thesis). Utrecht Netherlands Geographical Studies 390.

Vancutsem, C., Lubamba, J. P. K., Blaes, X., Pekel, J. F., de Wasseige, C., & Defourny, P. (2006). *The Land Cover Map of the Democratic Republic of Congo*. Louvain-la-Neuve: Leuven University Press, UCL-Geomatics.

Vorosmarty, C. J., Green, P., Salisbury, J., & Lammers, R. B. (2000). Global Water Resources: Vulnerability from Climate Change and Population Growth. *Science*, 289. PMID:10894773

Wallace, J. S. (2000). Increasing agricultural water use efficiency to meet future food production. *Agriculture, Ecosystems & Environment*, *82*(1-3), 105–119. doi:10.1016/S0167-8809(00)00220-6

Walli, A., Tøttrup, C., Naeimi, V., Bauer-Gottwein, P., Bila, M., Mufeti, P., & Koetz, B. (2013, September). TIGER-NET–enabling an Earth Observation capacity for Integrated Water Resource Management in Africa. In *ESA Living Planet Symp.*, Edinburgh, UK.

Walsh, A. (2012). The Commodification of the Public Service of Water: A Normative Perspective. *Public Reason*, *3*(2), 90–106.

Wambua, P. P., Luwesi, C. N., Njuguna, J. W., & Ngoma, J.-B. N. (2017). Financing water management and developmemt: An Introduction to Financial Instruments and Operations. In Innovative Water Finance in Africa: A guide for Water Managers – Vol. 2: Economics and Principles of Water Finance Innovations. Academic Press.

Wasige, J. E., Groen, T. A., Smaling, E., & Jetten, V. (2013). Monotoring Basin-Scale Land Cover Changes in Kagera Basin of Lake Victoria Using Ancillary Data and Remote Sensing. *International Journal of Applied Earth Observation and Geoinformation*, *21*, 32–42. doi:10.1016/j.jag.2012.08.005

Wass, P. (Ed.). (1995). *Kenya's Indigenous Forests: Status, Management and Conservation. IUCN*. Cambridge, UK: Gland.

Water Resources Management Authority. (2009). *Integrated Water Resources Management and Water Efficiency Plan for Kenya*. Author.

Waziri, M., & Ogugbuaja, V.O. (2010). Inter relationships between physico-chemical water pollution indicators: A case study of River Yobe-Nigeria. *AJSIR*, *1*(1), 76-80.

Waziri, M., & Ogugbuaja, O. V. (2012). Prediction of some water quality indices in River Yobe-Nigeria, through annual projections. *Frontiers in Science*, *2*(4), 58–61. doi:10.5923/j.fs.20120204.02

WHO. (1993). *Guidelines for drinking water quality* (2nd ed.). Geneva, Switzerland: WHO.

WHO. (2006). *WHO Guidelines for Drinking Water Quality* (3rd ed.; Vol. 1). WHO.

WHO. (2008). *Guidelines for Drinking-water Quality* (3rd ed.). Geneva, Switzerland: World Health Organization.

Wichelns, D. (2002). An economic perspective on the potential gains from improvements in irrigation water management. *Agricultural Water Management*, *52*(3), 233–248. doi:10.1016/S0378-3774(01)00134-2

Wilhite, D. A., & Glantz, M. H. (1985). Understanding the Drought Phenomenon: The role of definitions. *Water International*, *10*(3), 111–120. doi:10.1080/02508068508686328

Willems, P. (1998). Hydrological Applications of Extreme Value Analysis. In H. Wheater & C. Kirby (Eds.), *Hydrology in a Changing Environment* (Vol. 3, pp. 15–25). Chichester, UK: John Wiley & Sons.

Willems, P., Olsson, J., Arnbjerg-Nielsen, K., Beecham, S., Pathirana, A., Gregersen, I. B., ... Nguyen, V.-T.-V. (2012). *Impacts of Climate Change on Rainfall Extremes and Urban Drainage Systems*. London: IWA Publishing. doi:10.2166/9781780401263

Williams, W. D. (1966). Conductivity and the concentration of total dissolved solids in Australian lakes. *Australian Journal of Marine and Freshwater Research*, *17*(2), 169–176. doi:10.1071/MF9660169

Wimmer, D., Fürst, J., & Herrnegger, M. (2015). *MaMa-Hydro: Exploring Water Resources Planning and Management options in the Nyangores headwater catchment of the vulnerable Maasai Mara River Basin in Kenya*. Book of Maps: Technical Report No. 2. Vienna: Kommission für Entwicklungsforschung (KEF).

WLE. (2015). *Water, land and ecosystem in Africa (WLE in Africa) - 2015 project review and stakeholder engagement*. WLE workshop report. Available at: https://waterlandandecosystems.wikispaces.com/

WMO. (2008). Guide to Hydrological Practice Vol. I: Hydrology – From measurement to Hydrological information (6th ed.). WMO No. 168.

WMO. (2009). Guide to Hydrological Practice Vol. II: Management of water resources and application of hydrological practices (6th ed.). WMO No. 168.

WMO. (2012). Standardized Precipitation Index. WMO 1090.

Wolski, P., Savenije, H. H. G., Murray-Hudson, M., & Gumbricht, T. (2006). Modelling of the flooding in the Okavango Delta, Botswana, using a hybrid reservoir-GIS model. *Journal of Hydrology (Amsterdam), 331*(1-2), 58–72. doi:10.1016/j.jhydrol.2006.04.040

World Bank. (2000). World Development Report 2000/2001: Attacking Poverty. Washington, DC: Author.

WRMA & Mukurumudzi WRUA. (2012). *Sub-Catchment Management Plan for Mukurumudzi River*. Author.

WRMA. (2010). *Enforcement of Water Use Charges and Water Quality Thresholds in Kenya*. WRMA Evaluation Workshop, Meru.

Wu, J. S., Han, J., Annambhotla, S., & Bryant, S. (2005). Artificial neural networks for forecasting watershed runoff and stream flows. *Journal of Hydrologic Engineering, 10*(3), 216–222. doi:10.1061/(ASCE)1084-0699(2005)10:3(216)

WUR. (2017). *'KrishakBandhu' (KB) buckets and quarter acre kits*. Available at: www.wur.nl

WWAP. (2012). *The United Nations World Water Development Report 4: Managing Water under Uncertainty and Risk (Executive Summary)*. Paris, France: United Nations Educational, Scientific and Cultural Organization.

Xiao, L. (1999). *Flashfloods in arid and semi arid zones*. IHP Technical documents in Hydrology. No 23. UNESCO.

Yacouba, D., Hu, G., & Wen, X. (2009). Assessment Of Land Use Cover Changes Using Ndvi And Dem In Puer And Simao Counties, Yunnan Province, China. *Marsland Press World Rural Observations, 1*(2), 1–11.

Yang, C., Raskin, R., Goodchild, M., & Gahegan, M. (2010). Geospatial cyberinfrastructure: Past, present and future. *Computers, Environment and Urban Systems, 34*(4), 264–277. doi:10.1016/j.compenvurbsys.2010.04.001

Yates, D., Sieber, J., Purkey, D., & Huber-Lee, A. (2005). WEAP21 – A Demand, Priority, and Preference-Driven Water Planning Model: Part 1, Model Characteristics. *Water International, 30*(4), 487–500. doi:10.1080/02508060508691893

Yevjevich, V. (1967). *An objective approach to defitions and investigations of continental hydrological drought. Hydrology Paper No. 23*. Fort Collins, CO: Colorado State University.

Yevjevich, V. (1972). *Probability and Statistics in Hydrology*. Fort Collins, CO: Water Resources publications.

Yira, Y., Diekkrüger, B., Steup, G., & Bossa, A. Y. (2017). Impact of climate change on hydrological conditions in a tropical West African catchment using an ensemble of climate simulations. *Hydrology and Earth System Sciences, 21*(4), 2143–2161. doi:10.5194/hess-21-2143-2017

Yuan, X., Zhang, M., & Wang, L. (2016). *Understanding and seasonal forecasting of hydrological drought in the anthropocene*. Hydrology and Earth Science System Sciences; doi:10.5194/hess-2016-592

Zhan, Q., Wang, J., Peng, X., Gong, P., & Shi, P. (2002). Urban built-up land change detection with road density and spectral information from multi-temporal Landsat TM data. *International Journal of Remote Sensing*, *23*(15), 3057–3078. doi:10.1080/01431160110104728

Zlinszky, A., & Timár, G. (2013). Historic maps as a data source for socio-hydrology: A case study of the Lake Balaton wetland system, Hungary. *Hydrology and Earth System Sciences*. doi:10.5194/hess-17-4589-2013

About the Contributors

Christopher Misati Ondieki is an independent research hydrologist who has served as Senior Lecturer at the Department of Geography, Kenyatta University, Kenya (1994- 2015). He earned his PhD in Meteorology from Nairobi University in 1994, having obtained his Masters in Hydrology from the VUB Belgium in 1985. Between 1987 and 1993, he was Research Hydrologist, Laikipia Research Programme, Nanyuki, Kenya, under the auspices of University of Berne, Switzerland with focus on ASAL hydrology. He has published widely. His research interests are in Hydrological processes and watershed management of river basins. He is a member of Hydrological Society of Kenya (HSK).

Johnson U. Kitheka is an environmental consultant holding BSc, MSc and PhD degrees in Geography and Hydrology from the University of Nairobi, Kenya. He has more than 20 years' experience in the research, assessment and monitoring of both coastal marine and freshwater resources. Dr. Kitheka has also participated in the implementation of several projects such as the NUFFIC NICHE Capacity building project at the South Eastern Kenya University and the UNEP-GEF Project focused on addressing the impacts of land-based activities in the Western Indian Ocean Basin. He is also experienced in Transboundary Diagnostic Analysis (TDA), Integrated River Basin and Coastal Zone Management, Strategic Action Programmes on land based interventions and National Action Plans for coastal natural resources management. Dr. Kitheka has also published widely in refereed international scientific journals. He has previously worked with the Kenya Marine and Fisheries Research Institute as a Senior Research Officer and the United Nations Environment Programme as an Environmental Scientist and is currently the Dean of the School of Water Resources Science and Technology of South Eastern Kenya University.

* * *

Jacobine T. Amutenya is equipped with 6 years of hydrology technical knowledge combined with a BSc. (Computer Science) Degree, MSc.(Water Science engineering: Hydro informatics), and a certificate in Basic Hydrologic sciences for the African region. Her key experience involves providing hydrological advisory services, collecting processing and analysing of hydrological data, managing ICT infrastructures as well as using remote sensing, hydraulic and hydrologic modelling and GIS applications for water resource management. As an aspiring researcher, her key interests are in using advanced ICT technology as well as programming mathematical models and online hydro informatics management systems. She is currently researching on Monitoring drought dynamics based on observed and remote sensing data.

Gerald A. Corzo is a winner of the Tison award 2012 of the IAHS association in hydrology. Civil Engineer with Master and PhD in Hydroinformatics for solving hydrological extreme events. He has an extensive experience in modelling for different problems in Integrated water resources management. His high skills in ICT technology have been used to develop various types of scientific applications. As researcher and lecturer of different academic institutions, he has worked and coordinated academic programs. His expertise has been mainly developed around programming mathematical models and statistical models for hydroinformatics problems. Since the last six years, he has been working on online modelling paradigms and how data from global hydrological models and their uncertainty can be integrated and analysed online.

Gerd Förch graduated from Hannover University, Germany, in 1973 with a Dr.-Ing. (PhD) in Civil Engineering, specialised in Water Resources Management; worked as research fellow in water resources engineering including waste management; took over visiting professorships in Jordan (Amman) and Ethiopia (Addis Ababa); was appointed the chair of water resources management at Siegen University in 1991; served in the University Board of Governors; developed close working links and partnerships with universities in Eastern Africa and South East Asia concentrating on IWRM, Integrated Watershed Management and environmental management (IWMNet); developed means of cooperation between science and development; established various master's programmes at African Universities and supervised a number of master and PhD students; he is DAAD Alumnus and member of the Centre of International Capacity Development at Siegen University; he retired from teaching and works presently as consultant for German (Development) Cooperation in Africa and Asia; his present focus is Karamoja/Uganda.

Nele Förch is an advisor for institutional development, water governance and integrated water resources management and has been working in Eastern Africa for more than ten years. She was the regional coordinator for the Centre for International Capacity Development for the University of Siegen. She is currently a technical advisor with the GIZ Water Sector Reform Programme (WSRP) in Kenya and directly works with the Kenyan Water Resources Management Authority. Her research interests include issues in water governance and integrated watershed management relating to water security, stewardship, participatory processes and particularly conflict transformation.

Josef Fürst is an associate Professor the Institute of Water Management, Hydrology and Hydraulic Engineering (IWHW) at the University of Natural Resources and Life Sciences, Vienna (BOKU). His main fields of research include hydrology and hydrography, the use of Ground Penetrating Radar (GPR) in hydrology, groundwater modelling and management, and GIS in Hydrology and Water Management.

Mathew Herrnegger is a post-doctoral researcher at the Institute of Water Management, Hydrology and Hydraulic Engineering (IWHW) at the University of Natural Resources and Life Sciences, Vienna (BOKU). He holds a Diploma in Environmental Engineering and Water Management and a PhD from BOKU-University. His fields of interest include catchment hydrology and modelling, flood forecasting, spatio-temporal variability of evapotranspiration, uncertainties and inverse modelling of precipitation and GIS in Hydrology and Water Management.

Jacob Katuva holds a MSc degee in Environmental and Biosystems Engineering from the University of Nairobi. He is currently a Dphil researcher in the Smith School of Enterprise and Environment, University of Oxford, United Kingdom. He has experience in hydrological analyses, rainfall-runoff modelling and water balance models and water resources management, Management Information Systems (MIS) and Geographical Information Systems (GIS) and advanced geostatistics.

Mwadini J. Khatib received a BSc in Environmental Engineering (2011) from Ardhi University, Dar es Salaam, Tanzania and MSc in Integrated Watershed Management (2016) from Kenyatta University, Nairobi, Kenya. During his undergraduate studies, he has won several Academic Prizes for excellent performance from different organization in Tanzania. He is registered graduate engineer in the Engineers Registration Board (ERB) of Tanzania. Mr. Mwadini Khatib is also a member of International Association for Impact Assessment (IAIA). Currently, Mwadini is working with the Ministry of Lands, Water, Energy and Environment, Zanzibar-Tanzania as Environmental Engineer of the Department of Energy and Minerals. He is engaged in various water, energy and environmental projects within East Africa.

Cush Ngonzo Luwesi holds a PhD in the field of economics of water and climate change from Kenyatta University (Nairobi). He is an associate professor of economics and environment at the University of Kwango (DR Congo), and the Deputy Director General (Academic) at the Health College of Kenge (DR Congo). He is also a member of the scientific advisory committee of the Climate Research for Development (CR4D) program championed by the United Nations' Economic Commission for Africa (UNECA). Prior to these assignments, he was the Focal Region Manager of the CGIAR Research Program on Water, Land and Ecosystems (WLE) in West Africa, a program managed by the International Water Management Institute (IWMI) on behalf of the Consultative Group for International Agriculture Research (CGIAR). He taught economics and geography at Kenyatta University (Nairobi) and South Eastern Kenya University (Kitui), Kenya, and has authored over 50 publications, some presented during international conferences in 33 countries.

Dominic Mazvimavi is a Professor of Water and Environmental Sciences and Director of the Institute for Water Studies at the University of the Western Cape, South Africa. He obtained a PhD in Hydrology from the Wageningen University and the International Institute for Geo-information and Earth Observation in The Netherlands, an MSc in Hydrology from the Vrije Universiteit Brussel, Belgium, and a BSc Honours from the University of Zimbabwe. Dominic Mazvimavi started his career in 1981 as a Hydrologist in the Ministry of Water Resources and Development in Zimbabwe, and in 1987 joined the University of Zimbabwe teaching hydrology and environmental science. From 2004 to 2008 he was at the Okavango Research Institute of the University of Botswana where he led projects on the environmental flow assessment of the Okavango River Basin in Angola, Botswana, and Namibia, and environmental monitoring of tsetse fly eradication in northern Botswana. Dominic Mazvimavi serves/has served as an Associate Editor for Hydrology and Earth Systems Science Journal, International Journal of Applied Earth Observation and Geoinformation, Physics and Chemistry of the Earth Journal, African Journal of Aquatic Science.

Shadrack K. Murimi is a Senior Lecturer and Chairman of Geography Department, Kenyatta University. He obtained a Bachelors Degree (B.Ed. (Sc), Geochemistry major) from Kenyatta University in 1986, and awarded Master of Science (Hydro-Geomorphology) in 1990. He undertook his Doctorate degree through German Academic Exchange Services (DAAD), in-country Programme in Trier University,

Germany and Kenyatta University and awarded Ph.D. degree in 2005. His areas of specialization include numerical climatic and hydrogeomorphic data analysis using remotely sensed data products and GIS functionalities, as well as Participatory community integrated GIS (PGIS) modelling mapping environmental watershed resources. Shadrack teaches both bachelors and postgraduate students. He has participated in local and international conferences and workshops and participated in research projects under variable themes related to communities' watershed resource utilization and geomorphic aspects. He has published in regional and international journals and book chapters.

Anne Njeri Njoroge is a holder of Bsc. in Mathematics and Geography from Egerton University, Kenya and Masters of Science in Hydrology and Water Resources from Kenyatta University with special interest in watershed research.

Amos Yesutanbul Nkpeebo is a Project Coordinator at Friends of the Earth – Ghana. He is also a Research Fellow at Mutatio Institute in Accra, Ghana. He holds Msc degree in Integrated Watershed Management (IWM) from Kenyatta University in Nairobi, Kenya.

Joy Apiyo Obando is an Associate Professor at the Department of Geography, Kenyatta University specializing in Physical Geography, Geomorphology and Integrated Watershed Management. She works as an interface between physical geography and management in the watershed. She coordinated the IWM programme at the Geography Department. Currently her research focuses on groundwater dynamics and watershed management. She coordinates an international Masters programme hosted at Kenyatta University. She has won several national and international research grant awards and has published widely. She is a member of the steering Committee of the International Geographical Union Geomorphology and Society Commission; and also a member of the International Advisory Board of Water Resources Research Center.

Anthony Opiyo Ogutu is a Senior Lecturer at South Eastern Kenya University and holds a PhD in Philosophy, MA in Philosophy and BA Sacred Theology from Pontifical Gregorian University in Rome, BA in Philosophy from pontifical Gregorian University in Rome and a Diploma in Philosophy from Consolata Diocese Institute of Philosophy in Nairobi Kenya. He has also held various administrative positions. He has been chair of the department and Dean of school of humanities and social sciences at South East Kenya University. He has also published widely and supervised MA and PhD students. He is currently in the Department of Philosophy and Religious Studies.

Luke Olang is a Senior Lecturer at the Faculty of Engineering Sciences and Technology of the Technical University of Kenya. He holds a PhD from University of Natural Resources and Life Sciences, Vienna (BOKU), Austria. His research interests are in water resource management studies that attempt to understand hydrological variability within the coupled terrestrial water, energy and agricultural cycles. This includes the intersection of human activities; and translation of the derived knowledge to improve management of the concerned bio-geo-chemical systems. He is involved a number of research initiatives, including a project funded by the Austrian Partnership Programme in Higher Education & Research (Appear) on 'Capacity building on the Water-Energy-Food Security Nexus through Research and Training in Kenya and Uganda (CapNex). Dr Olang has published in the larger area of climate change adaptation within the water resources sector, and is presently leading regional initiatives towards enhanced and integrated drought risk management within the Greater Horn of Africa region.

Gerald (Augusto) Corzo Perez is research/lecturer and coordinator of the Hydroinformatics Laboratory at UNESCO-IHE. Dr Gerald A. Corzo has extensive experience in modelling water resources using advanced ICT technology. He is an expert in programming mathematical models and online hydroinformatics management systems. In 2012 he won the Tison award as young scientist from the IAHS association. He has worked on different international institution such as Wageningen University in the Netherlands and the Technologic of Monterrey in Mexico. He has coordinated the statistics of the Climate change inventory of adaptation and mitigation actions for Latin-America, presented at the WWF in 2012. The last two years he served as chair of the session on geo-statistics at the European Geoscience Union. From 2011 to 2012 he created and led the LatinAqua network for water research scientist in Latin-America.

Chris Shisanya is currently the Dean School of Humanities and Social Sciences and Full Professor of Agroclimatology in the Department of Geography at Kenyatta His current areas of research interest include: climate change and adaptation, Climate change and human development, climate change and security, integrated soil fertility management, integrated watershed management and systems analysis. He is a prolific researcher having won over 45 national and international research grant awards in the last twenty years. He has over 60 peer-reviewed Journal publications and 10 book titles to his credit. He serves on the Editorial Boards of various Academic Scientific Journals.

Doris Wimmer is a Master student BSc degree Institute of Water Management, Hydrology and Hydraulic Engineering of the University of Natural Resources and Life Sciences, Vienna, Austria. Her main research interest is on effects of land-use on Hydrology.

Index

H

I

K

L

M

N

O

P

R

S

T

V

W

Y

www.ingramcontent.com/pod-product-compliance
Lightning Source LLC
LaVergne TN
LVHW062216200926
848484LV00006B/1413
9781799801634